Further volumes in this series:

D. P. S. Verma, T. Hohn (eds.)
Genes Involved in Microbe–Plant Interactions

B. Hohn, E. S. Dennis (eds.)
Genetic Flux in Plants

A. D. Blonstein, P. J. King (eds.)
A Genetic Approach to Plant Biochemistry

T. Hohn, J. Schell (eds.)
Plant DNA Infectious Agents

D. P. S. Verma, R. B. Goldberg (eds.)
Temporal and Spatial Regulation of Plant Genes

E. S. Dennis, D. J. Llewellyn (eds.)
Molecular Approaches to Crop Improvement

R. G. Herrmann (ed.)
Cell Organelles

T. Boller, F. Meins (eds.)
Genes Involved in Plant Defense

Plant Gene Research
Basic Knowledge and Application

D. P. S. Verma (ed.)

Signal Transduction in Plant Growth and Development

SpringerWienNewYork

Dr. Desh Pal S. Verma

Department of Molecular Genetics and Plant Biotechnology Center
The Ohio State University
Columbus, Ohio

Softcover reprint of the hardcover 1st edition 1996

Printed by Novographic, Ing. Wolfgang Schmid, A-1230 Wien
Typeset by Indah Photosetting Centre Sdn. Bhd., Kuala Lumpur
Graphic design: Ecke Bonk
Printed on acid-free and chlorine-free bleached paper

With 49 Figures

Library of Congress Cataloging-in-Publication Data

Signal transduction in plant growth and development / edited by D.P.S. Verma.
p. cm. — (Plant gene research)
ISBN-13:978-3-7091-7476-0 e-ISBN-13:978-3-7091-7474-6
DOI: 10.1007/978-3-7091-7474-6

1. Growth (Plants). 2. Plants — Development. 3. Cellular signal transduction. I. Verma, D. P. S. (Desh Pal S.). 1944–
II. Series.
QK731.S555 1996
581.3 — dc20

95-53834
CIP

ISSN 0175-2073
ISBN-13:978-3-7091-7476-0

Preface

Growth and development of plants are controlled by various environmental cues that are sensed by the plant via various signal transduction pathways coupled to specific responses. The complexities of these pathways have begun to unfold as many genes have been identified and some of them isolated and characterized in recent years. Some of these pathways are conserved from yeast to plants, being regulated by various kinases and phosphatases. In addition, plants have many unique pathways that transduce specific signals such as light, phytohormones, and oligosaccharides. The latter act as elicitors of pathogenic responses or inducers of new organs such as flower and root nodules. This group of signal molecules are unique as they not only act as mitogenic regulators to initiate cell division but also control full organ formation and thus should be considered as morphogens. This suggests that plants have evolved some unique receptors and signal transduction pathways which allows these organisms to respond to specific stimuli from the environment or plant pathogens and symbionts.

This volume highlights some of the examples of the plant signal transduction machinery, opening new vistas in research on plant growth and development. This field of research is rapidly moving forward and many new insights are being revealed. The new technologies including the use of bacteria, yeast, and *Arabidopsis* as functional complementation systems are providing proof of function of many of the proteins that show homology to those from other organisms. Several new frontiers remain to be crossed in this field that may require development of new tools. For example, understanding of plant embryogenesis, organogenesis, flower initiation, and defense mechanisms cannot be studied in heterologous systems. Moreover, many basic processes in plants such as cytokinesis, cell wall formation, and lignification require further attention in light of the basic architecture of plants. These studies will eventually lead to the improvement of crops and the use of plants as new resources for producing desirable products to meet the growing needs of mankind.

Desh Pal S. Verma

Contents

List of Contributors

Assmann, Sarah M., Biology Department, 208 Mueller Laboratory, Pennsylvania State University, University Park, PA 16802, U.S.A.

Battey, Nicholas H., School of Plant Sciences, University of Reading, Whiteknights, Reading RG6 2AS, U.K.

Beachy, Roger N., Division of Plant Biology, Department of Cell Biology, The Scripps Research Institute, La Jolla, CA 92037, U.S.A.

Carré, Isabelle, NSF Center for Biological Timing, Department of Biology, Gilmer Hall, University of Virginia, Charlottesville, VA 22903, U.S.A.

Echevarria, Cristina, Laboratorio de Fisiologia Vegetal, Faculdad de Biologia, Universidad de Sevilla, E-41012 Sevilla, Spain.

Epel, Bernard L., Division of Plant Biology, Department of Cell Biology, The Scripps Research Institute, La Jolla, California, U.S.A. (permanent address: Botany Department, Tel Aviv University, 69978 Tel Aviv, Israel)

Fenczik, Csilla A., Division of Plant Biology, Department of Cell Biology, The Scripps Research Institute, La Jolla, CA 92037, U.S.A.

Goring, Daphne R., Biology Department, York University, North York, Ont. M3J IP3, Canada.

Hong, Zonglie, Department of Molecular Genetics and Plant Biotechnology Center, The Ohio State University, Columbus, OH 43210, U.S.A.

Huber, Steven C., Agricultural Research Service, U.S. Department of Agriculture and Department of Crop Science, North Carolina State University, Raleigh, NC 27695-7631, U.S.A.

Kaiser, Werner M., Julius-von-Sachs-Institut für Biowissenschaften, Universität Würzburg, D-97082 Würzburg, Federal Republic of Germany.

Kay, Steve A., NSF Center for Biological Timing, Department of Biology, Gilmer Hall, University of Virginia, Charlottesville, VA 22903, U.S.A.

Klessig, Daniel F., Waksman Institute and Department of Molecular Biology and Biochemistry, Rutgers University, Piscataway, NJ 08855, U.S.A.

Miao, Guo-Hua, Biotechnology Research, Agricultural Products Department, E.I. duPont de Nemours & Company, Inc., Experimental Station, Wilmington, DE 19880-0402, U.S.A.

Pierre, Jean-Noël, Institut de Biotechnologie des Plantes, UA CNRS D 1128, Université de Paris-Sud, Centre d'Orsay, F-91405 Orsay Cedex, France.

Ritchie, Siân M., School of Plant Sciences, University of Reading, Whiteknights, Reading RG6 2AS, U.K.

Rothstein, Steven J., Department of Molecular Biology and Genetics, University of Guelph, Guelph, Ont. NIG 2W1, Canada.

Scherer, Günther F. E., Botanisches Institut, Universität Bonn, D-53115 Bonn, Federal Republic of Germany.

Verma, Desh Pal S., Department of Molecular Genetics and Plant Biotechnology Center, The Ohio State University, Columbus, OH 43210, U.S.A.

Vidal, Jean, Institut de Biotechnologie des Plantes, UA CNRS D 1128, Université de Paris-Sud, Centre d'Orsay, F-91405 Orsay Cedex, France.

Wobbe, Kristin K., Waksman Institute, Rutgers University, Piscataway, NJ 08855, U.S.A.

Yang, Zhenbiao, Department of Plant Biology and Plant Biotechnology Center, The Ohio State University, Columbus, OH 43210, U.S.A.

Signal Transducing Proteins in Plants: an Overview

Zhenbiao Yang

Contents

I. Introduction

Living organisms have evolved various mechanisms for the modulation of specific cellular or metabolic processes in response to specific extracellular signals. These regulatory mechanisms generally involve a cascade of biochemical events called signal transduction pathways. Based on a number of well-characterized signal transduction pathways in animal systems, a paradigm of signaling mechanisms primarily consisting of transmembrane receptors, GTP-binding proteins, second messengers, and protein kinases has been established. Signal transduction pathways conforming to this paradigm are found in various eukaryotes as diverse as fungi and mammals. The conservation of these mechanisms raises the possibility that signal transduction pathways in plants may also follow this signaling paradigm. Indeed, recent studies using biochemical, molecular and genetic approaches have revealed the existence and physiological function of many conserved signaling proteins in plants. The genetic approach has also led to the discovery of several novel plant regulatory proteins. The purpose of this chapter is to provide a comprehensive review on structure and function of both conserved and novel plant signaling proteins with emphasis on their role in signal transduction controlling specific developmental processes in plants. This review is not intended to be exhaustive but rather to reflect the major themes of the exciting research in the area of plant signal transduction.

II. Signal Transduction Paradigm

Most of our knowledge of the mechanism for cellular communication comes from studies using animal model systems such as *C. elegans, Drosophila, Xenopus* oocytes and mammalian cell lines. The intricate cellular communication in animals involves numerous kinds of signals, including peptidyl hormones, soluble small molecules, gaseous chemicals, steroids and other lipids, extracellular matrix, and physical factors such as light and temperature. In spite of the diversity of stimuli and corresponding responses, most well-studied signal transduction pathways are consistent with a paradigm as illustrated in Fig. 1. Upon stimulation by an extracellular signal, a plasma membrane receptor activates a GTP-binding protein. The GTP-binding protein, existing as two interchangeable conformational states (GDP- and GTP-bound), serves as a molecular switch controlling the signal transduction pathway. It is turned on when the bound GDP is replaced with GTP and off when the bound GTP is hydrolyzed back to GDP by the intrinsic GTPase activity. The activated GTP-binding protein either directly regulates a cascade of protein kinases or indirectly modulates the activity of second messenger-dependent protein kinases. The latter involves one or more effectors of GTP-binding proteins, which typically is an enzyme responsible for the generation of second messengers. The protein kinases alter the phosphorylation status of a specific set of proteins such as transcriptional factors leading to specific changes in gene expression, metabolic and/or cellular processes.

Two best-characterized families of transmembrane receptors are transmembrane receptor kinases (TRKs) with a single transmembrane domain and an intracellular tyrosine kinase domain and G protein-coupled receptors (GCRs) that typically contain seven transmembrane domains but no kinase domain. GCRs directly modulate the heterotrimeric GTP-binding protein (commonly called G protein) composed of α, β, and γ subunits. The α subunit of G protein directly interacts with the receptor and contains GTP/GDP binding and GTPase activities (Hepler and Gilman, 1992). These activities are regulated by its interaction with the receptor as well as with β and γ subunits. The heterotrimeric G protein complex is inactive; its interaction with the receptor releases β and γ subunits from the α subunit, which in turn activates the

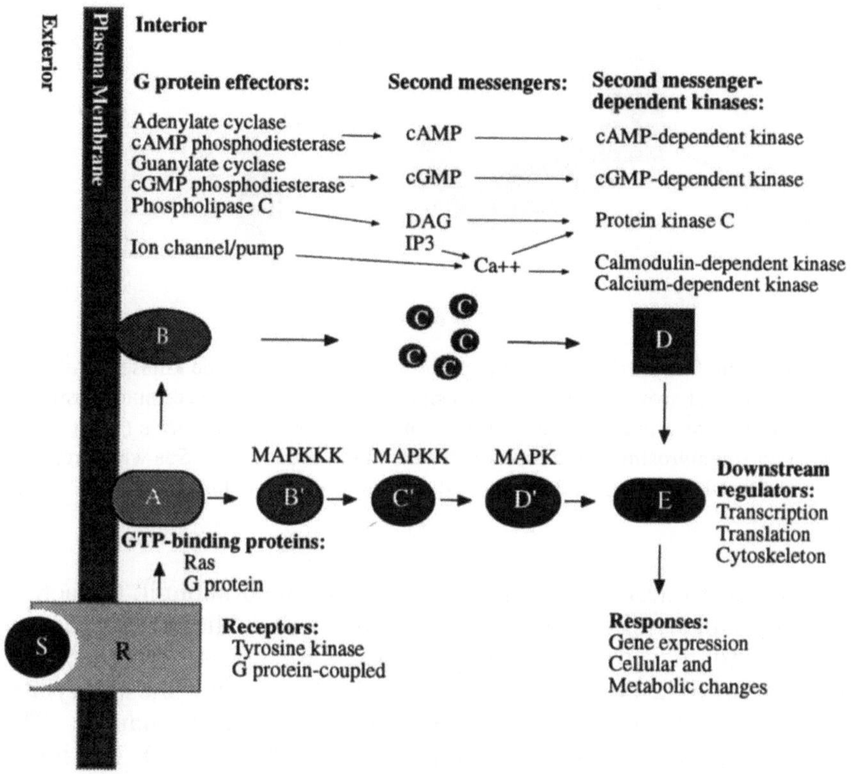

Fig. 1. Signal transduction paradigm. Schematical presentation of a general pattern of most signal transduction pathways found in animal and fungal systems. An extracellular signal (S) binds to a transmembrane receptor (R), which in turn activates a GTP-binding protein (A). The GTP-binding protein either regulates a cascade of protein kinases (B′–D′) or a G protein effector (B) leading to changes in levels of intracellular signals called second messengers (C). The second messenger is required for the activity of specific protein kinases (D). The protein kinases are responsible for the modulation of distinct regulators (E) controlling transcription and translation of specific proteins, cytoskeletal structures, and metabolic activities

G protein effector. Thus, the specificity of G protein activation and effector regulation primarily resides in the α subunit, consistent with the existence of a large family of diverse α subunits. However, recent work demonstrates that there are also multiple forms of β and γ subunits that may function to modulate the activity of effector proteins (Hepler and Gilman, 1992). Major G protein effectors include ion channels, phospholipases, and enzymes involved in the metabolism of cyclic nucleotides. The effectors are responsible for the regulation of intracellular concentration of second messengers including cAMP, cGMP, DAG, IP_3, and Ca^{++}. These second messengers then modulate the activity of specific protein kinases or phosphatases known as second messenger-dependent protein kinases or phosphatases.

TRKs often modulate the activity of the monomeric small GTP-binding protein called Ras. Ras is biochemically and structurally similar to the α

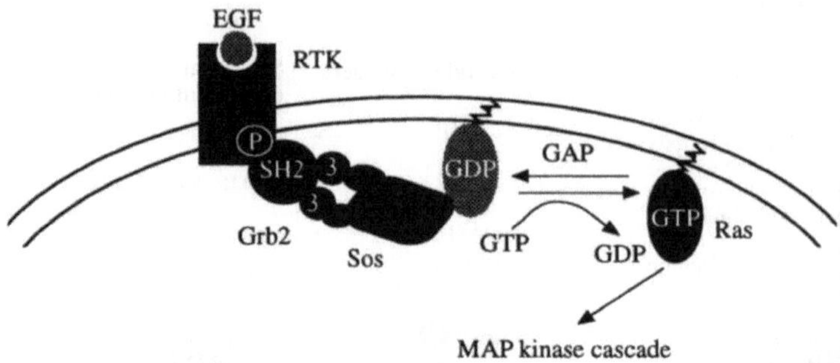

Fig. 2. Regulation of the Ras small GTP-binding protein by receptor tyrosine kinase. Binding of ligand (e.g., EGF, epidermal growth factor) to a tyrosine kinase receptor induces autophosphorylation on a tyrosine residue. Binding of the complex of adaptor protein Grb2/Sos (Ras GTP/GDP exchanger) to the phosphotyrosine of the receptor results in the activation of Sos which replaces GDP on inactive Ras with GTP. Active Ras then turns on the MAP kinase cascade

subunit of the heterotrimeric G protein. Compared to the α subunit, the activation of the Ras small GTP-binding protein requires two additional factors, an adaptor (Grb2) and a GDP/GTP exchanger (also called Ras activator). The adaptor contains an SH2 domain that interacts with the auto-phosphorylated receptor and two SH3 domains that bind the GDP/GTP exchanger. This exchanger activates Ras by replacing bound GDP with GTP (Fig. 2). The intrinsic GTPase activity in Ras is much weaker but can be activated by GTPase activating proteins (GAP). In animal systems, activated Ras directly turns on a cascade of protein kinases, typically composed of MAP kinase kinase kinase, MAP kinase kinase, and MAP kinase or their respective homologs (see below for more details). In the budding yeast, however, Ras mediates the activity of adenylate cyclase.

It should be pointed out that there are numerous signaling mechanisms that are exceptions to this paradigm. One extreme example is the signaling of steroid hormones. These hormones bind to intracellular receptors with a ligand binding domain and a transcriptional activation domain (Evans, 1988). Upon excitation, the receptor is translocated into the nucleus to directly regulate gene expression. Another example is the existence of alternative signaling pathways for the transmembrane receptors. The same tyrosine kinase receptors are also known to directly activate several other signaling proteins than Ras, including phospholipase C, phosphatidylinositol-3-kinase and phosphotyrosine phosphatase. This suggests that the TRK-mediated signaling is more complicated than just the linear Ras-modulated MAP kinase cascade. On the other hand, the MAP kinase cascade can also be mediated by other signalling proteins than Ras such as trimeric G proteins, protein kinase C and protein kinase A (Wilson, 1994). It is not uncommon that a specific signaling protein can mediate multiple signaling proteins and vice versa. More complexity of signal transduction mechanisms arises from the very key nature of signaling proteins per se—the ability to be down-regulated or negatively regulated. Protein kinases that phosphorylate and thus inactivate receptors, GAPs that activate Ras GTPase, cAMP or cGMP phosphodiesterases that lower the level of cAMP or cGMP, and protein phosphatases that dephosphorylate protein kinases are just a few examples of such negative regulators. This complex nature of signaling mechanisms provides opportunities for any given signal transduction pathway to cross-talk with other pathways. In conclusion, although the linear signal transduction pathway is the key to the specificity of stimulus-response, there exist mechanisms required for networking between different signaling pathways.

III. Conserved Signaling Proteins in Plants

In contrast to animal systems, signal transduction mechanisms in plants are poorly understood. No plant signaling pathway has been fully elucidated although increasing number of conserved signaling proteins have been identified in plants by various approaches. Based on the sequence conservation of their genes, many of these proteins have been identified by molecular strategies such as PCR amplification and DNA hybridization with animal or yeast DNA probes. However, the function for most of the conserved proteins thus identified remains elusive. The second most common approach relies on the biochemical study of specific activities of conserved signaling proteins. This approach, usually involving the direct determination of enzymatic activities or the use of specific agonists or antagonists, has revealed important functions for several key signaling proteins such as G proteins and calmodulins. Unfortunately, this biochemical approach had also resulted in some controversial conclusions regarding the existence and the function of certain signaling proteins in plants. The controversy primarily arose from inability to link changes in activity to specific plant signal transduction pathways as well as from difficulty in purifying such proteins. Finally, recent advances in functional cloning of *Arabidopsis* genes involved in signal transduction have significantly contributed to the discovery of conserved signaling proteins in plants. This approach establishes the biological function of the gene products homologous to signaling proteins found in other systems, but the biochemical activity of many of these genes remains to be determined. This section focuses on the discussion of key conserved signal transducing proteins found in plants using these three different approaches.

A. Plant Transmembrane Receptors

Like animal cells, plant cellular communication involves numerous extracellular signals such as growth regulators, light, peptides, gaseous chemicals, oligosaccharides, extracellular matrix, and mechanical stimuli. The molecular mechanism for the perception of most of these signals is poorly understood. Although considerable efforts have been devoted to the biochemical characterization of the receptors for several growth hormones and fungal elicitors, the identity of these receptors is largely unknown (Estelle, 1992). Perhaps the only well-established plant receptors are the photoreceptors, phytochromes (Bowler and Chua, 1994). The existence of the typical tyrosine receptor kinases and G protein-coupled receptors is still obscure in plants. Nonetheless, genes encoding putative receptors with structural features similar to the animal Ser/Thr receptor kinase and the prokaryotic two-component regulator have recently been cloned.

1. Receptor-like Ser/Thr Kinases

The first receptor-like Ser/Thr kinase (RLK) from plants, maize ZmPK1, was cloned by a PCR strategy based on conserved protein kinase domains (Walker and Zhang, 1990). Subsequently, a number of genes encoding protein kinases with structures similar to ZmPK1 have been isolated (Chang et al., 1992; Goring and Rothstein, 1992; Stein et al., 1991; Tobias et al., 1992). Although overall sequence similarities among different RLKs can vary considerably, they share many topological features that are characteristic of transmembrane receptor kinases. First, all have an N-terminal hydrophobic domain that is thought to function as signal peptide. Second, all RLKs contain a single internal putative transmembrane domain that is flanked by the N-terminal extracellular domain and the C-terminal cytoplasmic kinase domain. Third, the extracellular domain contains putative N-linked glycosylation sites like the extracellular domain of the animal kinase receptor. The glycosylation is thought to be important for ligand binding (Ulrich and Schlessinger, 1990). The extracellular domains of all plant RLKs can be classified into two groups, those that have similarity to SLG proteins involved in self-incompatibility and those that contain a leucine-rich repeat sequence (LRR). The SLG-homologous domain also contains a cysteine rich region possibly involved in the formation of secondary and tertiary structure required for ligand-binding. LRR are found in many proteins and play a central role in protein–protein interaction (Kobe and Deisenhofer, 1994). Thus, LRR may be involved in the oligomerization of receptor polypeptides, which is shown to be essential for ligand binding or receptor activation (Massagne et al., 1994; Ullrich and Schlessinger, 1990). Finally, they all have a cytoplasmic domain characteristic of Ser/Thr kinase. Ser/Thr kinase activity has been demonstrated for most of the plant RLKS (Chang et al., 1992; Goring and Rothstein, 1992; Horn and Walker, 1994; Stein and Nasrallah, 1993). This kinase domain is reminiscent of the Ser/Thr kinase domain of the animal TFG-β receptors (X. Lin et al., 1991; Massague et al., 1994). The animal receptor Ser/Thr kinases represent a growing family of kinase receptors that regulate many cellular processes including cell growth, differentiation, motility, organization, and death in diverse animals such as mammals, frogs and flies (Massague et al., 1994).

Among all cloned RLKs, several are involved in self-incompatibility in *Brassica*. These RLKs originally called S-proteins are closely related to each other within both the SLG domain and the kinase domain (Nasrallah and Nasrallah, 1993). Mutations in these RLKs do not affect other aspects of growth and development, suggesting that the *Brassica* RLKs have evolved specific function in the control of self-incompatibility. Consistent with their function, they are specifically expressed in stigma and pollen. Physiological function of other RLKs is not clear. Three *Arabidopsis* RLKs ARK1, ARK2, and ARK3 share striking sequence homology with S-proteins, but obviously have no self-incompatibility function since a self-incompatibility system does not exist in this plant. Promoter:GUS fusion expression indicates that ARK2 is exclusively expressed in above-ground tissues (cotyledons, leaves, and sepals) while ARK3 expression is limited to small groups of cells in the root-hypocotyl transition zone and at the base of lateral roots, auxiliary buds and pedicels (Dwyer et al., 1994). Two other *Arabiodopsis* RLKs (RLK1 and RLK4) related to SLGs show distinct expression patterns (Walker, 1993), RLK1 is specifically expressed in rosette leaves while expression levels for RLK4 are much higher in roots than in rosettes leaves. Two *Arabidopsis* RLKs (TMK1 and RLK5) with LRR sequences do not seem to have differential expression in different organs (Chang et al., 1992; Walker, 1993). The multiplicity, diversity and complex expression pattern of RLK genes suggest that members of the RLK family have diverse cellular functions.

The structural features described above suggest that the plant RLKs are involved in perception and transmission of extracellular signal but their specific function in signal transduction pathways has not been elucidated. Especially, lack of knowledge of the nature of ligands for these RLKs renders the study of their function in signal transduction particularly difficult. Self-incompatibility may provide a useful assay system for the identification of ligands involved in the activation of SLG receptor kinases. The presence of LRR sequence in the extracellular domain suggests that RLKs may function as dimers or oligomers. The animal TGF-β receptors are able to bind ligands as monomers but require heterodimer or heteromultimers for receptor activation. Molecular cloning of genes based on protein–protein interaction is a powerful technique for identifying signaling proteins and may be useful for identifying peptide ligands and downstream components of RLK-mediated signal transduction pathways. Recently, a type 2C protein phosphatase (KAPP) interacting with RLK5 has been cloned by screening an *E. coli* expression CDNA library using ^{32}P-labeled RLK5 kinase domain fused to glutathione-S-transferase (Stone et al., 1994). Whether this is a downstream component or an enzyme that desensitizes the receptor remains to be determined. It appears that the mechanism for activating downstream component of the TGF-β receptors is different from that for tyrosine kinase receptors (Massague et al., 1994; Wieser et al., 1993). The cytoplasmic Ser/Thr kinase activity is required for signaling of the TGF-β receptors, but no substrates of these Ser/Thr kinases have been identified (Massague et al., 1994; Wieser et al., 1993). Interestingly, KAPP has been shown to be phosphorylated by RLK5 although the physiological significance of this phosphorylation is not clear (Stone et al., 1994). Further studies of the RLK5-interacting protein phosphatase may provide important insight into the mechanism for the action of Ser/Thr receptor kinases.

2. A Two-Component Sensing System in Plants

Arabidopsis ETR1 protein involved in early steps of ethylene signal transduction resembles the bacterial two-component system (Chang et al., 1993). The two-component system, composed of a transmembrane sensor protein and a response regulator, is responsible for regulating adaptive responses to environmental and nutritional cues in prokaryotes. The sensor protein has topology analogous to the eukaryotic transmembrane receptor kinase in that it has an N-terminal variable sensor domain situated in periplasmic space and a cytoplasmic histidine kinase domain. Instead of a single transmembrane domain, the bacterial sensor protein has two transmembrane domains flanking the sensor domain. Activation of sensor domain leads to the autophosphorylation of histidine residue on kinase domain, which in turn activates the regulator protein by phosphorylation of a conserved asparate residue within the regulator. In several instances, the sensor and the regulator are contained in a single polypeptide (see references in Chang et al., 1993). The deduced protein sequence of the ETR1 gene cloned by chromosome walking has primary sequence and topology similar to the single polypeptide two-component system (Chang et al., 1993). ETR1 protein contains a variable N-terminal domain that has no apparent sequence similarity to any known proteins and might be involved in ethylene perception. This is followed by one or more potential transmembrane domains, a cytoplasmic histidine kinase-like domain and a C-terminal regulator-like domain. The putative kinase domain has 5 motifs including a histidine phosphorylation site that are characteristic of the bacterial histidine kinase domain. The putative regulator domain has all critical conserved residues including an aspartate residue (site of phosphorylation) and predicted three dimensional structure found in the bacterial regulator domain (Chang et al., 1993). Because of the topological similarity to the two-component regulator and of the involvement in early steps of ethylene signal transduction, it was suggested that ETR1 protein might be an ethylene receptor (Chang et al., 1993). Although this is an intriguing possibility, there is no direct evidence for its function as a receptor. Further studies on its biochemical and structural properties and downstream effector protein are necessary to assess the biochemical role of ETR1 in ethylene signal transduction.

B. GTP-binding Proteins

GTP-binding proteins belong to a large group of structurally related but functionally versatile regulatory proteins that are involved in virtually every key cellular process in eukaryotes. One of the most conspicuous functions of GTP-binding proteins is to serve as molecular switch in the signal transduction pathway. The role of both heterotrimeric G proteins and small GTP-binding proteins in signal transduction has been demonstrated essentially in all eukaryotic organisms studied including fungi, slime molds, worms, flies, frogs and mammals. There is a growing body of evidence that GTP-binding proteins are also key players in many signal transduction pathways in plants. Recently, genes encoding both groups of GTP-binding proteins have been identified in plants.

1. Heterotrimeric G Proteins

Biochemical evidence for role in signal transduction. Studies involving G protein agonists or antagonists indicate that G proteins are involved in light signal transduction (Neuhaus et al., 1993; Romero and Lam, 1993; Warpeha et al., 1991) and the regulation of K^+ channel opening (Farley-Grenot and Assmann, 1991; Li and Assmann, 1993) in plant cells. Understanding of the physiological role of G proteins in animal cells is facilitated by two bacterial toxins, cholera and pertussis toxins, that act as specific G protein agonist and antagonist, respectively. Cholera toxin ADP-ribosylates a cysteine residue of GTP-bound α subunit locking G protein in active state while pertussis toxin ADP-ribosylates an arginine residue of GDP-bound α subunit locking it in inactive state. Using isolated guard cell protoplasts and patch clamp techniques, Farley-Grenot and Assmann (1991) showed that G protein agonists (cholera toxin and GTPγS) decrease inward K^+ current in the protoplasts, suggesting G protein modulates the reduction of guard cell K^+ uptake and stomatal closure mediated by abscissic acid.

Several studies have implicated G protein involvement in light signal transduction (Neuhaus et al., 1993; Romero and Lam, 1993; Warpeha et al., 1991). A blue light-regulated G protein activity was observed on the plasma membrane isolated from apical buds of dark grown pea seedlings (Warpeha et al., 1991). A 40 kDa polypeptide of an α subunit of G protein in the membrane fraction was ADP-ribosylated by cholera toxin only in the excited state (i.e., in the presence of both GTP and blue light) but by pertussis toxin in absence of GTP and blue light. Photoaffinity labeling of this polypeptide by a nonhydrolyzable GTP analog was also regulated by blue light. These results indicate that the 40 kDa polypeptide is the α subunit of the blue light-mediated G protein associated with the plasma membrane (Warpeha et al., 1991). Blue light activation of the G protein in isolated membrane system implies that the membrane fraction contains both blue light receptor and G protein, thus suggesting that the G protein might be regulated through a G protein-coupled receptor.

Using microinjection techniques, Neuhaus et al. (1993) show that G protein antagonists (GDPβS and pertussis toxin) inhibit phytochrome A-mediated responses including activation of photoregulated gene expression, chloroplast development and anthocyanin biosynthesis in wild type tomato cells. Injection of agonists (cholera toxin, GTPγS, and GMP-PNP) into phytochrome A-deficient tomato cells restore all these phytochrome A-mediated responses, as does the introduction of tomato phyA protein. However, Ca^{++} or calmodulin only partially rescues phyA$^-$ phenotypes by stimulating photoregulated gene expression and partial chloroplast development (Neuhaus et al., 1993). A combination of cGMP and Ca^{++} resulted in fully developed chloroplasts and anthocyanin biosynthesis (Bowler et al., 1994). Based on these results, it was proposed that an early step of phyA-mediated signal transduction is heterotrimeric G protein(s) that acts upstream of two branch pathways respectively mediated by cGMP and Ca^{++}/calmodulin (Bowler et al., 1994). These studies did not distinguish if phyA regulates a single or multiple G proteins. G-protein-coupled receptors in animal cells typically belong to a family of transmembrane receptors that directly interact with G proteins (Boldwin, 1994). Assuming G protein is indeed an early step in the phytochrome signal transduction, the regulation of phytochrome-mediated G protein appears not to follow the classical example of receptor–G protein interactions in animal cells since there is no evidence that phytochrome is a transmembrane or membrane-associated protein (Quail, 1991). To address whether one or more G proteins are involved in the phytochrome signal transduction and how phytochrome mediates G protein activity, it is essential to identify specific G proteins involved in the phytochrome responses.

Molecular cloning and characterization of G proteins. The biochemical evidence for the involvement of G proteins in plant signal transduction is supported by cloning of genes encoding G protein subunits from plants. Genes encoding homologs of the G protein α subunit have been isolated from *Arabidopsis* (GPA1) and tomato (TGA1) (Ma et al., 1990, 1991). More recently, homologs of a G protein β subunit have also been cloned from both *Arabidopsis* (AGB1) and maize (ZGB1). Genes for γ subunit have not been identified from any higher plant species. In animal systems, a large number of G proteins have been described (Hepler and Gilman, 1992). Based on sequence comparison, these plant homologs of both α and β subunits are distinct from respective subunits of all known G proteins (Ma et al., 1990; Weiss et al., 1994). It is not clear whether the G protein-dependent signal transduction pathways in plants described above are regulated by a single or several different G proteins. Nonetheless, the G protein involving GPA1 or its homologs appears to be different from the light-regulated G proteins described above since GAP1 does not have a C-terminal cysteine residue required for ADP-ribosylation by cholera toxin, suggesting that multiple kinds of G proteins exist in plants (Ma et al., 1990; Weiss et al., 1994). Whether GPA1 and AGB1 belong to the same G protein is not clear, neither is their physiological role in plant signal transduction. Analysis of the temporal and spatial patterns of GPA1 expression during *Arabidopsis* life cycle shows that GPA1 is distributed in all organs except mature seeds (Weiss et al., 1993). Higher levels of expression were observed in meristematic tissues and elongating cells. GPA1 is more abundant in young tissues than mature tissues where GPA1 is primarily found in vascular system and specific cells of particular organs (e.g., mesophyll cells of leaves and epidermis of petals). This complex pattern of GPA1 suggests that GPA1 is likely to be involved in various signal transduction pathways during plant development (Weiss et al., 1993). To understand the role of GPA1, AGB1, and their homologs in signal transduction pathways, it is necessary to elucidate their physiological function using more direct approaches such as alteration of their expression in specific cells or tissues and to identify their effectors and the extracellular signals that regulate their activity.

2. The Ras Superfamily of Small GTP-binding Proteins

Structural and functional conservation. The ras superfamily of small GTP-binding proteins are key regulators of diverse cellular processes such as cell growth and differentiation, establishment of cell polarity, and intracellular trafficking. Based on amino acid sequence similarity, the Ras superfamily is divided into at least three major families: Ras, rho, and rab/YPT. Each family generally has distinct cellular functions (Hall, 1990, 1994; Kahn et al., 1992; Takai et al., 1992) (see Table 1). Among the three families, the regulatory functions of the Ras proteins are best understood. Ras functions as a molecular switch in signal transduction pathways controlling cell proliferation and differentiation in a number of divergent organisms including mammals, worms, insects, and fungi. The role of the rho proteins is distinct in that they are key components of the signal transduction pathway regulating the organization of the actin cytoskeleton. The rab/YPT proteins play an essential role in the intracellular vesicle trafficking. Additional families of small GTP-binding proteins have been described, including nucleus-localized Ran/TC4 involved in the checkpoint control of mitotic cell cycle (Lounsburry et al., 1994; Moore and Blobel, 1993) and Sar/Arf involved in the control of vesicle trafficking (Verma et al., 1994).

The small GTP-binding proteins contain conserved structural motifs required for GTP/GDP binding, GTPase activity, membrane anchoring (via isoprenylation) and interaction with downstream effector molecules (Bollag and McCormick, 1991; Downward, 1992; Takai et al., 1992). Point mutations within these motifs are known to affect signal transducing activity of the small GTP-binding proteins. Noticeably, mutations in residues corresponding to Gly-12, Ala-59, Gln-61 and Glu-63 of the mammalian $p21^{ras}$ result in impaired GTPase, and thus the signal transduction pathway is permanently on, even in the absence of

Table 1. Structural and functional comparisons between Ras, rho, and rab/YPT

Family	Function	C_3-ribosylation[a]	C-terminus[b]	Isoprenylation[c]
Ras	cell growth/ differentiation	no	CAAX	FTase
Rho	actin cytoskeletal organization	yes	CAAL	GGTase I
Rab/YPT	vesicle trafficking	no	CC CXC CCXX	GGTase II

[a] C_3 an exoenzyme from *Clostridium botulinum*

[b] C-terminal signature sequence for protein isoprenyltransferase. C, cysteine; A, aliphatic amino acid; X, any amino acids; L, leucine

[c] Enzymes responsible for C-terminal isoprenylation: FTase, protein farnesyltransferase; GGTase, protein geranylgeranyltransferase

signals (Bollag and McCormick, 1991; Downward, 1992; Takai et al., 1992). These so-called dominant positive mutations account for the oncogenicity of Ras oncoproteins. A second group of mutations (Gln-61 plus Cys-186, or Ser-17) are dominant inhibitory; they interfere with the function of endogenous wild type proteins by sequestering an upstream activator or downstream effector. These dominant mutations have served as instrumental tools for studying the functions of small GTP-binding proteins in signal transduction (Bollag and McCormick, 1991). The activities of these mutations are conserved in various eukaryotes including mammals, insects, worms, slime mold and fungi (Esch et al., 1993; Lev, 1993; Wigler, 1993).

Identification in plants. The rho, rab, and ran/TC4 small GTP-binding proteins have been identified in plants (Terryn et al., 1993; Verma et al., 1994; Z. Yang and Watson, 1993). To date, no plant Ras proteins have been found although they exist in all other eukaryotes including mammals, insects, worms, slime molds and fungi (Chardin, 1993). Members of the rab family of small GTP-binding proteins have been characterized from various plant species and there is evidence that they are involved in the control of vesicle transport pathways as the fungal and animal rab proteins (Bednarek et al., 1994; Cheon et al., 1993; Verma et al., 1994). The plant rab proteins have been extensively reviewed recently (Terryn et al., 1993; Verma et al., 1994). Only the rho family of small GTP-binding proteins from plants is discussed in this chapter.

The first plant rho gene, *Rop1Ps*, has recently been isolated from pea (Z. Yang and Watson, 1993). Rop1Ps contains all conserved structural features found in small GTP-binding proteins as well as those unique to the rho proteins. The presence of conserved domains for GDP/GTP binding and GTPase activity suggests that Rop1Ps is a GTP-binding protein. This is confirmed by the specific GTP-binding activity of the *E. coli*-expressed Rop1Ps protein (Z. Yang and Watson, 1993). The effector-binding domain in Rop1Ps is also conserved, suggesting that the plant rho effectors are similar to those present in other systems. Rop1Ps has a C-terminal CSIL conforming to the signature sequence CAAL of the rho proteins (C, cysteine; A, aliphatic amino acids; and L, leucine) required for posttranslational isoprenylation by protein geranylgeranyltransferase I (*GGTase I) (Reiss et al., 1991). In fact, *E. coli*-expressed Rop1Ps is geranylgeranylated in vitro (J. Zhu and Z. Yang, unpubl. results). In addition, Rop1Ps contains a rho-specific residue Asn-42, the site for rho-specific ADP-ribosylation by the C_3 exoenzyme toxin from *Clostridium botulinum* (Aktories et al., 1989; Sekine et al., 1989). The C_3 toxin has been a useful tool for studying the cellular function of the rho proteins in animal cells and thus may also be used to assess their functions in plants.

In plants, rho is encoded by a multiple gene family with at least six isogenes in pea (Z. Yang and Watson, 1993) and three in *Arabidopsis thaliana* (H. Li and Z. Yang, unpubl. data). cDNA clones for two *Arabidopsis* rho isogenes, *Rop1At* and *Rop2At*, have been isolated and sequenced (H. Li and Z. Yang, unpubl. data). Sequence comparison shows that Rop1At is 95% identical to Rop1Ps and 88% identical to Rop2At at amino acid level. Phylogenetic analysis suggests plants have evolved a unique subfamily of rho proteins; this may reflect their distinct role in controlling the organization of the unique actin arrays in plants. Plant cells have a wide range of distinct actin cytoskeletal arrays that vary with cell types and stages of cell cycle. These include nucleus-surrounding filaments (positioning the nucleus), subcortical actin cables (controlling cytoplasmic streaming), cortical fine filaments (interacting with microtubules), preprophase band filaments (controlling the site of cell division), and phragmoplast filaments (directing division plane formation). The existence of the multiple rho gene family raises an interesting possibility that each rho may have a distinct role in mediating one or more of these specific actin arrays. Multiple rho isoforms are also found in both mammals and yeast (Aktories et al., 1989; Matsui and Toh-e, 1992; Sekine et al., 1989). In both systems, different members of the rho family have distinct cellular functions. In human cells, Rac1 mediates actin filament accumulation on the plasma membrane leading to membrane ruffling in fibroblasts while rhoA regulates the assembly of actin stress fibers and focal adhesions induced by growth factors (Ridley and Hall, 1992; Ridley et al., 1992). In the budding yeast, CDC42Sc is essential for the asymmetrical distribution of actin filaments and thus controls the establishment of cell polarity whereas Rho3 and Rho4 regulate directional bud growth (Johnson and Pringle, 1990; Matsui and To-he, 1992; Ziman et al., 1990). Interestingly, both human and *C. elegans* homologs of CDC42 are able to complement temperature-sensitive mutation in yeast, indicating that the function of this protein in controlling actin organization and cell polarity is conserved in evolutionarily divergent organisms (Chen et al., 1993; Shinjo et al., 1990). CDC42Sc and Rho1 are colocalized with asymmetrically distributed actin filaments in the bud site, suggesting that the rho proteins may directly interact with or in proximity with the actin binding proteins to regulate the actin organization. Analysis of colocalization of the plant rho proteins with the plant actin arrays may provide important clues as to their function in regulating the organization of specific actin arrays.

Another possible reason for the existence of multiple rho isogenes is that each is involved in a specific developmental process as a result of developmentally regulated rho gene expression. RNA blot hybridization analysis indicates that Rop1Ps is preferentially expressed in apical buds and root tips of pea seedlings (Z. Yang and Watson, 1993) and Western blot analysis using polyclonal antibodies against Rop1Ps shows that rho proteins are more abundant in apical buds and root tips of seedlings (Y. Lin and Z. Yang, unpubl. results). Our results suggest that Rop1ps may play a role in regulating cell division in pea. In flowers, rho is most abundant in anthers and pollinated styles, raising an interesting possibility that rho is required for pollen germination (Y. Lin and Z. Yang, unpubl. data). Using indirect immunofluorescence involving Rop1Ps antibody, we show that rho is preferentially localized to the apex of pollen tubes, suggesting that it is involved in the regulation of the actin-dependent tip growth of pollen tubes. Since the polyclonal antibody is likely to react with different rho isoforms, the role of a specific isoform in the accumulation of rho proteins observed by Western blots is not clear. In *Arabidopsis*, Rop1At and Rop2At are differentially expressed in different organs and stages of development as indicated by RNA blot hybridization analysis (H. Li and Z. Yang, unpubl. results). Rop1At appears to be specifically expressed in mature flowers and siliques while Rop2At mRNA is found in all organs but more abundant in young flowers than other organs. More detailed analysis of temporal and spatial expression patterns for various rho isogenes may provide evidence for distinct developmental function of rho isoforms. Transgenic expression of rho isoforms containing dominant mutations described above may be a powerful tool for understanding the role of rho proteins in signal transduction controlling plant development. Specific function of rho isoforms in developmental control may also depend on specific actin cytoskeletal structures and cellular process(es) mediated by the respective rho isoforms. In conjunction with analyses of rho differential expression and subcellular localization, thorough analyses of such transgenic plants at molecular, cytoskeletal, cellular, and organismal levels may provide important insight into the signal transduction mechanisms that coordinate the control of actin cytoskeletal organization with the spatial control of cell growth and division throughout plant development.

3. Membrane Association of GTP-binding Proteins via Isoprenyl Moieties

Membrane association is a prerequisite for signaling activity of several proteins such as GTP-binding proteins in the early events of signal transduction pathways. Thus, the mechanism controlling membrane association may play an important role in signal transduction. Membrane anchoring of both heterotrimeric and small GTP-binding proteins involves posttranslational modification through covalent attachment of one or more hydrophobic moieties such as isoprenyl, myristoyl and palmitoyl groups (Casey, 1994). The α subunit of G proteins is myristoylated while the γ subunit is isoprenylated. All the three major families of small GTP-binding proteins are also isoprenylated (Glomset and Farnsworth, 1994). Protein isoprenylation, first discovered in yeast and animals, is covalent attachment of an isoprenyl group (either farnesyl or geranylgeranyl) to the cysteine residue of the C-terminal consensus sequence (-CXXX) of specific proteins. Isoprenylation is not only required for membrane association of GTP-binding proteins by increasing their hydrophobicity but also for interaction with G protein-coupled receptor or Ras regulators (Kisselev et al., 1994; Orita et al., 1993). Three distinct enzymes catalyzing isoprenylation reactions have been identified in fungal and animal systems (see Table 1). In general, each enzyme modifies a specific group of proteins with unique C-terminal signature sequence CXXX recognized by respective transferases (see Table 1).

Recent studies demonstrate that protein isoprenylation also occurs in plants (Biermann et al., 1994; Randall et al., 1993; Z. Yang et al., 1993; Zhu et al., 1993). Using yeast Ras1 protein containing C-terminal CAIM or CAIL for an in vitro enzyme assay, it was shown that plant cells have both FTase and GGTase I (Randall et al., 1993). As mentioned above, the plant rho proteins contain C-terminal signature sequence CAAL for GGTase I and are specifically geranylgeranylated in an in vitro assay, suggesting that isoprenylation may also play a role in plant signal transduction involving the rho proteins. To understand the role of protein isoprenylation in plant signal transduction, we have cloned genes encoding protein farnesyltransferase subunits from plants. A cDNA encoding a pea homolog of the FTase β subunit was isolated using a RT-PCR strategy based on the conserved amino acid sequences found in yeast and mammalian FTase β subunits (Z. Yang et al., 1993). Using a similar PCR approach, we have cloned a cDNA encoding a pea homolog of the FTase α subunit (D. Qian et al., unpubl. results). Different FTases from yeast, mammals and plants have different substrate specificity (Randall et al., 1993). Although the pea homologs of FTase α and β subunits are more similar to mammalian FTase sequences than to the yeast sequences, the plant β subunit is quite unique in that it has an additional sequence of about 50 amino acid residues that is absent in both yeast and mammalian enzyme. The plant FTase appears to have higher specificity for substrates with CAIM than with CVLS (Randall et al., 1993). It would be interesting to determine if this unique sequence contributes to the substrate specificity of plant FTase or unique regulatory mechanisms that may be present in plant FTase. Plant FTase may be involved in the cell cycle control since our preliminary studies suggest that FTase mRNA is preferentially expressed in cells actively dividing both in tobacco BY-2 suspension cell culture and in pea seedlings (J. Katsuta et al., unpubl. results). Identification of FTase substrate proteins may facilitate our understanding of the physiological function of FTase. The only known substrates of plant FTase are homologs of a chaperonine DnaJ and a family of novel proteins whose biochemical and physiological functions are unknown (Biermann et al., 1994; Randall et al., 1993; Zhu et al., 1993). In animal and yeast systems, γ subunits of G proteins and the Ras small GTP-binding proteins are specifically farnesylated, however, these proteins have not been discovered in plants.

C. G Protein Effectors

There is biochemical evidence that the three major classes of G protein effectors (ion channels, phospholipases, and enzymes that metabolize cyclic nucleotides) are present in plants. Several different ion channels including Ca^{++}, K, and Na^{+} channels are known to be directly regulated by G proteins in animal cells. Using whole-cell patch clamp techniques, it was shown that the inward K^{+} current in guard cells and outward K^{+} in mesophyll cells are regulated by G proteins (Farley-Grenot and Assmann, 1991; Li and Assmann, 1993). However, it is not clear if these channels are direct effectors of G proteins or are indirectly regulated by G proteins through other second messengers. Evidence for G protein-regulated activity for other plant ion channels is lacking.

Phosphoinositide-specific phospholipase C (PLC) plays a key role in animal and fungal signal transduction pathways (Berridge, 1993; Rhee et al., 1989; Yoko-o et al., 1993). PLC hydrolyzes phosphoinositide into diacylglycerol (DAG), an activator of protein kinase C (PKC), and inositol trisphosphate (IP_3), a key second messenger that mobilizes intracellular Ca^{++} storage. Very little is known about the role of DAG in plant signal transduction although there is evidence that it may be involved in the regulation of mitotic progression (Larsen and Wolniak, 1990). Further, typical protein kinase C found in animals and yeast has not been identified in plants. However, IP_3 releases Ca^{++} from vacuoles of plant cells through Ca^{++} channels as shown by patch clamp studies and photolysis of caged IP_3 induces transient increases in Ca^{++} levels leading to inactivation of K^+ channel and stomatal closure in guard cells (reviewed in Coté and Crain, 1993). Although these studies using exogenous IP_3 indicate the involvement of IP_3 in plant signal transduction, its physiological role has not been established. Nonetheless, the existence of PLC activities has been shown in several plant species (reviewed in Coté and Crain, 1993). As in animal cells, both membrane-associated and soluble PLC activities are found in plant cells. Based on biochemical properties of membrane-associated PLC activities, it appears that there exist multiple forms of membrane-associated PLC as well. In mammals, there are four PLC subfamilies, PLC-α, PLC-β, PLC-γ, and PLC-δ (Berridge, 1993; Rhee et al., 1989). Only members of PLC-β subfamily are shown to be regulated by G proteins (Berridge, 1993; Thomas et al., 1991). Recently, it was found that PLC activity increases transiently prior to H_2O_2 production in response to polygalacturonic acid, an elicitor of the oxidative burst in cultured soybean cells (Legendre et al., 1993). This transient increase in PLC activity can also be promoted by the G protein agonist, mastoparan, which also induces the oxidative burst in soybean cells. These results suggest that a G protein regulated PLC may be involved in the signal transduction leading to the oxidative burst in plant cells. In spite of these biochemical studies indicating the existence of multiple forms of PLC including a G protein-regulated PLC, establishing their role in plant signal transduction awaits molecular identification of plant PLCs.

The most common G protein effectors are adenylate or granulate cyclases and cyclic nucleotide phosphodiesterases that mediate the level of intracellular cAMP or cGMP. The established role of cAMP in signal transduction is found in a diversity of organisms ranging from bacteria through fungi and slime molds to animals (Barritt, 1992). In *E. coli* and other bacteria, cAMP is involved in metabolic regulation. In yeast, level of cAMP, a key second messenger in the signal transduction pathways controlling growth and differentiation, is mediated by cAMP adenylate cyclase, the effector of the Ras small GTP-binding proteins. cAMP works both as extracellular signal and intracellular messenger in the regulation of slime mold development. In animals, cAMP is present in almost all cell types and functions as a key second messenger in a great number of cellular processes. Direct regulation of adenylate cyclase by G proteins is a principal mechanism for mediating the level of cAMP in animals. In some cases, cAMP phosphodiesterase can be direct effector of G proteins (Barritt, 1992).

There is considerable evidence for the existence of components of cAMP-mediated signaling system in plants although its role in plant signal transduction has not been unequivocally resolved. cAMP, adenylate cyclase, and cAMP phosphodiesterase have been detected in plants (Carriate et al., 1988; Lusini et al., 1991). A kinase that has similar properties to animal cAMP-dependent protein kinase was partially purified from petunia, suggesting cAMP may act as a second messenger in plant cells (Polya et al., 1991). The role of cAMP regulatory system in plant signal transduction is further supported by recent studies by Li and colleagues (1994a). Using whole cell patch clamp techniques, they showed that introduction of 1 to 5 mM of cAMP increases outward K^+ current in *Vicia faba* mesophyll cells. The outward current is reduced by specific inhibitors of cAMP-dependent protein kinase (PKA) but enhanced by the catalytic subunit of PKA from bovine brain. One should be cautious with these results since the effective concentrations of exogenously applied agents are much greater than physiological concentrations found in animal cells (Li et al., 1994a). Unfortunately, neither the physiological significance of cAMP level and activity of adenylate cyclase and cAMP phosphodiesterase nor the molecular identify of these enzymes has been demonstrated in plants. One difficulty in determining the physiological role of cAMP, as suggested by Trewavas and Gilroy (1992), could be that cAMP is confined to only specific cell types such as guard cells. Direct demonstration of cAMP-mediated signal transduction in plants may have to rely on molecular and genetic approaches. Cloning of adenylate cyclase and cAMP phosphodiesterase based on sequence conservation may prove to be fruitful, as shown for protein kinases discussed below. Genetic approaches in *Arabidopsis* and other plants might lead to identification and subsequent cloning of genes involved in cAMP regulation.

In contrast, much less effort has been devoted to understanding the role of cGMP in plant signal transduction. cGMP was detected in plants but no work on its role in signal transduction was conducted until recent elegant experiments by Chua's group (Bowler et al., 1994). By using microinjection techniques, they found that cGMP promotes the phytochrome-regulated anthocyanin biosynthesis in tomato cells (*aurea* mutant) deficient in phytochrome A while cGMP antagonists inhibit these processes in wild type cells. The effective cGMP concentrations were between 30 and 80 µM, similar to the physiological concentrations found in animal cells. More interestingly, their results also indicate that cGMP acts downstream of G proteins and that cGMP together with Ca^{++} induces a full spectrum of processes including chloroplast development and anthocyanin synthesis that they previously showed to be mediated by G protein alone (Neuhaus et al., 1993). By analogy to G protein-mediated cGMP signaling in animal photoperception systems, these results suggest that either guanylate cyclase or cGMP phosphodiesterase may be the effector of a phytochrome-activated G protein. In mammalian retina, cGMP phosphodiesterase is the effector of transducin (G protein). Transducin, upon activation by rhodopsin (photoreceptor protein), in turn activates cGMP phosphodiesterase leading to the reduction of cGMP levels (Barritt, 1992). In *Drosophila* photoreceptor cells, activation of transducin results in an increase in cGMP levels. The increase in cGMP levels as a result of G protein-mediated activation of guanylate cyclase is found in several other G protein-cGMP signaling systems (Barritt, 1992). Thus, the phytochrome-activated cGMP signaling might involve guanylate cyclase as G protein effector. To date, the molecular identities of guanylate cyclase and cGMP phosphodiesterase from plants remain obscure.

D. Calmodulin and Related Ca^{++}-binding Proteins

The role of Ca^{++} as a key second messenger in plant cells is well established. Ca^{++} signaling is implicated in a vast number of cellular and developmental processes in plants, including polarized cell growth, mitosis, cytoplasmic streaming, and processes mediated by light, growth regulators, mechanical stimuli and gravity force (Gehring et al., 1990; Hepler and Wayne, 1985; Knight et al., 1992; Trewavas and Gilroy, 1992). The ability of this universal molecule to regulate such a diversity of specific responses is probably, to a great extent, attributed to spatiotemporal distribution of specific Ca^{++} effectors. Ca^{++} can directly activate protein kinases or calmodulins that in turns activate protein kinases or other regulatory proteins. Calmodulin (CaM) is a ubiquitous Ca^{++}-binding protein that activates CaM/Ca^{++}-dependent protein kinases and phosphatases. Typical CaMs contain four EF hand Ca^{++}-binding domains whose sequences are highly conserved. Many CaMs and related proteins have been identified through molecular cloning from plants yet their role in specific signal transduction pathways is largely unknown (Braam and Davis, 1990; Jena et al., 1989; Ling and Zienlinski, 1989; Perera and Zienlinski, 1992; Toda et al., 1985). A CaM gene (TCH1) and two CaM-related genes (TCH2 and TCH3) were isolated from *Arabidopsis* based on the induction of their mRNAs by touch or other mechanical stimuli such as wind and rain (Braam and Davis, 1990). Five additional CaM genes and 6 more CaM-related genes were cloned from *Arabidopsis*, of which at least three were touch-inducible (reviewed in Sistrunk et al., 1994). Interestingly, it was shown that wind also induces a rapid increase in Ca^{++} levels in tobacco (Knight et al., 1992). These results suggest that Ca^{++}/CaM signaling has an important role in the regulation of plant responses to mechanical stresses. It is not clear why multiple CaM and related proteins are induced by these stresses and what function each of these inducible Ca^{++}-binding proteins has in stress responses. Ca^{++}/CaM has also been implicated in other signal transduction pathways including phytochrome-mediated signal transduction in tomato cells (Neuhaus et al., 1993) and regulation of K^+ activity in guard cells (Luan et al., 1993). However, it is not known which CaM isoforms are involved in these pathways, neither is the effector of plant CaMs and related proteins.

E. Second Messenger-dependent Protein Kinases

In animal cells, protein kinases are the most common targets of second messengers. These include cAMP-dependent protein kinase (PKA), cGMP-dependent protein kinase, Ca^{++}/DAG-regulated protein kinase (PKC) and Ca^{++}/CaM-dependent kinases. Although conclusive evidence for the existence of these kinases in plants is lacking, the presence of the second messengers and their implication in plant signal transduction as discussed above suggests these kinases may also be key signal transducing proteins in plants. This is supported by the discovery of a rising number of plant genes encoding protein kinases with structural similarity to the animal second messenger-dependent protein kinases.

While the classical second messenger-dependent protein kinases have not been clearly demonstrated, a novel family of second messenger-dependent protein kinases, Ca^{++}-dependent protein kinases, are shown to be ubiquitous in plants (D. Roberts, 1993; D. Roberts and Harmon, 1992).

1. Ca^{++}-dependent Protein Kinases

Ca^{++}-dependent protein kinases (CDPK), first discovered in plants, is different from Ca^{++}/CaM-dependent kinase and PKC in that its activity is independent of CaM or phospholipids (Harper et al., 1991; Putnam-Evans et al., 1990). CDPKs contain a calmodulin-like Ca^{++}-binding domain that is linked to the N-terminal kinase domain by a junction domain (Harper et al., 1991, 1994). It is shown that the junction domain acts as pseudosubstrate-type autoinhibitor that probably binds to the kinase active site of CDPK in the absence of the activator, Ca^{++} (Harper et al., 1994). Upon binding of Ca^{++} to the calmodulin-like domain, the pseudosubstrate could be released from kinase active site, leading to the activation of kinase. The analogous autoinhibitory regulatory mechanism is common in all the animal second messenger-dependent protein kinases and phosphatases (Kemp and Person, 1991; Newton, 1993). The similarity of the regulatory mechanism between this novel Ca^{++}-dependent protein kinase and the animal second messenger-dependent protein kinases suggests a mechanistic conservation for diverse second messenger-dependent protein kinases in plants and animals.

CDPKs are widespread in both lower and higher plants and have recently been found in protists (see references in Harper et al., 1994; D. Roberts and Harmon, 1992). Plants contain a large family of CDPK genes (Harper et al., 1994; D. Roberts and Harmon, 1992). Several CDPK genes have been cloned from various plant species including soybean, *Arabidopsis*, carrot, rice and corn (see references in Harper et al., 1994; D. Roberts and Harmon, 1992). In spite of their widespread distribution and well-characterized biochemical properties, their physiological function and role in signal transduction are still not well understood. Identification of CDPK substrates may provide clues to its function. To date, the only known CDPK substrate is nodulin 26, a transmembrane protein that is proposed to play a role in metabolite transport across the symbiosome membrane in legume root nodules (D. Roberts, 1993). Another putative substrate is the plasma membrane H^{+}-ATPase pump but its in vivo phosphorylation by CDPK has not been reported (D. Roberts, 1993; D. Roberts and Harmon, 1992). Perhaps, the most significant finding related to the function of plant CDPKs is its association and colocalization with actin cytoskeleton. Immunocytochemical studies show that CDPK is colocalized with actin microfilaments in *Tradescantia* pollen tubes and internodal cells of *Chara* (McCurdy and Harmon, 1992; Putnam-Evans et al., 1989). In both cells a Ca^{++}-dependent actomyosin is implicated in the directional movement of organelles or vesicles (D. Roberts and Harmon, 1992). In vitro phosphorylation of the regulatory light chain of smooth muscle myosin by plant CDPKs was observed (D. Roberts, 1993; D. Roberts and Harmon, 1992). Moreover, a recent study indicates that a maize pollen-specific CDPK is required for pollen germination and tube growth (Estruck et al., 1994). These studies together suggest that CDPK plays a role in the regulation of actinomyosin-mediated vesicle transport required for cell growth.

2. Protein Kinases with Similarity to Second Messenger-dependent Kinases

Using PCR or homologous probing strategies, several homologs of non-receptor Ser/Thr protein kinases have been cloned from different plant species (Biermann et al., 1990; Lawton et al., 1989; X. Lin and Watson, 1992; X. Lin et al., 1991). These kinase homologs show strongest sequence similarity to the animal second messenger-dependent protein kinase, especially cyclic nucleotide-dependent protein kinases and PKCs. For instance, a pea homolog, PsPK5, has 36% and 31% sequence identity to bovine CAPK and PCK, respectively. PsPK5 has higher sequence similarity to other plant homologs of similar kinases (46% identity to bean PVPK-1), suggesting that plants have evolved a family of protein kinases related to the second messenger dependent-protein kinase. In pea, a suite of five transcripts encoding these kinase homologs (PsPK1 to PsPK5) were identified by RT-PCR and expression of these transcripts is differentially regulated by light (X. Lin et al., 1993). During light-mediated greening of etiolated seedlings, PsPK3 and PsPK5 transcripts are down-regulated while PsPK2 and PsPK4 transcript are up-regulated. The accumulation of PsPK1 and transcripts remains constant. Of these light-regulated PsPK genes, PsPK5 is most interesting (R. Khanna and J. C. Watson, pers. comm.) since PsPK5 mRNA is more abundant in hooks than in apical buds but undetectable in other parts of dark-grown seedlings including stems, roots and root tips. The hook region is the most photoresponsive portion of an etiolated seedling and contains highest levels of phytochrome. Furthermore, PsPK5 mRNA decreases rapidly in response to continuous white light (detectable within 20 min) but does not respond to pulses of red or blue light. However, decreases in PsPK5 mRNA can be induced by continuous red or blue light, suggesting that PsPK5 gene regulation follows a high irradiance response regulated by phytochrome B. Although the function of these kinase homologs is not clear, the complex patterns of PsPK expression suggest that multiple kinase may be involved in the light-mediated developmental processes. Whether any of these pea putative kinases play a role in light signal transduction remains to be determined. Delineation of biochemical properties of these kinases such as the regulation of kinase activity may provide important functional information as to whether they are true second messenger-dependent protein kinases in plants.

F. MAP Kinase Cascade

In yeast and animal systems, a cascade of protein kinases called the MAP (mitogen-activated protein) kinase cascade plays a central role in signal transduction pathways (Crews and Erikson, 1993; Nishida and Gotoh, 1993). As shown in Fig. 1, this cascade involves three key protein kinases, MAP kinase kinase kinase (also called Raf or Mek kinase), MAP kinase kinase (or Mek), and MAP kinase. The MAP kinase cascade is responsible for controlling an array of diverse growth and developmental processes in various eukaryotic organisms, including growth factor-mediated processes in mammals, eye development, and differentiation of embryo terminal system in *Drosophila*, vulval development in *C. elegans*, and osmosensing, pheromone-mediated mating process and cell enlargement in yeast (Cooper, 1994; Nishida and Gotoh, 1993; Wilson, 1994). In animal systems, the MAP kinase cascade is modulated by receptor tyrosine kinase and the Ras small GTP-binding protein. Genetic studies show that this linear signaling pathway simply functions as a common switch to control processes mediated by specific ligands in different organisms or cells (Cooper, 1994; Nishida and Gotoh, 1993; Wilson, 1994). However, biochemical studies indicate that the MAP kinase cascade cross-talks with other signaling proteins including G proteins, protein kinase C, and protein kinase A (Wilson, 1994). In yeast, three distinct MAP kinase cascades (HOG1, FUS3/KSS1, and MPK1) respectively control the osmosensing, pheromone-mediated mating process and cell enlargement (Cooper et al., 1994). Each cascade is distinctly regulated: the HOG1 pathway by a prokaryote-type two-component regulator, FUS3/KSS1 by receptor-linked G protein, and MPK1 by protein kinase C. G protein- and protein kinase C-mediated MAP kinase cascade are also found in mammals (Nishida and Gotoh, 1993).

The striking conservation and versatility of the MAP kinase cascade suggest that this signaling mechanism also operates in plants. Indeed, all components of this kinase cascade have been discovered in plants (reviewed in Heberle-Bors and Hirt, 1994). Genes encoding MAP kinase have been cloned from alfalfa, *Arabidopsis*, pea, and tobacco. Plants contain a family of at least five different MAP kinases (Heberle-Bors and Hirt, 1994), analogous to the multiplicity of MAP kinase in yeast and animals. The function of these MAP kinases in plants is not clear. The transcript level of an alfalfa MAP kinase fluctuates in a cell cycle-dependent manner (Heberle-Bors and Hirt, 1994). Further, the activation of *Arabidopsis* MAP kinase is induced by auxin and is correlated with the auxin-induced activation of cell division in tobacco cell cultures (Mizoguchi et al., 1994). These results together suggest a plant MAP kinase cascade may be involved in the control of cell proliferation, similar to one of the roles of the mammalian MAP kinase cascade. MAPKK homologs have also been found in tobacco but their function is unknown. A tobacco MAPKKK, NPK1, is able to specifically complement a yeast mutant defective in BCK1, a MAPKKK of the MPK1 cascade controlling cell wall construction and proliferation (Banno et al., 1993). This suggests that the mechanism regulating the activity of NPK1 and BCK1 might be conserved. As discussed below, a MAPKKK homolog CTR1 has been shown to be negative regulator of ethylene responses (Kieber et al., 1993). Therefore, evidence is accumulating that the MAP kinase cascade is also a key signal transduction mechanism in plants.

Identification of an *Arabidopsis* gene, *CTR1*, encoding a MAPKKK in the ethylene signal transduction pathway indicates that the MAP kinase cascade may be central to this pathway (Kieber et al., 1993). CTR1 has a C-terminal kinase domain that shows strong sequence similarity to a mammalian Raf MAPKKK, Raf. The N-terminal region of CTR1 does not have the typical motif found in the N-terminal regulatory domain of Raf, although sequence similarity also extends to the N-terminal regulatory domain. The Raf N-terminal domain functions as negative regulator; it binds to activated Ras leading to activation of Raf. The component immediately upstream of CTR1 is unknown. Recent studies with animal Raf indicate that the biochemical significance of Ras–Raf interaction is to recruit Raf to the intracellular side of the plasma membrane since a modified Raf that is able to associate with the plasma membrane through isoprenylation has constitutive Raf activity independent of Ras (Leaves et al., 1994; Stokoe et al., 1994). Although the universality of such Raf regulatory mechanism has yet to be shown, these studies suggest that direct regulators of Raf can be diverse, consistent with the fact that several distinct regulators mediate the MAP kinase cascade both in fungi and animals. There is no evidence that G protein is involved in the ethylene signal transduction pathway, and Ras and protein kinase C have not yet been identified in plants. Chang et al. (1993) suggest that ETR1 (a putative two-component regulator, see above) may be the direct regulator of CTR1 since genetic studies place ETR1 upstream of CTR1 and ETR1 has a domain that shows topological similarity to Ras. In fact, a MAP kinase cascade regulated by a two-component regulatory system has recently been found for yeast osmosensing (Maeda et al., 1994). In the osmosensing system, low osmotic conditions activate phosphorylation of histidine kinase within the osmotic sensor SNL1, resulting in phosphorylation of aspartate residue of the regulator domain of SNL1 and a second separate regulator SSK1. The phosphorylation of SSK1 causes inhibition of the osmosensing MAP kinase, HOG1, leading to reduced glycerol synthesis. It is possible that a MAP kinase cascade involving CTR1 is similarly regulated by ETR1 either directly or indirectly through a separate regulator like SSK1. A model for ETR1/CTR1-mediated ethylene signal transduction pathway has recently been proposed (Bowler and Chua, 1994). An alternative model, as illustrated in Fig. 3, is the one analogous to the yeast osmosensing system. This model may explain how the negative regulator CTR1 is involved in signal transduction controlling the inhibition of cell expansion and elongation mediated by ethylene. According to this model, MAPKKK activity of CTR1 is negatively regulated by ethylene as a result of phosphorylation induced by ETR1. This is consistent with the fact that CTR1 works as a negative regulator of ethylene responses. This model also predicts that the activation of CTR1 induces cell expansion or elongation which is dependent on the remodeling of cell wall, that appears to be suppressed by ethylene (Kieber et al., 1993). Interestingly, the yeast MPK1 kinase cascade is known to regulate cell wall remodeling and cell enlargement (Cooper et al., 1994). The CTR1 kinase cascade might be a central regulatory mechanism controlling cell growth modulated by various signals in plants. For instance, a stimulatory growth hormone such as auxin might activate the kinase cascade resulting in increased cell expansion or elongation. Auxin is shown to activate an *Arabidopsis* MAP

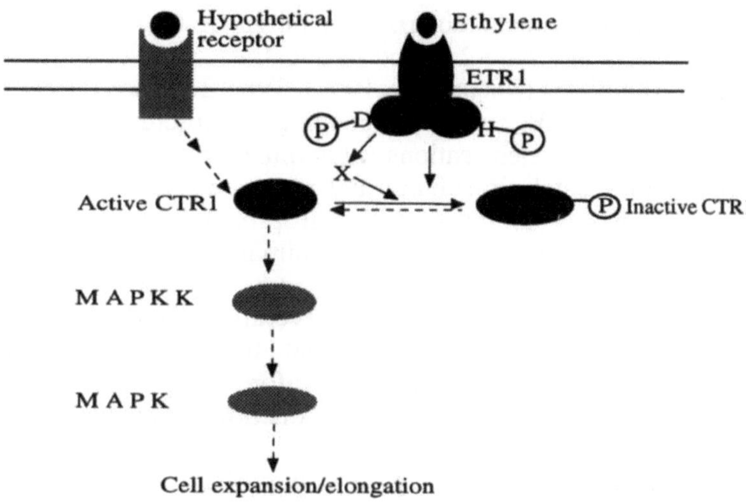

Fig. 3. A possible model for ethylene signal transduction pathway regulating the inhibition of cell growth and elongation. Ethylene is perceived by a prokaryote-type two-component system ETR1; binding of ethylene to sensor domain of ETR1 causes phosphorylation of histidine residue (H) of the cytoplasmic histidine kinase domain. The activation of histidine kinase domain results in the phosphorylation of a aspartate residue (D) on the regulator domain, which either directly or indirectly through an unknown regulator (X) inhibits CTR1 activity possibly through phosphorylation. This inhibition turns off the CTR1-mediated MAP kinase cascade required for cell expansion and elongation. Other signals such as auxin stimulate cell expansion and elongation by positively regulating the MAP kinase cascade. Solid symbols and arrows indicate steps that have been identified. Dotted symbols and arrows indicate hypothetical pathway

kinase (Mizoguchi et al., 1994), but whether it would be the one invovled in the CTR1-mediated cascade has yet to be determined. In addition to the Ras-mediated activation, the mammalian MAPKKK is also positively regulated by PKC or negatively by PKA, suggesting that phosphorylation of MAPKKK plays an important role in coordinating cross-talks of the Ras-mediated pathway with other pathways. The phosphorylation state of the MAPKKK (CTR1) might have a similar role in the networking of various signaling pathways in the regulation of cell expansion or elongation. Moreover, CTR1, when inactive as a regulator of the kinase cascade, might have function(s) in the regulation of other ethylene responses unrelated to cell growth. Although highly speculative, this alternative model shown in Fig. 3 provides a possible mechanism for opposing effects of certain plant growth regulators on cell growth.

G. Protein Phosphatases

Since most signal transduction pathways involve activation of protein kinases (either receptor kinases or second messenger-dependent kinases), protein phosphatases have been traditionally thought to play a key role in down-regulation or desensitization of signal transduction. A large number of protein phosphatases (PPs) have been characterized in animal cells. They are classified into two groups, type-1 and type-2, based on substrate specificity and sensitivity to endogenous inhibitors (Shenolikar, 1994). Type-2 PPs are further divided into three subgroups, 2A, 2B, and 2C, depending on structure, substrate spectrum, and regulation by divalent cations (Shenolikar, 1994). All four types of PPs are composed of a catalytic subunit and regulatory subunit and are conserved in all eukaryotes investigated. Although protein kinases are important negative regulators in signal transduction in animal cells, recent studies have revealed a number of phosphatases that are key positive regulators (Shenolikar, 1994). Further, the activity of several PPs is also dependent on second messengers. For instance, PP2B type phosphatase (calcineurin) is specifically activated by Ca^{++}/calmodulin. Therefore, PPs provide an important revenue for cross-talks between signal transduction pathways (Shenolikar, 1994).

Protein phosphatases are also emerging as important components of signal transduction pathways in plants. All four types of PPs have been identified and genes encoding PP-1, PP2A, and PP2C cloned from several plant species (Ferreira et al., 1993; Luan et al., 1993; MacKintosh et al., 1990, 1991; Smith and Walker, 1991, 1993). *Arabidopsis* contains a family of five genes encoding the catalytic subunit of PP-1; at least one of these, like yeast PP-1, appears to be involved in the cell cycle control (Ferreita et al., 1993; Smith and Walker, 1993). Studies using PP inhibitors suggest that several different PPs including PP-1, PP2A and PP2B are involved in the regulation of K^+ influx in guard cell (Li et al., 1994b; Luan et al., 1993). Cyclophilin-cyclosporin A, specific inhibitor of calcineurin, blocks Ca^{++}-induced inactivation of K^+ influx channels (Luan et al., 1993), suggesting that a calcineurin-like, Ca^{++}-dependent PP plays a role in the regulation of Ca^{++}-mediated K^+ influx and stomatal closure in guard cells. Stomatal closure is triggered by elevated ABA levels but ABA also regulates many other aspects of developmental processes and stress responses. ABA signal transducing mutants that have pleotropic effects on the ABA-controlled processes have been identified and one gene corresponding to such a mutant, ABI1, encodes a PP2C-like protein (Leung et al., 1994; Meyer et al., 1994). Interestingly, ABI1 has some novel structural features that are not present in typical PP2C protein. Although the C-terminal part shows strong sequence similarity (35% identity, 55% similarity) to the catalytic subunit of the yeast and rat PP2C, the N-terminal portion contains an EF hand motif-Ca^{++}-binding site. The unique structural features suggest that the N-terminal domain of ABI1 may have a regulatory role possibly involving Ca^{++}. A similar PP (*rgdC*) with a Ca^{++}-binding domain fused to catalytic domain is found in *Drosophila* and plays a role in phototransduction (Steele et al., 1992). Whether ABI1 corresponds to the calcineurin-like activity found in *Vicia* guard cells has not been investigated.

The fact that ABI1 has pleotropic effect indicates that ABI1 takes part in an early step in the ABA signal transduction pathway. Evidence for the role of PPs in the early steps of signal transduction is also found in animal cells (Shenolikar, 1994). Recently, another PP2C-like protein, KAPP, was identified in *Arabidopsis* by cloning of cDNAs encoding protein that interacts with an *Arabidopsis* receptor-like protein kinase, RLK5 (Stone et al., 1994). Similarly, KAPP contains a PP2C catalytic domain fused to an N-terminal RLK5-binding domain (KI domain) that has no apparent homology to known proteins. The binding of KI domain to RLK5 depends on the autophosphorylation of RLK5. This is analogous to the animal protein tyrosine phosphatases with SH2 domains that bind to phosphotyrosine of tyrosine kinase receptor. Such tyrosine phosphatases are implicated in signaling from the tyrosine kinase receptor, suggesting that KAPP might also play an important role in an early step of signaling from RLK5 (Stone et al., 1994). Elucidation of the physiological function of RLK5 may facilitate the understanding of the role of KAPP in signal transduction, and vice versa.

H. Phosphatidylinositol Kinases

Phosphatidylinositol kinases (PIKs) phosphorylate hydroxyl groups on the inositol ring of phosphotidylinositides. Several PIKs phosphorylating specific positions on the inositol ring have been characterized, including phosphatidylinositol-3-kinase (PI3K), PI4K, and PI5K. Among these, the role of PI3K in signal transduction is best understood. PI3K, composed of a catalytic and a regulatory subunit, is involved in signal transduction pathways mediated by several extracellular signals such as insulin, IGF, PDGF and EGF, and lysophosphatidic acid in mammalian cells (see references in Hong and Verma, 1994). In yeast, PI3K is involved in the regulation of protein targeting to vacuoles (see references in Hong and Verma, 1994). The specific function of PI3K is reflected in its association with various cellular components. PI3K is known to directly interact with tyrosine kinase receptors through SH2 domain in the regulatory subunit (Kotani et al., 1994). PI3K is also associated with the actin cytoskeleton (Payastre et al., 1991) and appears to be a downstream component of rho in the signal transduction pathway controlling actin cytoskeletal organization (Kumagai et al., 1993; Zhang et al., 1993). There is evidence that PI5K is also downstream of rho in the signaling pathway controlling PDGF and thrombin-mediated cell adhesion (Chong et al., 1994).

Both PI3K and PI4K have been identified in plants (Hong and Verma, 1994; Xu et al., 1992; W. Yang et al., 1993). PI4K is associated with cytoskeleton in plant cells (Xu et al., 1992) and a PI4K activator is also found to directly interact with polymerized actin (W. Yang et al., 1993), suggesting that PI4K may play an important role in the signal transduction modulating the cytoskeletal structure in plant cells. A gene encoding the catalytic subunit of PI3K was recently cloned from soybean (Hong and Verma, 1994). The soybean PI3K is encoded by a small gene family and the expression of a distinct isoform of PI3K is activated during membrane proliferation in young nodules, suggesting that this PI3K plays a role in the development of the peribacteroid membrane (Hong and Verma, 1994). PI3K activity is predominantly present in membrane fraction but it is not known if it is also associated with cytoskeleton. The fact that the cytoskeleton dynamically interacts with membrane systems through phosphoinositide association and association of PI kinases with membrane and cytoskeleton suggests that PI kinases may provide an important link to signal transduction pathways controlling cytoskeleton.

IV. Novel Plant Signaling Mechanisms

A. *A Lesson from Light-mediated Signal Transduction*

Plant growth and development fundamentally differ from animals in many aspects. One distinct feature of plant development is plasticity, typified by its ability to regenerate from a somatic cell of most parts of the plant and to alter developmental course in response to environment cues. Among various environmental factors, light has the most dramatic effect on plant development. Several photoreceptor systems including phytochrome and blue light receptors are known to modulate various developmental processes. Phytochromes appear to be unique to plants since similar signaling molecules have not been discovered in other organisms and the reversibility of red and far-red absorbing forms of phytochrome is unparalleled in any other photoreceptor systems. In spite of the distinct properties of the photoreceptor in plants, current evidence indicates that early steps of signal transduction pathways mediated by phytochrome bear striking similarity to the signaling paradigm found in animals. Nevertheless, genetic studies in phytochrome signal transduction have also led to identification of novel proteins (Det1 and Cop9, Cop11) regulating phytochrome-mediated morphogenesis (Deng, 1994).

A series of genes (*cop* and *det* genes) involved in photomorphogenesis were identified by screening for mutants that show constitutive photomorphogenesis (COP) or de-etiolated (DET) phenotypes (Chory, 1993; Deng, 1994). Several of these genes, including *cop1*, *cop9*, *cop11* and *det1* genes, have been cloned (Castle and Meinke, 1994; Deng et al., 1992; Pepper et al., 1994; Wei et al., 1994). All except *cop1* encode novel proteins that show no significant sequence similarity to any known proteins. Cop1 is a novel fusion of a Zn^{++}-binding domain and a WD-40 repeat domain known to be involved in protein–protein interaction (Deng et al., 1992). The novelty of these signaling proteins suggests that plants may have evolved unique regulatory mechanisms for the control of plastic developmental programs. However, it remains to be seen whether or not similar proteins are present in non-plant systems. The biochemical mechanism for the action of these proteins in signal transduction is not clear. Det1 is a nuclear-localized protein but does not directly bind DNA, suggesting that it may interact with transcriptional factors in regulating gene expression (Pepper et al., 1994). Cop9 is a component of a large functional complex whose formation is probably mediated by light (Wei et al., 1994). Interestingly, the formation of this complex also requires Cop8 and Cop11. Such a signaling complex may provide a mechanism for the cross-talk between light signal and other signal transduction pathways. Cop1 contains at least three recognizable structural domains: an N-terminal Zn^{++}-binding motif followed by a putative coiled-coil region and multiple WD-40 repeats homologous to the β subunit of the G protein. Genetic analyses indicate all three domains are important for the function of Cop1 (McNeillis et al., 1994). Interestingly, a recent study indicates that Cop1 nucleocytoplasmic partitioning is regulated by light (von Arnim and Deng, 1994). In both hypocotyl cells of *Arabidopsis* seedlings and epidermal cells of onion bulbs, GUS-Cop1 fusion protein is only present in the cytoplasm in the light but enriched in the nucleus in the dark. The level of the nuclear GUS-Cop1 is correlated with the extent of repression of photomorphogenic development. In root cells, Cop1 is constitutively nuclear, independent of light. These results indicate that Cop1 acts in the nucleus, possibly as a transcriptional repressor, to suppress photomorphogenesis in the dark and in roots.

The tissue-specific regulation of Cop1 nucleocytoplasmic partitioning is consistent with the hypothesis that light-mediated signal transduction integrates with signaling pathways mediated by internal developmental signals. Further support for this hypothesis comes from the analysis of other *cop* and *det* mutants. A series of *fusca* genes involved in embryo development have been identified by screening for mutants that have excess anthocyanin production in embryo. These mutants also exhibit many other phenotypes including

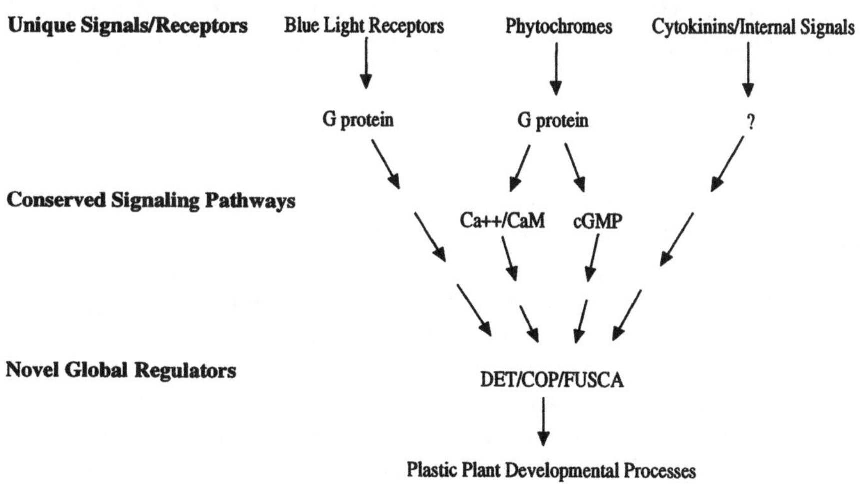

Fig. 4. A model for the integration of signaling pathways mediated by different light receptors and internal signals. Plants have unique extracellular signals or receptors that modulate conserved signaling pathways involving G proteins, second messengers and second messenger-dependent protein kinases. The conserved signaling mechanism regulates a battery of novel global regulators that integrate information from different upstream pathways to control developmental processes. Such an integrative mechanism allows plant development to be plastic. The global regulators may either form a complex or a cascade in which at least one regulator receives signals from upstream pathways and probably more than one regulators act as transcriptional regulators to control expression of genes involved in developmental processes mediated by light and cytokinin

constitutive photomorphogenesis, limited cell expansion, ectopic expression of trichomes and chloroplasts, and adult lethality (Castle and Meinke, 1994; Misera et al., 1994). In fact, many of the *fusca* genes are allelic to the *cop* and *det* genes (Castle and Meinke, 1994; Misera et al., 1994; Pepper et al., 1994; Wei et al., 1994). For instance, *det1, cop1, cop9*, and *cop11* are allelic to *fus2, fus1, fus7,* and *fus6,* respectively. Epistasis analysis indicates that these gene products are common downstream components of different signal transduction pathways mediated by various photoreceptors including phytochromes and blue light receptors (Castle and Meinke, 1994; Misera et al., 1994; Pepper et al., 1994; Wei et al., 1994). In addition, it is shown that *det* mutants have enhanced sensitivity to cytokinin and that the *det1* phenotype can be mimicked by addition of cytokinin to wild type seedlings, suggesting that cytokinin may be an internal signal regulating *det* activity (Chory et al., 1994). On the basis of these observations, it has been suggested that the *cop/det/fusca* gene products define a global regulatory mechanism through which various signal transduction pathways mediated by internal and environmental signals converge in the regulation of development (Castle and Meinke, 1994; Misera et al., 1994; Wei et al., 1994). In the case of light signal transduction, upstream of this novel global regulatory system are likely to be the conserved signaling molecules such as G proteins, cGMP, and Ca^{++}/CaM (Bowler and Chua, 1994). Although how the conserved signaling proteins interact with the global regulators remains to be determined, such integrative signaling system might be a key molecular mechanism controlling developmental plasticity in plants (see Fig. 4).

B. Extracellular Matrix in Signal Transduction

Another significant difference in plant development from animals is the lack of cell movement and the critical role of cell position rather than lineage in the modulation of development. This underscores the importance of the mechanism controlling the direction of cell growth and division throughout plant development. Apparently, the regulation of cytoskeletal structures is a key to the spatial control of cell growth and division. The conserved elements such as the rho GTP-binding proteins and microtubule- and actin-binding proteins are expected to be crucial components of the signal transduction pathways controlling the cytoskeletal organization in plant cells. However, the mechanism that defines the positional information is likely to be unique to plants. Evidence is emerging that the plant extracellular matrix may play a pivotal role in conveying such spatial information (K. Roberts, 1994). In particular, extracellular arabinogalactan-proteins (AGPs) may be critical regulators of cell growth and division, especially the directional growth and division, as indicated by several recent studies. First of all, secreted AGPs from embryonic carrot cell line can convert a non-embryonic to an embryonic line (Krenger and van Holst, 1993). Further, using antibody against specific epitopes of AGPs, it was shown that these epitopes are expressed in specific spatial patterns in the root and that these AGPs are the markers for cell positions rather than cell types (Knox et al., 1989, 1991). Finally, preliminary studies using AGP ligands (Yariv phenylglycosides) indicate that AGP is involved in controlling the direction of cell expansion in *Arabidopsis* root epidermal cells (J. Zhu, pers. comm.). The mechanism of action for AGP is totally unknown. In animals, extracellular matrix plays an important role in cell signaling through a family of transmembrane proteins called integrins that provide linkage between ECM, actin cytoskeleton, and signal transduction pathways mediated by other extracellular signals. There is evidence for the existence of integrins in plant cells (see review in Shibaoka and Nagai, 1994), suggesting that integrins may also be involved in the signaling of plant ECM such as AGPs. Recent cloning of genes encoding AGP protein backbone (Du et al., 1994) may provide a useful tool for understanding the role of AGPs in plant signaling and biochemical mechanism of the ECM-mediated signaling processes.

V. Conclusions and Future Challenges

Plant signal transduction mechanisms bear striking similarity to those in animal and fungal systems. A battery of conserved signaling molecules are found in plants, including receptor-like protein kinases, heterotrimeric G proteins, small GTP-binding proteins, phospholipase C, cyclic nucleotides, Ca^{++}/calmodulin, and various signal transducing protein kinases and phosphatases. The role of some of these molecules in plant signal transduction has been demonstrated by biochemical or genetic studies. However, several key conserved signaling proteins are still obscure in plants. Further, plant signal transduction pathways involving the conserved signaling proteins have yet to be elucidated.

Conserved signaling proteins not yet identified in plants include G protein-coupled transmembrane receptors, receptor tyrosine kinase, the Ras family of small GTP-binding proteins and their regulators (GTP/GDP exchangers and GTPase activating proteins), adenylate and guanylate cyclases and cAMP and cGMP phosphodiesterases, and typical PCK, PKA and Ca^{++}/CaM-dependent protein kinases. Cloning of genes involved in signal transduction using molecular genetic approaches in *Arabidopsis* is expected to expand the list of the conserved signaling proteins in plants. Molecular techniques such as homologous probing and PCR and biochemical approaches such as the use of specific antagonists and agonists will continue to play an important role in discovering additional conserved regulatory proteins from plants.

One major challenge has been and will remain to be the elucidation of the function of conserved regulatory proteins identified in plants. The biochemical activity for most of the signaling proteins discovered by genetic approaches has not been studied. Furthermore, numerous genes encoding the conserved regulatory proteins have been isolated using molecular approaches, but few have known function in plant signal transduction. Although antisense RNA technology has been used to study the function of non-regulatory proteins in plants, it has not been particularly useful for signaling proteins probably because of their instability and low abundance. Nonetheless, several other genetic approaches are potentially more powerful. These include transgenic expression of genes containing dominant positive or inhibitory mutations, identification of *Arabidopsis* lines tagged with T-DNA inserted into genes of interests and gene knock-out by homologous recombination.

Delineation of plant signal transduction pathways involving conserved signaling proteins will not only provide insight into the evolution of signal transduction mechanisms but the molecular and biochemical mechanisms of plant growth and development. Since many signaling proteins physically interact with upstream or downstream components, genes encoding new components may be cloned using recently developed strategies based on in vivo or in vitro protein interaction (e.g., using the yeast two hybrid system). Such strategies are also useful for identifying factors involved in cross-talks between signaling pathways. Genetic studies such as screening for novel mutants in signal transduction and epistasis analysis of signal transduction mutants will continue to be a pivotal tool in furthering our understanding of plant signal transduction. These molecular and genetic approaches together with biochemical analyses of catalytic or regulatory features of signaling proteins will be necessary for deciphering a clear picture of signal transduction pathways. In addition to the molecular and genetic strategies, microinjection techniques, widely used to study signal transduction in animal systems, has also proven to be useful in plant systems (Bowler et al., 1994; Neuhaus et al., 1993). These techniques are especially powerful when used in junction with a well defined genetic system such as plant mutants that are deficient in specific signaling proteins.

Perhaps the most interesting and challenging question is how and why the conserved signaling proteins or pathways are involved in the modulation of the unique developmental processes in plants. Recent studies of light signal transduction have provided some hints for the answer to this question. These studies are consistent with the model that the conserved signaling molecules (e.g., G proteins, cGMP and Ca^{++}) function as intermediate steps of the plant signal pathways to integrate unique early and downstream signaling components in controlling specific plant processes. The fact that the novel signal proteins (Cop/Det/Fusca) are required for transduction of intrinsic signal(s) besides light suggests these novel proteins may provide key mechanisms for networking between signaling pathways. Such networking may be fundamental for the regulation of plasticity in plant development. Future research directed at understanding how the conserved pathways interface with the unique signaling mechanism will provide important insight into the mechanism controlling developmental plasticity in plants.

Acknowledgements

The work in author's laboratory is supported by grants from U.S. Department of Agriculture (NRICGP # 94-37304-1111) and National Science Foundation (MCB-9496224).

VI. References

Aktories K, Braun U, Rosener S, Just I, Hall A (1989) The rho gene product expressed in *E. coli* is substrate of botulinum ADP-ribosyltransferase C3. Biochem Biophys Res Commun 158: 209–213

Balwin JM (1994) Structure and function of receptors coupled to G proteins. Curr Opin Cell Biol 6: 180–190

Banno H, Hirano K, Nakamura T, Irie K, Nomoto S, Matsumoto K, Machida Y (1993) NPK1, a tobacco gene that encodes a protein with a domain homologous to yeast BCK1, STE11, byr2 protein kinases. Mol Cell Biol 13: 4745–4752

Barritt GJ (1992) Communication within animal cells. Oxford University Press, Oxford

Bednarek SY, Reynolds TL, Schroeder M, Grabowski R, Hengst L, Gallwitz D, Raikhel NV (1994) A small GTP-binding protein from *Arabidopsis thaliana* functionally complements the yeast *YPT6* null mutants. Plant Physiol 104: 591–596.

Berridge MJ (1993) Inositol triphosphate and calcium signalling. Nature 361: 315–325

Biermann B, Johnson EM, Feldman LJ (1990) Characterization and distribution of a maize cDNA encoding a peptide similar to the catalytic region of second messenger dependent protein kinases. Plant Physiol 94: 1609–1615

Biermann BJ, Morehead TA, Tate SE, Price JR, Randall SK, Crowell DN (1994) Novel isoprenylated proteins identified by an expression library screen. J Biol Chem 269: 25251–25254.

Bollag G, McCormick F (1991) Regulators and effectors of *ras* proteins. Annu Rev Cell Biol 7: 601–632

Bowler C, Chua N-H (1994) Emerging themes of plant signal transduction. Plant Cell 6: 1529–1541

Bowler C, Neuhaus G, Yamagata H, Chua N-H (1994) Cyclic GMP and calcium mediate phytochrome phototransduction. Cell 77: 73–81

Braam J, Davis RW (1990) Rain-, wind-, and touch-induced expression of calmodulin and calmodulin-related genes in *Arabidopsis*. Cell 60: 357–364

Carricarte VC, Bianchini GM, Muschietti JP, Tellez-Inon MT, Perticari A, Torres N, Flawia MM (1988) Adenylate cyclase activity in a higher plant, alfalfa (*Medicago sativa*). Biochem J 249: 807–811

Casey PJ (1994) Lipid modifications of G proteins. Curr Opin Cell Biol 6: 219–225

Castle LA, Meinke DW (1994) A *FUSCA* gene of *Arabidopsis* encodes a novel protein essential for plant development. Plant Cell 6: 25–41

Chang C, Schaller GE, Patterson SE, Kwok SF, Meyerowitz EM, Bleecker AB (1992) The TMK1 gene from *Arabidopsis* codes for protein with structural and biochemical characteristics of a receptor protein kinase. Plant Cell 4: 1263–1271

Chang C, Kwok SF, Bleecker AB, Meyerozitz EM (1993) *Arabidopsis* ethylene-response gene *ETR1*: similarity of product to two-component regulators. Science 262: 539–544

Chardin P (1993) *ras* homologs: a comparison of primary structures. In: Lacal JC, McCormick F (eds) The Ras superfamily of GTPases. CRC Press, Boca Raton, pp 203–229

Chen W, Lim HH, Lim L (1993) The *CDC42* homologue from *Caernorhabditis elegans*, complementation of yeast mutation. J Biol Chem 268: 13280–13285

Cheon C-I, Lee N-G, Siddique A-BM, Verma DPS (1993) Roles of plant homologs of rab1p and rab7p in the biogenesis of the peribacteroid membrane, a subcellular compartment formed de novo during root nodule symbiosis. EMBO J 12: 4125–4135

Chong LD, Trynor-Kaplan A, Bokoch GM, Schwartz MA (1994) The small GTP-binding protein rho regulates a phosphatidylinositol 4-phosphate 5-kinase in mammalian cells. Cell 79: 507–513

Chory J (1993) Out of darkness: mutants reveal pathways controlling light-regulated development. Trends Genet 9: 167–172

Chory J, Reinecke D, Sim S, Washburn T, Brenner M (1994) A role of cytokinins in de-etiolation of *Arabidopsis det* mutants have an altered response to cytokinins. Plant Physiol 104: 339–347

Cooper JA (1994) Straight and narrow or tortuous and intersecting? Curr Biol 4: 1118–1121

Coté GG, Crain RC (1993) Biochemistry of phosphoinositides. Annu Rev Plant Physiol Plant Mol Biol 44: 333–356

Crews CM, Erikson RL (1993) Extracellular signals and reversible protein phosphorylation: what to Mek of it all. Cell 74: 215–217

Deng X-W (1994) Fresh view on light signal transduction in plants. Cell 76: 423–426

Deng X-W, Matsui M, Wei N, Wagner D, Chu A-M, Feldmann KA, Qauil PH (1992) *COP1*, an *Arabidopsis* regulatory gene, encodes a protein with both a zinc-binding motif and a Gβ homologous domain. Cell 71: 791–801.

Downward J (1992) Regulatory mechanisms for Ras proteins. BioEssays 14: 177–184

Du H, Simpson RJ, Moritz RL, Clarke AE, Bacic A (1994) Isolation of the protein backbone of an arabinogalactan-protein from the styles of *Nicotiana alata* and characterization of a corresponding cDNA. Plant Cell 6: 1643–1653

Dwyer KG, Kandasamy MKM, Acciai SIJ, Kudish B, Miller JE, Nasrallah ME, Nasrallah JB (1994) The superfamily of S locus-related sequences in *Arabidopsis*: diverse structures and expression patterns. Plant Cell 6: 1829–1843

Esch RK, Weeks G, Firtel RA (1993) Regulation and function of two developmentally regulated *ras* genes in *Dictyostelium*. In: Lacal JC, McCormick F (eds) The Ras superfamily of GTPases. CRC Press, Boca Raton, pp 173–186

Estelle M (1992) The plant hormone auxin: insight in sight. BioAssays 14: 439–444

Estruch J, Kadwell S, Merlin E, Crossland L (1994) Cloning and characterization of a maize pollen-specific calcium-dependent calmodulin-independent protein kinase. Proc Natl Acad Sci USA 91: 8837–8841

Evans RM (1988) The steroid and thyroid hormone receptor superfamily. Science 240: 889–895

Farley-Grenot K, Assmann SM (1991) Evidence for G-protein regulation of inward K^+ channel current in guard cells of fava bean. Plant Cell 3: 1037–1044

Ferreira PCG, Hemerly AS, Montagu MV, Inze D (1993) A protein phosphatase 1 from *Arabidopsis thaliana* restores temperature sensitivity of a *Schizosaccharomyces pombe* $cdc25^{ts}/wee1^{-}$ double mutant. Plant J 4: 81–87

Gehring CA, Williams DA, Cody SH, Parish RW (1990) Phototropism and geotropism in maize coleotiles are spatially correlated with increases in cytosolic free calcium. Nature 345: 528–530

Glomset JA, Farnsworth CC (1994) Role of protein modification reactions in programming interactions between ras-related GTPases and cell membranes. Annu Rev Cell Biol 10: 181–205

Goring DR, Rothstein SJ (1992) The S-locus receptor kinase gene in a self-incompatible *Brassica napus* line encodes a functional serine/threonine kinase. Plant Cell 4: 1273–1281

Hall A (1990) The cellular function of small GTP-binding proteins. Science 249: 635–640.

Hall A (1994) Small GTP-binding proteins and the regulation of the actin cytoskeleton. Annu Rev Cell Biol 10: 31–54

Harper JF, Sussman MR, Schaller GE, Putnam-Evans C, Charboneau H, Harmon AC (1991) A calcium-dependent protein kinase with a regulatory domain similar to calmodulin. Science 252: 951–954

Harper JF, Huang J-F, Lloyd SJ (1994) Genetic identification of an autoinhibitor in CDPK, a protein kinase with a calmodulin-like domain. Biochemistry 33: 7267–7277

Heberle-Bors CJC, Hirt H (1994) MAP kinases: universal multipurpose signaling tools. Plant Mol Biol 24: 407–416

Hepler JR, Gilman AG (1992) G proteins. Trends Biochem Sci 17: 383–387

Hepler PK, Wayne RO (1985) Calcium and plant development. Annu Rev Plant Physiol Plant Mol Biol 36: 397–439

Hong Z, Verma DPS (1994) Phosphatidylinositol 3-kinase is induced during soybean nodule organogenesis and is associated with membrane proliferation. Proc Natl Acad Sci USA 91: 9617–9621

Horn MA, Walker JC (1994) Biochemical properties of the autophosphorylation of RLK5, a receptor-like protein kinase from *Arabidopsis thaliana*. Biochim Biophys Acta 1208: 65–74

Jena PK, Reddy ASN, Poovaiah BW (1989) Molecular cloning and sequencing of a cDNA for plant calmodulin: signal-induced changes in the expression of calmodulin. Proc Natl Acad Sci USA 86: 3644–3648

Johnson DI, Pringle JR (1990) Molecular characterization of *CDC42*, a *Saccharomyces cerevisiae* gene involved in the development of cell polarity. J Cell Biol 111: 143–152

Kahn RA, Der CJ, Bokoch GM (1992) The ras superfamily of GTP-binding proteins: guidelines on nomenclature. FASEB J 6: 2512–2513

Kemp KE, Pearson RB (1991) Intrasteric regulation of protein kinases and phosphatases. Biochim Biophys Acta 1094: 67–76

Kieber JJ, Rothenberg M, Roman G, Feldman KA, Ecker JR (1993) *CTR1*, a negative regulator of the ethylene response pathway in *Arabidopsis*, encodes a member of the Raf family of protein kinases. Cell 72: 427–441

Kisselve OG, Ermolaeva M, Gautam N (1994) A farnesylated domain in the G protein γ subunit is a specific determinant of receptor coupling. J Biol Chem 269: 21399–21402

Knight MR, Smith SM, Trewavas AJ (1992) Wind-induced plant motion immediately increases cytosolic calcium. Proc Natl Acad Sci USA 89: 4967–4971

Knox JP, Day S, Robert K (1989) A set of cell surface glycoproteins forms an early marker of cell position, but not cell type, in the root apical meristem of *Daucus carota* L. Development 106: 47–56

Knox JP, Linstead PH, Peart J, Cooper C, Roberts K (1991) Developmentally regulated epitopes of cell surface arabinogalactan proteins and their relation to root tissue pattern formation. Plant J 1: 317–326

Kobe B, Deisenhofer J (1994) The leucine-rich repeat: a versatile binding motif. Trends Biochem Sci 19: 415–421

Kotani K, Yonezawa K, Hara K, Ueda H, Kitamura Y, Sakaue H, Ando A, Chavanieu A, Calas B, Grigorescu F, Nishiyama M, Waterfield MD, Kasuga M (1994) Involvement of phosphoinositide 3-kinase in insulin- or IGF-1-induced membrane ruffling. EMBO J 13: 2313–2321

Kreuger M, van Holst G-J (1993) Arabinogalactan proteins are essential in somatic embryogenesis of *Daucus carota* L. Planta 189: 243–248

Kumagai N, Morii N, Fujisawa K, Nemoto Y, Narumiya S (1993) ADP-ribosylation of *rho* p21 inhibits lysophosphatidic acid-induced protein tyrosine phosphorylation and phosphatidylinositol 3-kinase activation in cultured Swiss 3T3 cells. J Biol Chem 268: 24535–24538

Larsen PM, Wolniak SM (1990) 1,2-dioctanoylglycerol accelerates or retards mitotic progression in *Tradescantia* stamen hair cells as a function of the time of its addition. Cell Motil Cytoskeleton 16: 190–203

Lawton MA, Yamamoto RT, Hanks SK, Lamb CJ (1989) Molecular cloning of plant transcripts encoding protein kinase homologs. Proc Natl Acad Sci USA 86: 3140–3144

Leaves SJ, Paterson HF, Marshall CJ (1994) Requirements for Ras in Raf activation is overcome by targeting Raf to the plasma membrane. Nature 369: 411–414

Legendre L, Yueh YG, Crain R, Haddock N, Heinstein PF, Low PS (1993) Phospholipase C activation during elicitation of the oxidative burst in cultured plant cells. J Biol Chem 268: 24559–24563

Leung J, Bouvier-Durand M, Morris P-C, Guerrier D, Chefdor F, Giaraudat J (1994) *Arabidopsis* ABA response gene *ABI1*: features of a calcium-modulated protein kinase. Science 264: 1448–1452

Lev Z (1993) *ras* genes in *Drosophila melanogaster*. In: Lacal JC, McCormick F (eds) The Ras superfamily of GTPases. CRC Press, Boca Raton, pp 187–220

Li W, Assmann SM (1993) Characterization of a G-protein-regulated outward K^+ current in mesophyll cells of *Vicia faba* L. Proc Natl Acad Sci USA 90: 262–266

Li W, Luan S, Schreiber SL, Assmann SM (1994a) Cyclic AMP stimulates K^+ channel activity in mesophyll cells of *Vicia faba* L. Plant Physiol 106: 957–961

Li W, Luan S, Schreiber SL, Assmann SM (1994b) Evidence for protein phosphatase 1 and 2A regulation of K^+ channels in two types of leaf cells. Plant Physiol 106: 963–970

Lin HY, Wang X-F, Ng-Eaton E, Weinberg RA, Lodish HF (1992) Expression cloning of a TGF-type II receptor, a functional transmembrane serine/threonine kinase. Cell 68: 775–785

Lin X, Watson JC (1992) cDNA sequence of PsPK5, a protein kinase homolog form *Pisum sativum* L. Plant Physiol 100: 1072–1074

Lin X, Feng X-H, Watson JC (1991) Differential accumulation of transcripts encoding protein kinase homologs in greening pea seedlings. Proc Natl Acad Sci USA 88: 6951–6955

Ling V, Zienlinski RE (1989) Cloning of cDNA sequences encoding the calcium-binding protein, calmodulin, from barley (*Hordeum vulgare* L.). Plant Physiol 90: 714–719

Lounsburry K, Beddow A, Macara I (1994) A family of proteins that stabilize the Ran/TC4 GTPase in its GTP-bound conformation. J Biol Chem 269: 11285–11290

Luan S, Li W, Rusnak F, Assmann SM, Schreiber SL (1993) Immunosuppressants implicate protein phosphatase regulation of K^+ channels in guard cells. Proc Natl Acad Sci USA 90: 2202–2206

Lusini P, Trabalzini L, Franchi GG, Bovalini L, Martelli P (1991) Adenylate cyclase in roots of *Ricinus communis*; stimulation by GTP and Mn^+. Phytochemistry 30: 109–111

Ma H, Yanofsky MF, Meyerowitz EM (1990) Molecular cloning and characterization of *GPA1*, a G protein a subunit gene from *Arabidopsis thaliana*. Proc Natl Acad Sci USA 87: 3821–3825

Ma H, Yanofsky MF, Huang H (1991) Isolation and sequence analysis of *TGA1* cDNAs encoding a tomato G protein α subunit. Gene 107: 189–195

McCrudy DW, Harmon AC (1992) Calcium-dependent protein kinase in the green alga *Chara*. Planta 188: 54–61

McNeillis TW, von Arnim AG, Araki T, Komeda Y, Misera S, Deng X-W (1994) Genetic and molecular analysis of an allelic series of cop1 mutants suggests functional roles for the multiple protein domains. Plant Cell 6: 487–500

Mackintosh RW, Haycox G, Hardie DG, Cohen PTW (1990) Identification by molecular cloning of two cDNA sequences from the plant *Brassica napus* which are very similar to mammalian protein phosphatases-1 and -2A. FEBS Lett 276: 156–160

MacKintosh C, Coggins J, Cohen P (1991) Plant protein phosphatases: subcellular distribution, detection of protein phosphatase 2C and identification of protein phosphatase 2A as the major quinate dehydrogenase phosphatase. Biochem J 273: 733–738

Maeda T, Wurgler-Murphy SM, Salto H (1994) A two-component system that regulates an osmosensing MAP kinase cascade in yeast. Nature 369: 242–245

Massague J, Attisano L, Wrana J (1994) The TGF-β family and its composite receptors. Trends Cell Biol 4: 172–178

Matsui Y, Toh-e A (1992) Yeast RHO3 and RHO4 ras superfamily genes are necessary for bud growth, their defect is suppressed by a high dose of bud formation genes CDC42 and BEM1. Mol Cell Biol 12: 5690–5699

Meyer K, Leube MP, Grill E (1994) A protein phosphatase 2C involved in ABA signal transduction in *Arabidopsis thaliana*. Science 264: 1452–1458

Misera S, Muller AJ, Weiland-Heidecker U, Jurgens G (1994) The *FUSCA* genes of *Arabidopsis*: negative regulators of light responses. Mol Gen Genet 244, 242–252

Mizoguchi T, Gotoh Y, Nishida E, Yanaguchi-Shinozaki K, Hayashida N, Iwasaki T, Kamada H, Shinozaki K (1994) Characterization of two cDNAs that encode MAP kinase homologues in *Arabidopsis thaliana* and analysis of the possible role of auxin in activating such kinase activities in cultured cells. Plant J 5: 111–122

Moore MS, Blobel G (1993) The GTP-binding protein Ran/TC4 is required for protein import into the nucleus. Nature 365: 460–463

Nasrallah JB, Nasrallah ME (1993) Pollen-stigma signaling in the sporophytic self-incompatibility. Plant Cell 5: 1325–1335

Neuhaus G, Bowler C, Kern R, Chua NH (1993) Calcium/calmodulin-dependent and -independent phytochrome signal transduction pathways. Cell 73: 937–952

Newton AC (1993) Interaction of proteins with lipid headgroups: lessons from protein kinase C. Annu Rev Biophys Biomol Struct 22: 1–25

Nishida E, Gotoh U (1993) The MAP kinase cascade is essential for diverse signal transduction pathways. Trends Biochem Sci 18: 118–121

Orita S, Kaibuchi K, Kuroda S, Shimizu K, Nakanishi H, Takai Y (1993) Comparison of kinetic properties between two mammalian GTP/GDP exchange proteins, guanine nucleotide-releasing factor and *smg* GDP dissociation stimulator. J Biol Chem 268: 25542–25546

Payrastre B, van Bergen en Henegouwen PMP, Breton M, den Hartigh JC, Plantavid M, Verkleij AJ, Boonstra J (1991) Phosphoinositide kinase, diacylglycerol kinase, phospholipase C activities associated to the cytoskeleton: effect of epidermal growth factor. J Cell Biol 115: 121–128

Pepper A, Delaney T, Washburn T, Poole D, Chory J (1994) *DET1*, a negative regulator of light-mediated development and gene expression in *Arabidopsis*, encodes a novel nuclear-localized protein. Cell 78: 109–116

Perera IY, Zielinski RE (1992) Structure and expression of the *Arabidopsis* CaM-3 calmodulin gene. Plant Mol Biol 19: 649–664

Polya GM, Chung R, Menting J (1991) Resolution of a higher plant protein kinase similar to the catalytic subunit of cAMP-dependent protein kinase. Plant Sci 79: 37–45

Putnam-Evans CL, Harmon AC, Palevitz BA, Fechheimer M, Cormier MJ (1989) Calcium-dependent protein kinase is localized with F-actin in plant cells. Cell Motil Cytoskeleton 12: 12–22

Putnam-Evans CL, Harmon AC, Cormier MJ (1990) Purification and characterization of a novel calcium-dependent protein kinase from soybean. Biochemistry 29: 2488–2495

Quail PH (1991) Phytochrome: a light-activated molecular switch that regulates gene expression. Annu Rev Genet 25: 389–409

Randall SK, Marshall MS, Crowell DN (1993) Protein isoprenylation in suspension-cultured tobacco cells. Plant Cell 5: 433–442

Reiss Y, Stradley SJ, Gierasch LM, Brown MS, Goldstein JL (1991) Sequence requirement for peptide recognition by rat brain $p21^{ras}$ protein farnesyltransferase. Proc Natl Acad Sci USA 88: 732–736

Rhee SG, Suh P-G, Ryu S-H, Lee S-Y (1989) Studies of inositol phospholipid-specific phospholipase C. Science 244: 546–550

Ridley AJ, Hall A (1992) The small GTP-binding protein rho regulates the assembly of focal adhesions and actin stress fibers in response to growth factors. Cell 70: 389–399

Ridley AJ, Paterson HF, Johnston CL, Diekmann D, Hall A (1992) The small GTP-binding protein rac regulates growth factor-induced membrane ruffling. Cell 70: 401–410

Roberts DM (1993) Protein kinases with calmodulin-like domains: novel targets of calcium signals in plants. Curr Opin Cell Biol 5: 242–246

Roberts DM, Harmon AC (1992) Calcium-modulated proteins: targets of intracellular calcium signals in higher plants. Annu Rev Plant Physiol Plant Mol Biol 43: 375–414

Roberts K (1994) The plant extracellular matrix: in a new expansive mood. Curr Opin Cell Biol 6: 688–694

Romero LC, Lam E (1993) Guanine nucleotide binding protein involvement in early steps of phytochrome-regulated gene expression. Proc Natl Acad Sci USA 90: 1465–1469

Sekine A, Fujiwarna M, Narumiya S (1989) Asparagine residue in the rho gene product is the modification site for botulinum ADP-ribosyltransferase. J Biol Chem 264: 8602–8605

Shenolikar S (1994) Protein serine/threonine phosphatases—new avenues for cell regulation. Annu Rev Cell Biol 10: 55–86

Shibaoka H, Nagai R (1994) The plant cytoskeleton. Curr Opin Cell Biol 6: 10–15

Shinjo K, Kolan JG, Hart MJ, Narashimhan V, Johnson DI, Evans T, Cerione RA (1990) Molecular cloning of the gene for the human placental GTP-binding protein Gp (G25K): identification of the GTP-binding protein as the human homolog of the yeast cell-division-cycle protein CDC42. Proc Natl Acad Sci USA 87: 9853–9857

Sistrunk M, Antosiewicz DM, Purugganan MM, Braam J (1994) *Arabidopsis TCH3* encodes a novel Ca^{2+} binding protein and shows environmentally induced and tissue-specific regulation. Plant Cell 6: 1553–1565

Smith RD, Walker JC (1991) Isolation and expression of a maize type 1 protein phosphatase. Plant Physiol 97: 677–683

Smith RD, Walker JC (1993) Expression of multiple type 1 phosphoprotein phosphatases in *Arabidopsis thaliana*. Plant Mol Biol 21: 307–316

Steele FR, Wasjburn T, Rieger R, O'Tousa JE (1992) *Drosophila* retinal degeneration C (rgdC) encodes a novel serine/threonine protein phosphatase. Cell 69: 669–676

Stein JC, Nasrallah JB (1993) A plant receptor-like gene, the S-locus receptor kinase of *Brassica oleracea* L., encodes a functional serine/threonine kinase. Plant Physiol 101: 1103–1106

Stein JC, Howlett B, Boyes DC, Nasrallah ME, Nasrallah JB (1991) Molecular cloning of a putative receptor protein kinase gene encoded at the self-incompatibility locus of *Brassica oleracea*. Proc Natl Acad Sci USA 88: 8816–8820

Stokoe D, Macdonalds G, Cadwallader K, Symons M, Hancok JF (1994) Activation of raf as a result of recruitment to the plasma membrane. Science 264: 1463–1467

Stone JM, Collinge MA, Smith RD, Horn MA, Walker JC (1994) Interaction of a protein phosphatase with an *Arabidopsis* serine-threonine receptor kinase. Science 266: 793–795

Takai Y, Kaibuchi K, Kikuchi A, Kawata M (1992) Small GTP-binding proteins. Rev Cytol 133: 187–230

Terryn N, Van Montagu M, Inze D (1993) GTP-binding proteins in plants. Plant Mol Biol 22: 143–152

Thomas FMH, Geny B, Cockcroft S (1991) Identification of a novel cytosolic poly-phosphoinositide-specific phospholipase C (PLC-86) as the major G-protein-regulated enzyme. EMBO J 10: 2507–2512

Tobias CM, Howlett B, Nasrallah JB (1992) An *Arabidopsis thaliana* gene with sequence similarity to the S-locus, receptor kinase of *Brassica oleracea*. Plant Physiol 99: 284–290

Toda H, Yazawa M, Sakiyama F, Yagi K (1985) Amino acid sequence of calmodulin from wheat germ. J Biochem 98: 833–842

Trewavas A, Gilroy S (1992) Signal transduction in plant cells. Trends Genet 7: 356–361

Ullrich A, Schlessinger J (1990) Signal transduction by receptors with tyrosine kinase activity. Cell 61: 203–212

Verma DPS, Cheon C-I, Hong Z (1994) Small GTP-binding proteins and membrane biogenesis in plants. Plant Physiol 106: 1–6

von Arnim AG, Deng X-W (1994) Light activation of *Arabidopsis* photomorphogenic repressor COP1 involves a cell-specific regulation of its nucleocytoplasmic partitioning. Cell 79: 1035–1045

Walker JC (1993) Receptor-like protein kinase genes of *Arabidopsis thaliana*. Plant J 3: 451–456

Walker JC, Zhang R (1990) Relationship of a putative receptor protein kinase from maize to the S-locus glycoproteins of *Brassica*. Nature 345: 743–746

Warpeha KMF, Hamm HE, Rasenick MM, Kaufman LS (1991) A blue-light activated GTP-binding protein in the plasma membranes of etiolated peas. Proc Natl Acad Sci USA 88: 8925–8929

Wei N, Chamovitz DA, Deng X-W (1994) *Arabidopsis* COP9 is a component of a novel signaling complex mediating light control of development. Cell 78: 117–124

Weiss C, Huang H, Ma H (1993) Immunolocalization of the G protein α subunit encoded by the *GPA1* gene in *Arabidopsis*. Plant Cell 5: 1513–1528

Weiss CA, Garnaat CW, Mukai K, Yu Y, Ma H (1994) Isolation of cDNAs encoding guanine nucleotide-binding protein β-subunit homologues from maize (ZGB1) and *Arabidopsis* (AGB1). Proc Natl Acad Sci USA 91: 9554–9558

Wieser R, Attisano L, Wrano JL, Massague J (1993) Signaling activity of transforming growth factor β type II receptors lacking specific domains in the cytoplasmic region. Mol Cell Biol 13: 7239–7247

Wigler MH (1993) The ras system in yeasts. In: Lacal JC, McCormick F (eds) The Ras superfamily of GTPase. CRC Press, Boca Raton, pp. 155–172

Wilson C (1994) Receptor tyrosine kinase signalling: not so complex after all? Trends Cell Biol 4: 309–414

Xu P, Llyod C, Stager CJ, Drobak BK (1992) Association of phophatidylinositol 4-kinase with the plant cytoskeleton. Plant Cell 4: 941–951

Yang W, Burkhard W, Cavallius J, Merrick WC, Boss WF (1993) Purification and characterization of a phosphatidylinositol 4-kinase activator in carrot cells. J Biol Chem 268: 392–398

Yang Z, Watson JC (1993) Molecular cloning and characterization of rho, a ras-related small GTP-binding protein from the garden pea. Proc Natl Acad Sci USA 90: 8732–8736

Yang Z, Cramer CL, Watson JC (1993) Protein farnesyltransferase in plants: molecular cloning and expression of a homolog of the β subunit from the garden pea. Plant Physiol 101: 667–674

Yoko-o T, Matsui Y, Yagisawa H, Nojima H, Uno I, Toh-e A (1993) The putative phosphoinositide-specific phospholipase C gene, *PLC1*, of the yeast *Saccharomyces cerevisiae* is important for cell growth. Proc Natl Acad Sci USA 90: 1804–1808

Zhang J, King WG, Dillon S, Hall A, Feig L, Rittenhouse SE (1993) Activation of platelet phosphatidylinositide 3-kinase requires the small GTP-binding protein Rho. J Biol Chem 268: 22251–11154

Zhu J-K, Bressan RA, Hasegawa PM (1993) Isoprenylation of the plant molecular chaperone ANJ1 facilitates membrane association and function at high temperature. Proc Natl Acad Sci USA 90: 8557–8561

Ziman M, O'Brien JM, Ouellette LA, Church WR, Johnson DI (1991) Mutational analysis of *CDC42Sc*, a *Saccharomyces cerevisiae* gene that encodes a putative GTP-binding protein involved in the control of cell polarity. Mol Cell Biol 11: 3537–3544

G-Protein Regulation of Plant K^+ Channels

Sarah M. Assmann

Contents

I. Introduction

A diversity of GTP-binding proteins is found in living organisms. These include small G proteins involved in vesicle trafficking and secretion (e.g., rab proteins), cytoskeleton organization (e.g., rho and rac proteins), and cell growth and differentiation (e.g., the proto-oncogenic ras proteins), as well as tubulins, signal-recognition particles (SRPs), translation factors, and the signal-transducing heterotrimeric G proteins (Kaziro et al., 1991; Marsh and Goode, 1993; Boguski and McCormick, 1993). Common to all of these proteins is a mechanism in which the protein is activated upon binding of GTP and inactivated by subsequent GTP hydrolysis. This chapter focuses on the signal-transducing G proteins (reviewed in Gilman, 1987; Neer and Clapham, 1988; Simon et al., 1991) and their roles as regulators of plant ion channels. Other plant members of the GTPase superfamily are discussed in detail by Yang (1996) as well as in recent reviews (Kaufman, 1994; Ma, 1994; Terryn et al., 1993).

Signal-transducing heterotrimeric G proteins are composed of α, β, and γ subunits, with respective molecular weights of approximately 35–55 kDa, 35–40 kDa, and 7–10 kDa (Hepler and Gilman, 1992; Kaufman, 1994). These G proteins are associated with receptor proteins containing seven membrane spanning domains. Activation, e.g., by light or a hormone, of the relevant receptor molecule in turn activates the G protein, resulting in the exchange of GTP for GDP at a site on the α subunit of the G protein. The α subunit dissociates from the $\beta\gamma$ subunit pair and interacts with a downstream effector protein, thereby activating it. In unusual cases the $\beta\gamma$ subunit pair may be the activator; for example, a class of cardiac K^+ channels is regulated by the $\beta\gamma$ subunit pair

(Logothetis et al., 1987). Endogenous GTPase activity of the α subunit eventually hydrolyses the bound GTP to GDP. GTP hydrolysis returns the α subunit to its inactive, GDP-bound form and results in reassociation of the α and the βγ subunits. This reassociation also terminates any activity of the βγ dimer. Because the G protein remains active as long as GTP is bound, G-protein activation can result in amplification of the original signal, with the rapidity of GTP hydrolysis determining the extent of signal amplification. In some cases, GTPase-activating proteins (GAPS) interact with the G protein, resulting in an accelerated rate of GTP hydrolysis (McCormick, 1993; Wittinghofer et al., 1993). The GAP is often an ancillary protein (Boguski and McCormick, 1993), but there are also cases where the GAP and the effector protein are one and the same entity. For example, the L-type Ca^{2+} channel of rat frontal cortex appears to accelerate GTPase activity of its regulatory G protein, G_o (Sweeny and Dolphin, 1992). Some G proteins, notably ras proteins, also have associated with them other regulatory proteins (Boguski and McCormick, 1993): guanine-nucleotide-releasing proteins (GNRPS), guanine-nucleotide-dissociation inhibitors (GDIS), and guanine-nucleotide-dissociation stimulators (GDSS).

The complexity of G-protein signalling pathways is still being unravelled. There are examples of different receptors activating the same G protein, and of one receptor regulating more than one G protein (Birnbaumer, 1992, and references therein). Different subtypes of an effector enzyme may be activated, inhibited, or unaffected by βγ dimers (Birnbaumer, 1992; Camps et al., 1992a). In some cases, specific βγ dimers are required for the activity of a particular Gα. For example, microinjection of antisense oligonucleotides to two specific β subunits interfered with G_o-mediated Ca^{2+} channel modulation in a rat pituitary cell line, but antisense oligonucleotides to two other β subunits had no effect (Kleuss et al., 1992). In other cases βγ subunits of different G proteins appear functionally interchangeable (Gautam and Simon, 1990; Camps et al., 1992b; Katz et al., 1992; Birnbaumer, 1992). How all of this crosstalk occurs while still maintaining the specificity of stimulus-response coupling is one of the major questions in G-protein research (Neer and Clapham, 1988; Birnbaumer, 1992; Hepler and Gilman, 1992).

II. G-Protein Regulation of Animal Ion Channels: Background and Methodology

Ion channels are one of the primary targets of G-protein action in animal cells (Dunlap et al., 1987; Brown and Birnbaumer, 1990). G-protein regulation of ion channels comprises part of numerous signal transduction pathways, including those mediating vision, taste, and smell (Stryer, 1986; Dhallan et al., 1990; McLaughlin et al., 1992; Hepler and Gilman, 1992). Ion channels are distinct from many of the targets of G-protein action (the classic example is adenylate cyclase, the enzyme which catalyses production of CAMP) in that their activity is not assayed by standard enzymological methods. Assessment of whether an ion channel is regulated by G proteins typically requires the application of electrophysiological techniques. Since the study of G-protein regulation of ion channels in plant cells employs methods and criteria developed from studies on animal systems, it is of use to begin by briefly describing that methodology, and the functional definitions of G-protein regulation that arise from it.

Ion channels are regulated proteinaceous pores which span the cellular and organellar membranes of cells (Hille, 1992). Ions move through channels down their electrochemical gradient, which is determined according to the Nernst equation by two factors: the concentration difference for that particular ion across the membrane and the membrane potential difference or membrane voltage. To a first approximation, ion channels exist in either an open conformation, which allows the passage of ions, or a closed conformation, in which ion flux is curtailed. The majority of ion channels exhibit a selectivity for ionic species which is confered by the molecular structure of the channel polypeptide(s), and channels are typically "named" after their permeant ion, e.g., "K^+ channels" or "Ca^{2+} channels."

Since the early 1980s, the premier electrophysiological approach for the study of ion channels has been the patch clamp technique (Hamill et al., 1981; Neher and Sakmann, 1992). A glass microelectrode or "micropipette" with a tip aperture of about 1 µm is filled with an ionic solution as is required for electrical conductivity and brought in contact with the cell membrane by use of a micromanipulator. In the case of plant cells, this requirement for membrane contact necessitates the use of protoplasts. Gentle suction is applied to the back of the pipette through a port in the pipette holder, resulting in the formation of a tight seal between the rim of the pipette and the cell membrane. This tight seal dramatically reduces electrical noise, and ensures that ionic currents flowing through open ion channels in the enclosed patch of membrane are detected by the patch clamp amplifier. The amplifier converts the current signal to a voltage which can be displayed on an oscilloscope or digitized by an analog to digital converter for storage on computer disk.

Patch clamping can be performed in several different configurations (Fig. 1). The configuration initially obtained is called the "cell attached patch." If additional suction or high voltage is applied to the patch it can be ruptured, allowing diffusional equilibration between the solution in the patch pipette and the cell cytoplasm. Because the volume of pipette solution far exceeds that of the cytosol, the ionic composition of the cytosol in this "whole cell configuration" is largely determined by the pipette solution, with the exception of ionic species for which the cell may have strong or localized buffering capacities. Electrical data recorded in the whole cell configuration represent the net current flowing through all ion channels and carriers of the cell membrane. Thus, to separate out and observe current through a particular type of channel requires a judicious choice of solutions. For example, in order to observe current through K^+ channels (Fig. 2), one might employ relatively high concentrations of K^+ as a salt of an impermeant (and therefore non-current producing) anion.

Studies seeking to document G-protein action on ion channels typically involve the introduction of G-protein regulators to the cytosol via the pipette

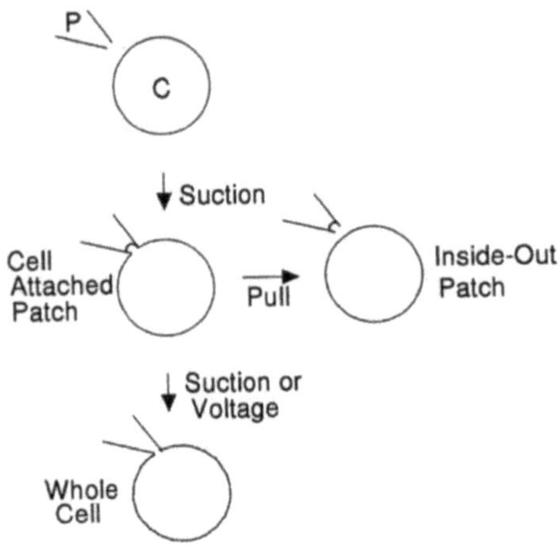

Fig. 1

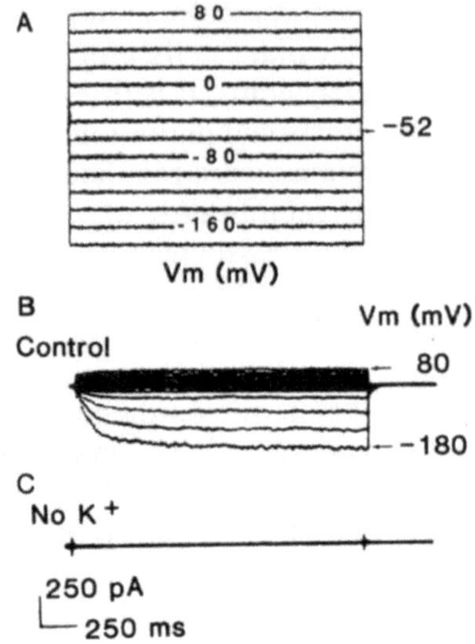

Fig. 2

Fig. 1. Membrane configurations used in patch clamping. P, micropipette; C, cell

Fig. 2. Whole cell K^+ currents from a guard cell protoplast of *Vicia faba*. A, Voltage protocol applied to the cell. Voltage pulses were applied sequentially in time and are superimposed for purposes of illustration. B, Inward (downward) and outward (upward) K^+ currents from the protoplast in response to the voltage protocol shown in A. Current responses were obtained sequentially in time and are superimposed for purposes of illustration. The external K^+ concentration was 13.8 mM and the internal (pipette and cytosol) concentration was 107 mM. C, Absence of K^+ currents when K^+ in the bath and pipette solutions is removed and replaced with the impermeant cation N-methyl glucamine. Reproduced from Fairley-Grenot and Assmann (1991)

solution, and the monitoring of subsequent current response. For example, GTPγS and GDPβS, hydrolysis-resistant analogs of GTP and GDP, can be introduced to the cytosol via the whole cell configuration. GTPγS locks G proteins into their active state, while GDPβS locks G proteins into their inactive state. As described in a recent methodological review (McFadzean and Brown, 1992), these nucleotides should be included in the pipette solution at concentrations of 100–500 μM for whole cell experiments, in order to achieve successful competition against endogenous guanine nucleotides. Because GDPβS is competing with endogenous GTP for binding, and does not bind irreversibly, it is sometimes necessary to apply even higher (mM) concentrations of this nucleotide in the whole cell configuration to see an effect (Brown, 1990; McFadzean and Brown, 1992). If guanine nucleotide analogs do affect the current response, it is important to determine the nucleotide specificity of this response to verify G-protein involvement. Whereas other enzymes such as nucleoside diphosphate kinases can utilize GTP or ATP as substrates (Heidbuchel et al., 1992), G proteins are typically highly selective for GTP over ATP.

For example, the affinity of ras for GTP is six orders of magnitude higher than its affinity for ATP (Wittinghofer et al., 1993).

Two natural G-protein regulators, the bacterial toxins cholera toxin (CTX) and pertussis toxin (PTX), are also useful as experimental tools which can be introduced via the patch pipette. Both of these toxins catalyse the ADP-ribosylation of G proteins (Gierschik, 1992). A given G protein may be ADP-ribosylated by one, both, or neither of the toxins; for example, transducin and tubulin can be ADP-ribosylated by both toxins (Hepler and Gilman, 1992; Burns et al., 1993). The site of ADP-ribosylation promoted by CTX is an arginine residue that appears to be present in all Gα's (Gierschik and Jakobs, 1990). However, only some G proteins appear susceptible to ADP-ribosylation by this toxin (Ui, 1990). Ribosylation at this site blocks GTPase activity, leaving the G protein in a permanently activated state (Cassel and Selinger, 1977; Hepler and Gilman, 1992). PTX catalyzes ADP-ribosylation at a cysteine site (West et al., 1985) that is present near the C-terminus of some G proteins (Ui, 1990). Ribosylation at this site results in a G protein that is no longer responsive to activation by the receptor protein (Gilman, 1987; Hepler and Gilman, 1992). Although there are cellular effects of PTX other than ADP-ribosylation of G proteins, these effects are observed at PTX concentrations of 1–10 μg/ml (Ui, 1990). Therefore, effects of PTX occurring at lower concentrations provide good evidence of G-protein involvement (McFadzean and Brown, 1992).

Observation of opposite effects of GTPγS vs. GDPβS on channel activity, absence of effects of corresponding adenine analogues, and regulation by low concentrations of one or both of the bacterial toxins is good initial evidence for G-protein involvement, which can then be further confirmed by biochemical dissection of the signal transduction pathway. Such dissection may also involve electrophysiological techniques. For example, one can administer via the patch pipette an antibody to the implicated G protein, and determine if the signalling response is thereby nullified (Yatani et al., 1988; Brown, 1990). Alternatively, G proteins may be first implicated in ion flux regulation by in vitro biochemical techniques, and electrophysiological techniques may subsequently be employed to study the dynamics and kinetics of such regulation in the living cell.

Whole cell analyses with non-hydrolysable nucleotide analogues alone can give indication of G-protein involvement, but give relatively little information on the pathway of G-protein action. Additional information can be garnered by repeating such studies on isolated membrane patches. If G-protein regulators alter ion channel function in the whole cell configuration but have no effect in the isolated patch configuration, then the implication is that G-protein regulation of channel behavior is *cytosol-dependent*, that is, involves a signal transduction cascade that includes cytosolic partners. For example, in retinal rod cells photon absorption by rhodopsin activates the G protein transducin. Activated transducin activates cGMP phosphodiesterase which decreases cGMP levels and thereby triggers cation channel closure, thus eliminating a depolarizing cation current and promoting cell hyperpolarization. Hyperpolarization reduces the amount of the neurotransmitter glutamate, that is released to downstream retinal cells (Stryer, 1986; Pugh and Miller, 1987; Chabre et al., 1993). All of this occurs in less than half a second (Chabre et al., 1993), emphasizing the exquisite temporal control that can be exerted by G-protein signalling pathways. The phosphatidylinositol pathway (Coté and Crain, 1993) is another example of a G-protein-mediated signal transduction cascade that is cytosol-dependent (Smrcka et al., 1991). In various cell types, hormonal signals such as bradykinin, angiotensin and vasopressin (Sternweis and Smrcka, 1993; Gutowski et al, 1991) activate G proteins of the G_q subfamily, which in turn activate the β isozyme of phospholipase C (PLC). The β isozyme can be activated by both α and βγ subunits (Sternweis and Smrcka, 1993; Camps et al., 1992a; Katz et al., 1992). PLC catalyses production of the two signalling molecules, diacylglycerol and inositol trisphosphate (IP_3), from phosphatidylinositol bisphosphate. IP_3-stimulated Ca^{2+} release from intracellular stores elevates cytosolic Ca^{2+} concentrations (Ca_i) and thus affects numerous processes, including ion channel behavior (e.g., Higashida and Brown, 1986).

Cytosol-dependent pathways may be elucidated not only by "test-tube" biochemistry but also by "pipette" biochemistry. For example, in one study (Kozlowski et al., 1991) of G-protein regulation of L-type Ca^{2+} currents in heart muscle cells, chemically caged GTPγS was introduced into the cells via the patch pipette. Upon "uncaging" (activation) of the GTPγS by UV light energy, Ca^{2+} currents increased, implicating G-protein activation of Ca^{2+} channels. The effectiveness of freed GTPγS was reduced in the presence of Rp-cAMP, an antagonist of cAMP-dependent protein kinase (PKA). This result suggested that Ca^{2+} channel opening resulted in part from GTPγS-activation of a G protein that stimulated adenylate cyclase activity, which in turn activated PKA.

Alternatively, G-protein regulators may be found to be effective in the isolated patch configuration, in which case it is said that the G-protein regulation is *membrane-delimited*. This is a functional definition, and the meaning is that G-protein regulation of the channel involves only components that are associated with the isolated patch. This pathway is presumed to exclude soluble signals and most, but perhaps not all, cytoskeletal-associated signalling molecules. A membrane-delimited pathway of G-protein action has been documented in animal systems for at least 11 K^+ channels, 3 Ca^{2+} channels, and 2 Na^+ channels (reviewed in Brown, 1993).

A membrane-delimited pathway may result from G-protein activation of membrane-associated enzymes whose substrates are ion channels. For example, G-protein activation of membrane bound phospholipase A2 results in arachidonic acid production, which has been shown to activate cardiac K^+ channels in isolated membrane patches (Kurachi et al., 1989; Kim et al., 1989). Alternatively, a membrane-delimited response may result from *direct* action of the G protein on the channel polypeptide. Proof of the latter mechanism requires purification of both the G protein and the channel protein(s), reconstitution of an active channel in an artificial membrane system, and demonstration that the purified G protein, when activated, does modulate channel activity (Brown, 1993). This technically demanding experiment has to date been performed with positive results for one animal channel, the Ca^{2+} channel from the T-tubules of skeletal muscle (Hamilton et al., 1991).

III. G-Protein Regulation of Plant Ion Channels

Evidence for G-protein involvement in ion channel regulation has been found in two types of plant cells to date: guard cells and mesophyll cells of *Vicia faba*. Below, the evidence for G protein-mediated channel regulation and the implications of such regulation for cell function are described.

A. Guard Cell Channels

In guard cells of *Vicia faba* patch clamped in the whole cell configuration, application of 500 μM GTPγS to the cytosol reduced the magnitude of inward K^+ currents, while application of 500 μM GDPβS increased inward K^+ current magnitude (Fairley-Grenot and Assmann, 1991) (Fig. 3 A). These concentrations are within the range commonly employed for elucidation of G-protein regulation of ion channels in animal systems (e.g., Kurachi et al., 1986; Breitwieser and Szabo, 1988; Dolphin and Scott, 1987; Brown, 1990; McFadzean and Brown, 1992). Enhancement of the guard cell K^+ current was seen with as little as 50 μM GDPβS; a dose–response curve for GTPγS has not been published. The fact that effects on K^+ currents of both GTPγS and GDPβS were observed suggests that the patch-clamped guard cell "rests" in an intermediate state which allows both up and down-regulation by G proteins (Brown, 1993). ATPγS and ADPβS had no effect on the K^+ currents, demonstrating that there is nucleotide specificity of the responses (Fairley-Grenot and Assmann, 1991). This specificity is in contrast to results obtained with anion

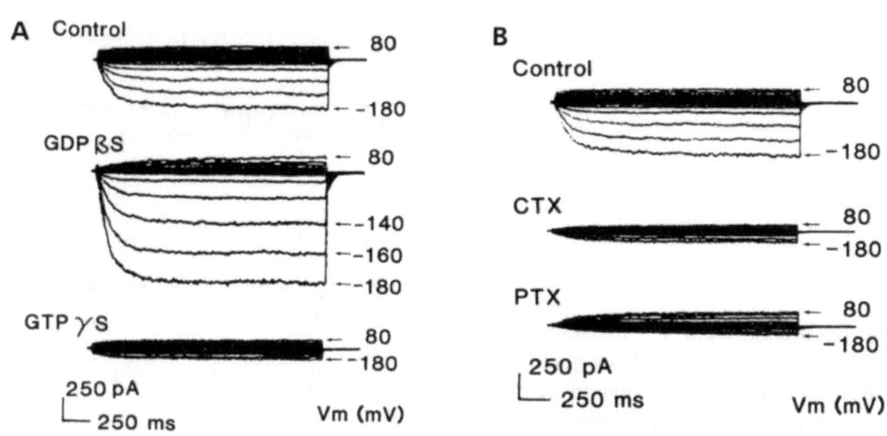

Fig. 3. Inhibition of inward K^+ current from a *Vicia faba* guard cell protoplast by application of GTPγS (500 μM) and by 200 ng/ml of cholera toxin (CTX) and pertussis toxin (PTX); enhancement of inward K^+ current by GDPβS (500 μM). Reproduced from Fairley-Grenot and Assmann (1991)

channels of guard cells, where both GTP and ATP appear to serve as activators (Hedrich et al., 1990). The bacterial toxins PTX and CTX, administered at 200 ng/ml, also decreased inward K^+ current magnitude (Fairley-Grenot and Assmann, 1991) (Fig. 3 B). Taken together, the nucleotide and toxin data indicate that a G protein regulates K^+ influx into guard cells.

As described above, in animal systems CTX typically activates G proteins while PTX inactivates G proteins. In contrast, in the guard cell system both toxins had the same, inhibitory (GTPγS-like) effects on K^+ currents. These results suggest either that G proteins are regulated differently by toxins in plant vs. animal cells or that there exists in guard cells one G protein that inhibits K^+ channels and is activated by CTX and another G protein that activates K^+ channels and is inhibited by PTX. It is interesting to note that the two toxins also have the same effect in another plant system: CTX and PTX both stimulate *cab* gene expression in soybean culture cells (Romero and Lam, 1993). One possible clue to the mechanism by which PTX might, contrary to expectation, activate a G-protein response comes from the observation that in a few cultured animal cell lines, PTX reduces basal, that is, receptor-independent, GTPase activity (Gierschik and Jakobs, 1990; Gierschik, 1992).

In the guard cell system, it was further observed that high cytosolic concentrations of the Ca^{2+} chelator, BAPTA (5 mM BAPTA provided in addition to standard buffering with EGTA), prevented K^+ current inhibition by GTPγS. BAPTA added to the control solution had no effect (Fairley-Grenot and

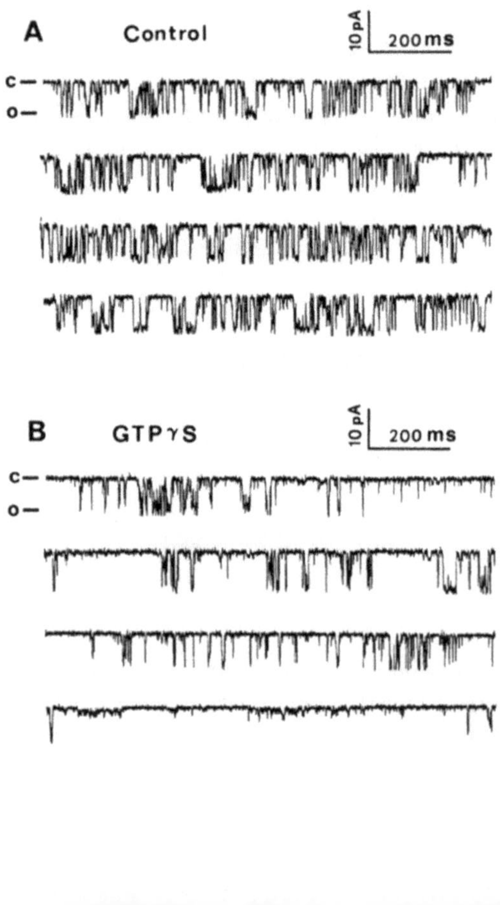

Fig. 4. Recording from an inside-out membrane patch of a *Vicia faba* guard cell, showing inhibition of inward K^+ channel activity after application of GTPγS (500 μM). c, current baseline when the channel was closed; o, current level when the channel was open. Reproduced from Assmann and Wu (1994)

Assmann, 1991). G-protein activation may therefore elevate Ca_i by triggering Ca^{2+} release from intracellular stores or by activating Ca^{2+} influx channels at the plasma membrane. G-protein elevation of Ca_i would provide a mechanism for the observed inhibition of K^+ currents, since these currents are known to be inhibited by experimental elevation of Ca_i (Schroeder and Hagiwara, 1989; Fairley-Grenot and Assmann, 1992; Lemtiri-Chlieh and MacRobbie, 1994). However, these data are also consistent with the possibility that G-protein regulation requires not Ca_i elevation but simply a permissive level of Ca_i that is not reached when high concentrations of Ca^{2+} chelators are present.

Given that BAPTA counteracts the effect of GTPγS, one might be tempted to conclude that G proteins regulate inward K^+ currents solely via a cytosolic cascade. However, there is a direct way to test this hypothesis, namely by determining whether G-protein regulators have any effect on K^+ channel activity in isolated membrane patches. Wu and Assmann (1994) tested the effects of GTPγS, GDPβS, PTX, and CTX on K^+ channels in inside-out membrane patches, using solutions with the same Ca^{2+} buffering as in the whole cell assays. They found that all of these regulators had effects on the isolated patches consistent with the effects observed in whole cell assays. GTPγS (Fig. 4), PTX, and CTX all inhibited channel activity (decreased open probability) while GDPβS enhanced channel activity (increased open probability). The control nucleotide, ATPγS, had no effect on open probability. The magnitude of the GDPβS effect was relatively small, possibly because in the isolated patch conditions (absence of exogenous GTP) most of the G proteins were already inactivated. Just as in the whole cell experiments, CTX and PTX both inhibited inward K^+ channel activity.

The whole cell and single channel data, taken together, suggest the following possible scenarios: (1) there is more than one pathway of G-protein action (possibly involving more than one G protein) in guard cells. The cytosol-dependent pathway requires *elevation* of Ca_i while the membrane-delimited pathway does not. (2) There are both cytosol-dependent and membrane-delimited pathways of G-protein action in guard cells. One or both of these pathways requires a *permissive* level of Ca^{2+} (present in both the isolated patch experiments and the whole cell experiments without BAPTA, but absent upon increasing the Ca^{2+} buffering strength by addition of BAPTA), without a requirement for elevation of Ca_i. (3) There is *only* a membrane-delimited pathway of G-protein action, which requires a permissive level of Ca_i. For example, G-protein activation might increase the affinity for Ca^{2+} of the channel protein itself or of other membrane-associated messengers. The discussion below assumes that the first of the above explanations is correct, because with that assumption we are best able to place G proteins in an integrated signal transduction scheme. However, as will be reiterated subsequently, this is still an assumption rather than a proven fact.

Role of G-protein regulation of K^+ channels in stomatal function. During stomatal opening (reviewed in Assmann, 1993; Kearns and Assmann, 1993), H^+ extrusion by a plasma membrane H^+ ATPase in response to stimuli such as light (Assmann et al., 1985; Serrano et al., 1988) results in a hyperpolarization of the membrane potential (i.e., "makes the membrane potential more negative"). Anion channel closure which prevents anion efflux may also contribute to hyperpolarization (Schroeder et al., 1993; Schwartz et al., 1995). This hyperpolarization both creates a driving force for K^+ uptake, and stimulates opening of the inward K^+ channels (Schroeder, 1988), which are voltage-regulated (Schroeder et al., 1987). K^+ enters the guard cell through these channels and, along with active Cl^- uptake, malate synthesis from starch breakdown, and, under some conditions, some contribution from photosynthetically produced solutes (Tallman and Zeiger, 1988; Reckmann et al., 1990; Poffenroth et al., 1992), decreases the osmotic potential of the guard cell. Decreased osmotic potential results in water influx, a separation and bowing out of the two guard cells, and a consequent increase in aperture of the stomatal pore. Since inward K^+ channels are the major conduits through which K^+ enters guard cells, effectors which inhibit these channels may slow or prevent stomatal opening. Therefore, it is anticipated that the G protein which is activated experimentally by GTPγS is activated in vivo by stimuli which prevent, reduce, or retard stomatal opening. One example of such a stimulus is the plant hormone abscisic acid (ABA). The following observations suggest that G-protein involvement in the ABA response is a reasonable working hypothesis:

1. ABA application inhibits inward K^+ currents in guard cells (Blatt, 1990; Schwartz et al., 1994; Lemtiri-Chlieh and MacRobbie, 1994).
2. G-protein activation (via GTPγS) similarly inhibits inward K^+ currents in guard cells (Fairley-Grenot and Assmann, 1991).
3. Under some conditions, ABA elevates Ca_i (McAinsh et al., 1990, 1992; Gilroy et al., 1991; Irving et al., 1992).
4. There is some evidence for ABA-activation of Ca^{2+} permeable channels (Schroeder and Hagiwara, 1990) and some evidence consistent with ABA-triggered release of Ca^{2+} from intracellular stores via a mechanism involving phosphoinositide turnover (MacRobbie, 1992; Parmar and Brearley, 1993; Coté and Crain, 1994). In animal cells, distinct G proteins do mediate both Ca^{2+} channel activation and release of stored Ca_i via the phosphatidylinositol pathway (Hepler and Gilman, 1992).
5. When Ca_i is deliberately buffered to low concentrations with high buffering strength, inhibition of whole-cell inward K^+ currents by GTPγS is prevented (Fairley-Grenot and Assmann, 1991).

Thus it is reasonable to hypothesize the following pathway:

ABA $\rightarrow$ G-protein activation $\rightarrow$ elevation of Ca_i $\rightarrow$ inhibition of K_{in} channels

If we hypothesize that one mechanism by which G proteins inhibit inward K^+ channels is by elevation of Ca_i, then what is the mechanism by which Ca_i inhibition occurs? One Ca^{2+}-dependent process in animal systems is activation of protein phosphatase 2B (PP2B). PP2B is activated by binding of Ca^{2+}-activated calmodulin (Klee et al., 1979, 1987; Liu et al., 1991), and, to a much lesser extent, by direct binding of Ca^{2+} (Stewart et al., 1982). We therefore investigated the possibility that Ca^{2+}-inactivation of inward K^+ channels is mediated by a plant PP2B homolog. Consistent with this hypothesis, introduction of constitutively-active (Ca^{2+}/calmodulin independent) bovine brain PP2B to the guard cell cytosol inhibited inward K^+ currents, even at low Ca_i. Conversely, the immunosuppressant/immunophilin complexes cyclosporin A/cyclophilin and FK506/FKBP, which are known specific inhibitors of PP2B (Liu et al., 1991, 1992), prevented the inhibition of inward K^+ currents by elevated Ca_i. These data indicate that Ca^{2+}-inhibition of inward K^+ currents is mediated by a phosphatase with homology to PP2B (Luan et al., 1993). Note that both calmodulin and PP2B are found not only in the cytosol, but also associated with the plasma membrane (Chantler, 1985; Collinge and Trewavas, 1989; Kincaid et al., 1987) as would be required to hypothesize their involvement in the isolated patch response.

Of great relevance to the idea that ABA acts in guard cells via a Ca^{2+}-regulated phosphatase is the recent deduction that an ABA-insensitive mutant of *Arabidopsis, abil*, which exhibits abnormally large stomatal apertures, is mutant in a gene which encodes a phosphatase with a Ca^{2+}-binding domain (Leung et al., 1994; Meyer et al., 1994). The gene has higher sequence similarity to another animal phosphatase, PP2C, than to PP2B, but also has homology over a limited region to the Ca^{2+}-binding subunit of PP2B. Only biochemical experimentation which tests whether this gene product is Ca^{2+}/CaM-dependent and sensitive to inhibition by immunosuppressant/immunophilin complexes will reveal whether the phosphatase involved in K^+ channel regulation is functionally of the 2B or 2C variety.

A more complete ABA signal transduction pathway might then look like this:

ABA → G-protein activation → elevation of Ca_i → activation of Ca^{2+}-dependent phosphatase → ?? → inhibition of K_{in} channels

The question marks indicate that it is not currently known whether the phosphatase substrate is the channel protein itself or another intermediate messenger.

There are also observations which suggest that the true picture of ABA-mediated regulation of the inward K^+ channel is significantly more complicated than the simple linear scheme outlined above. One caveat is that recent data indicate that it is intracellular ABA which is responsible for triggering K^+ channel inactivation (Schwartz et al., 1994). While externally applied ABA (10 μM) did inhibit inward K^+ channels and stomatal opening, this inhibition was dependent on external pH. Essentially no inhibition was observed at an external pH (pH_o) of 8.0, and increasing inhibition was observed as pH_o was lowered, a manipulation which promotes uptake of uncharged ABAH through the lipid bilayer (Kaiser and Hartung, 1981). In control experiments, changing pH_o alone had no effect on the currents (or on stomatal aperture in *Vicia*). The hypothesis that intracellular ABA that had been taken up by the guard cells was responsible for mediating K^+ channel inhibition was directly tested by applying ABA to the cytosol via the patch pipette solution in whole-cell configuration. Regardless of pH_o, cytosolic ABA (1 μM) was highly effective in inhibiting the inward K^+ currents. These data do not demonstrate that all ABA effects on guard cells are mediated by internal ABA (see Anderson et al., 1994), but they do suggest that we should revise our general paradigm of ABA response to include such possibilities as intracellular ABA receptors.

The implications of these results for G-protein involvement in the ABA response are unclear. Classically, heterotrimeric G proteins transduce signals from membrane-spanning receptors with seven transmembrane domains, suggesting that a G protein should not be the second messenger in this ABA response. However, recent evidence that the phytochrome signal is G protein mediated (see below) suggests that, in the case of plants, one should keep an open mind to the possibility that G proteins may transduce signals from cytosolic receptors. G proteins have also been implicated in organellar processes, including the regulation of a thylakoid-localized kinase (Millner, 1987), phosphorylation of proteins in isolated oat nuclei (Romero et al., 1991a), and vesicular transport through the Golgi in kidney cells (Hepler and Gilman, 1992). These results indicate that heterotrimeric G proteins, and presumably their receptors, can certainly be associated with intracellular membranes as well as with the cell membrane.

A second, important, caveat to the idea that a G protein serves as the intermediary between ABA and Ca_i elevation is that, as already described above, data are not yet available as to whether or not G protein activation does in fact result in elevation of Ca_i.

Another twist to the ABA-signal transduction pathway is that ABA itself does not always elevate Ca_i. Ca^{2+}-imaging experiments have shown that photolytic release of caged ABA always results in elevation of Ca_i only in guard cells of *Commelina* plants that have been grown at high temperatures (e.g., 40 °C). Plants that have been grown at 17 °C or lower do not show Ca_i elevation (Allan et al., 1993). The important but as yet unanswered questions are, do these low-temperature plants still show ABA-inhibition of inward K^+ channels? And do these plants still show K^+ channel inactivation in response to GTPγS? One study has shown that very high concentrations of EGTA or BAPTA prevent ABA-inhibition of inward K^+ channels in plants grown at 16–20 °C, and thus suggests that the ABA-Ca_i link is obligatory for the regulation of inward K^+ channels (Lemtiri-Chlieh and MacRobbie, 1994). It would be useful to further test this hypothesis with plants grown at different temperatures.

A third caveat is provided by the data of Lee et al. (1993), who observed a small enhancement of stomatal opening after treatment of epidermal peels with PTX or release of caged GTPγS microinjected into *Commelina* guard cells. This effect is opposite to that which would be predicted from the inhibition of inward K^+ channels observed in patch clamp studies. There are several possible ways to reconcile the two sets of results. In neither microinjection nor patch clamp studies can nucleotides be targeted by the experimenter to a particular G protein. Experimental conditions which differed in the two studies may therefore have promoted the activation of different G proteins, with opposite effects, in the two experiments. For example, it is known from studies on animal systems that adenylate cyclase is regulated by both stimulatory and inhibitory G proteins (Gilman, 1987). Alternatively, the impact of G protein-mediated inward K^+ channel inhibition on stomatal opening might not have been apparent in the study of Lee et al. (1993) because very high K^+ concentrations (100 mM KCl, 10 mM K^+-MES) were provided in the incubation medium. The impact of channel inhibition may, however, be significant in vivo, where apoplastic K^+ concentrations are thought to be in the low mM range (Bowling, 1987; Grignon and Sentenac, 1991). Finally, in Lee et al.'s study, caged GTPγS was more effective than non-caged GTPγS, (non-caged) GDPβS had no effect, and caged ATPγS was not tested as a control. Therefore, the effects of caged GTPγS in their study might be due to a product of cage lysis rather than to GTPγS itself.

A fourth caveat to the involvement of G proteins in the ABA response is that, even if future data do place a G protein between ABA and Ca_i elevation, the upstream and downstream components of the *membrane-delimited* pathway of G-protein action in guard cells remain to be determined. One step to addressing this issue is to ask whether or not ABA also has a membrane-delimited pathway of action; that is, whether ABA inhibits inward K^+ channels in the isolated patch. For both the membrane-delimited pathway and the cytosol-dependent pathway, it must also be kept in mind that a variety of environmental signals prevent stomatal opening, among them low ambient humidity, high intercellular concentrations of CO_2, darkness, and certain concentrations of the phytohormone auxin (reviewed in Zeiger, 1983; Zeiger et al., 1987; Assmann, 1993). The effects of these signals on guard cell ion transport have been little investigated and it is certainly possible that one or more of these signals will be found to be G protein-mediated in a membrane-delimited or cytosol-dependent fashion.

B. Mesophyll Cell Channels

The same approaches described in Fairley-Grenot and Assmann (1991) were utilized to investigate G-protein regulation of K^+ channels in mesophyll cells of *Vicia faba* (Li and Assmann, 1993). Results indicate that *outward* K^+ channels in mesophyll cells are G protein regulated (Table 1). Thus, GTPγS reduced outward K^+ current, while GDPβS enhanced outward K^+ currents. Just as in guard cells, strong buffering of Ca_i to low levels prevented the effect of GTPγS. CTX inhibited the outward K^+ current. However, unlike the case in guard cells, PTX had no effect, suggesting that the mesophyll G protein involved in regulation of the outward K^+ channel is not identical to the guard-cell G protein which regulates the inward K^+ channel.

Role in mesophyll cell function. Adenylate cyclase is the classic example of a G protein-regulated enzyme in animal cells. Various K^+ and Ca^{2+} channels in animal systems (Brown, 1990; Dolphin, 1990) are regulated by G protein-containing signal transduction cascades that include cAMP and cAMP-activated

Table 1. Comparisons of current magnitude and kinetics between control mesophyll protoplasts and those treated with nucleotides, Ca^{2+}, or bacterial toxins (Li and Assmann, 1993)

Treatment	Current density[a] (μA/pF)	Probability[b]	Activation half-time (ms)
Control	28.5 ± 3.3		205 ± 27
ADP[βS] (500 μM)	30.6 ± 3.2	0.620	215 ± 35
GDP[βS] (500 μM)	44.4 ± 7.8	0.008	205 ± 29
GTP[γS] (500 μM)	18.4 ± 3.0	0.005	172 ± 21
GTP[γS] (1 mM)	10.7 ± 2.2	0.001	210 ± 29
Ca^{2+} (20 μM)[c]	16.7 ± 2.1	0.005	168 ± 40
BAPTA (5 mM)	30.6 ± 5.8	0.660	203 ± 35
GTP[γS] (500 μM) + BAPTA (5 mM)	33.4 ± 6.4	0.330	236 ± 27
GTP[γS] (1 mM) + BAPTA (5 mM)	29.2 ± 4.8	0.890	220 ± 31
CTX (500 ng/ml)	17.5 ± 1.3	0.003	185 ± 24
PTX (500 ng/ml)	27.7 ± 5.5	0.850	204 ± 20

[a] Time-activated current density after depolarization to +85 mV

[b] Statistical comparison of I_{out} (pA/pF) at membrane potential of +85 mV between control and various treatments. Significance criterion used is $P \leqslant 0.01$

[c] Free cytoplasmic Ca^{2+} concentration obtained by including 2 mM $CaCl_2$ in pipette solution was calculated as 20 μM with the BUFFA program (from G. Ryall, Flinders Medical Centre, South Australia)

kinase (protein kinase A or PKA). However, the question of whether cAMP regulates plant ion channels has only recently been investigated. Effects of cAMP on mesophyll K^+ currents were assayed according to the hypothesis that if adenylate cyclase acted downstream from the G protein responsible for the inhibitory effects of GTPγS observed, then cAMP should also reduce current magnitude. Instead, cAMP significantly enhanced current magnitude (Li et al., 1994a). This enhancement could be observed at physiological (μM) cAMP concentrations, provided that an inhibitor of phosphodiesterase activity was also supplied. These results suggest the presence in plants of high levels of phosphodiesterase activity, and may help to explain why measurements of cAMP levels in plants have been variable and quite controversial (Trewavas and Gilroy, 1991).

In the study by Li and colleagues, it was further observed that introduction of bovine brain PKA also enhanced current magnitude, consistent with the possibility that a PKA-type kinase mediates the cAMP effect in vivo. Conversely, inhibitors of protein phosphatases 1 and 2A (PP1 and PP2A) were also applied to test for opposite effects stemming from phosphatase activity. Both phosphatase inhibitors enhanced current magnitude, suggesting that PP1 and/or PP2A-like proteins inhibit channel activity in vivo, opposite to the effects of PKA. These results are consistent with, but not proof of, a pathway in which G-protein activation results in phosphatase activation and K^+ current inhibition (Li et al., 1994b).

Although patch clamp studies are leading to hypotheses concerning the intracellular messengers in a G protein-containing signal transduction scheme, the initial signal(s) which activates such a pathway in mesophyll cells remain(s) unknown. K^+ fluxes in mesophyll cells do not have the clearly defined specialized functions which are obvious in guard cell function. We can speculate that K^+ fluxes in mesophyll cells may be of importance in controlling cell turgor. Regulation of these fluxes may be important in the control of cell expansion during development. Closure of outward K^+ channels will prevent K^+ efflux and turgor loss, particularly under conditions where the membrane potential is depolarized. In addition, many enzymes are known to require K^+ as a co-factor (Bhandal and Malik, 1988), and it is possible that one mechanism of controlling enzyme activity is through the control of cellular K^+ concentrations.

IV. Future Prospects

A. Linking Electrophysiology and Biochemistry/Molecular Biology

While electrophysiological data give strong indication of ion channel regulation mediated by G proteins in guard cells and mesophyll cells, the specific G proteins responsible for these functions have yet to be identified. Thus, while we can say with confidence that a GTP-binding protein is involved in K^+ channel regulation, it remains possible that this protein will have features that are unique to plants and distinct from the classic structure of animal G proteins. In fact, at this point in time it would be premature to conclude that channel regulation is necessarily carried out by a heterotrimeric G protein as opposed to a new class of G proteins, specific to plants. The distinctness of plant signalling systems is well illustrated by the CDPKs (Harper et al., 1991), kinases first identified in plants which have an intrinsic calmodulin-like domain that sets them apart from the Ca^{2+}/CaM-activated protein kinases familiar to animal biochemists. The point is further emphasized by the fact that only one gene/gene product with high homology to animal α subunits has been identified to date, despite extensive efforts (Terryn et al., 1993). The identification of only one true plant Gα gene indicates that there may not be high sequence conservation of the signal-transducing G proteins between plants and animals. The Gα gene that has been identified in *Arabidopsis* (*GPA1*) encodes a deduced amino acid sequence that is 36% identical to the sequence of bovine rod transducin (Ma et al., 1990). A similar gene has been identified in tomato (*TGA1*) (Ma et al., 1991). Immunolocalization studies of the *GPA1* gene product, GPα1, show that this protein is present in all parts of the plant, except mature seed (Weiss et al., 1993; Huang et al., 1994). GPα1 is found at higher levels in meristematic regions and in the vascular system, but is also found in other cell types, including mesophyll cells (Weiss et al., 1993; Huang et al., 1994). The function of GPα1 is unknown, and it will be of interest to analyze whether plants that over- or under-express this protein exhibit alterations in ion channel behavior. Such studies will provide approaches toward attributing specific physiological functions to specific GTP-binding polypeptides.

Proteins that hydrolyse GTP, or bind GTP, GTPγS, or antibodies to mammalian Gα subunits have been reported in the higher plants *Lemna* (Hasunuma and Fundera, 1987; Hasunuma et al., 1987a), pea (Hasunuma et al., 1987b), zucchini (Jacobs et al., 1988; Drobak et al., 1988; Perdue and Lomax, 1992), oat (Romero et al., 1991b), soybean (Legendre et al., 1992), maize (Bilushi et al., 1991), rice (Zaina et al., 1990), *Arabidopsis thaliana, Vicia faba*, and *Commelina communis* (Blum et al., 1988). GTP-binding proteins have also been detected in barley aleurone protoplasts (Wang et al., 1993). In many cases, the proteins identified have lower molecular weights than predicted from animal α subunits (Terryn et al., 1993), again suggesting the possibility of unique aspects of plant G proteins. Two genes have been cloned which contain some homology to Gβ subunits (Deng et al., 1992; Ishida et al., 1993). Whether the gene products function in vivo as Gβ subunits has yet to be established. Polypeptides that show cross-reactivity to a Gβ antibody raised against the βγ dimer of transducin have also been reported (Warpeha et al., 1990). None of the putative Gα or Gβ proteins has been fully purified, with the result that no function can be specifically assigned to any of them, although there are, in some cases, clues from physiological studies (see below). If purification is accomplished or, alternatively, if the relevant genes can be cloned and expressed in an in vitro system, it will then be possible to utilize these proteins in patch clamp assays, to determine whether these G proteins (or antibodies to them) alter ion channel function. Given the observation of a membrane-delimited pathway of K^+ channel regulation in guard cells, it will also be of interest to reconstitute plant K^+ channels in a lipid bilayer and observe whether these channel polypeptides are directly regulated by G proteins.

B. Other Signal Transduction Systems

In addition to stomatal function and mesophyll K^+ transport, G proteins have been implicated in a variety of other signalling processes. Levels of auxin binding and GTPγS binding change in parallel in membrane vesicles of oat, suggesting G-protein involvement in transduction of the auxin signal (Zaina et al., 1990). GTPγS stimulation of phospholipid turnover in cultured cells of *Acer pseudoplatanus* may be further indication of G-protein involvement in a plant phosphatidylinositol pathway (Dillenschneider et al., 1986), as also surmised from the guard cell studies. G-protein involvement has also been proposed in defense responses of soybean (Legendre et al., 1992) and tomato (Vera-Estrella et al., 1994).

Several studies implicate G proteins in transduction of the red/far-red signals for which phytochrome is the photoreceptor. Bossen et al. (1990) showed that GDPβS introduced by electroporation inhibited red light-induced swelling of wheat mesophyll protoplasts, while GTPγS could induce swelling. Romero and colleagues (1991b) showed that red light administered in vivo to etiolated oat seedlings slightly increased GTP binding in subsequently harvested extracts, consistent with red light-activation of a G protein. Cholera toxin, which classically activates G proteins, had the same effect as red light on phytochrome-regulated gene expression, namely expression of the *cab* genes was upregulated while expression of the *phy* genes was down-regulated (Romero et al., 1991b). Recent data from Chua's laboratory further implicates G proteins in phytochrome-regulation of gene expression (Neuhaus et al., 1993). It was also recently demonstrated that red light enhances GTP binding by three nuclear envelope proteins (Clark et al., 1993). The methodology was selective for monomeric G proteins, and the authors suggest that monomeric G proteins may be translocated to the nucleus as a result of red-light activation of phytochrome. This interpretation highlights the possibility that classic heterotrimeric G proteins may not be found to be the primary signal-transducing G proteins in plant systems.

A comprehensive study by Warpeha et al. (1990) also documents G-protein involvement in transduction of the signal from a blue light photoreceptor. In that study, GTP binding of a 40 kDa protein in pea plasma membrane was specifically regulated by blue light. ADP-ribosylation by CTX and PTX and cross-reactivity to α subunit antibodies similarly identified an approximately 40 kDa putative GTP-binding protein.

Ion fluxes are certain to be involved in many of the light responses of plant cells as well as in plant signal transduction processes involving hormones and elicitor compounds. It will therefore be of interest in the future to apply to these systems the powerful electrophysiological techniques that are now available for the analysis of G-protein involvement in ion transport regulation.

Note added in proof. Two recent papers, by Armstrong and Blatt (1995) and Kelly et al. (1995), provide important new information on the role of G proteins in the regulation of guard cell K^+ channels.

V. References

Allan AC, Trewavas AJ, Fricker MD (1993) Elevation of cytoplasmic calcium in *Commelina* guard cells after flash photolysis of caged-ABA is affected by growth temperature. In: Society of Experimental Biology Symposium: Membrane Transport in Plants and Fungi, p 8

Anderson BE, Ward JM, Schroeder JI (1994) Evidence for an extracellular reception site for abscisic acid in *Commelina* guard cells. Plant Physiol 104: 1177–1183

Armstrong F, Blatt MR (1995) Evidence for K^+ channel control in *Vicia* guard cells coupled by G-proteins to a 7TMS receptor mimetic. Plant J 8: 187–198

Assmann SM (1993) Signal transduction in guard cells. Annu Rev Cell Biol 9: 345–375

Assmann SM, Simoncini L, Schroeder JI (1985) Blue light activates electrogenic ion pumping in guard cell protoplasts of *Vicia faba*. Nature 318: 285–287

Bhandal IS, Malik CP (1988) Potassium estimation, uptake, and its role in the physiology and metabolism of flowering plants. Int Rev Cytol 110: 205–254

Bilushi SV, Shebunin AG, Babakov AV (1991) Purification and subunit composition of a GTP-binding protein from maize root plasma membranes. FEBS Lett 291: 219–221

Birnbaumer L (1992) Receptor-to-effector signaling through G proteins: roles for βγ dimers as well as α subunits. Cell 71: 1069–1072

Blatt MR (1990) Potassium channel currents in intact stomatal guard cells: rapid enhancement by abscisic acid. Planta 180: 445–455

Blum W, Hinsch K-D, Schultz G, Weiler EW (1988) Identification of GTP-binding proteins in the plasma membrane of higher plants. Biochem Biophys Res Commun 156: 954–959

Boguski MS, McCormick F (1993) Proteins regulating Ras and its relatives. Nature 366: 643–654

Bossen ME, Kendrick RE, Vredenberg WJ (1990) The involvement of a G-protein in phytochrome-regulated, Ca^{2+}-dependent swelling of etiolated wheat protoplasts. Physiol Plant 80: 55–62

Bowling DJF (1987) Measurement of the apoplastic activity of K^+ and Cl^- in the leaf epidermis of *Commelina communis* in relation to stomatal activity. J Exp Bot 38: 1351–1355

Breitwieser GE, Szabo G (1988) Mechanism of muscarinic receptor-induced K^+ channel activation as revealed by hydrolysis-resistant GTP analogues. J Gen Physiol 91: 469–493

Brown AM (1993) Membrane-delimited cell signaling complexes: direct ion channel regulation by G-proteins. J Membr Biol 131: 93–104

Brown AM, Birnbaumer L (1990) Ionic channels and their regulation by G protein subunits. Annu Rev Physiol 52: 197–213

Brown DA (1990) G-proteins and potassium currents in neurons. Annu Rev Physiol 52: 215–242

Burns RG, Farrell KW, Surridge CD (1993) Should the tubulins be members of the GTPase superfamily? In: Marsh J, Goode J (eds) The GTPase superfamily. Wiley, Chichester, pp 248–267

Camps M, Carozzi A, Schnabel P, Scheer A, Parker PJ, Gierschik P (1992a) Isozyme-selective stimulation of phospholipase C-β2 by G protein βγ-subunits. Nature 360: 684–686

Camps M, Hou C, Sidiropoulos D, Stock JB, Jakobs KH, Gierschik P (1992b) Stimulation of phospholipase C by guanine-nucleotide-binding protein βγ subunits. Eur J Biochem 206: 821–831

Cassel D, Selinger Z (1977) Mechanism of adenylate cyclase activation by cholera toxin: inhibition of GTP hydrolysis at the regulatory site. Proc Natl Acad Sci USA 74: 3307–3311

Chabre M, Antonny B, Bruckert F, Vuong TM (1993) The G protein cascade of visual transduction: kinetics and regulation. In: Marsh J, Goode J (eds) The GTPase superfamily. Wiley, Chichester, pp 112–124

Chantler PD (1985) Calcium-dependent association of a protein complex with the lymphocyte plasma membrane: probable identity with calmodulin-calcineurin. J Cell Biol 101: 207–216

Clark GB, Memon AR, Tong C-G, Thompson GA Jr, Roux SJ (1993) Phytochrome regulates GTP-binding protein activity in the envelope of pea nuclei. Plant J 4: 399–402

Collinge M, Trewavas AJ (1989) The location of calmodulin in the pea plasma membrane. J Biol Chem 264: 8865–8872

Coté GG, Crain RC (1993) Biochemistry of phosphoinositides. Annu Rev Plant Physiol Plant Mol Biol 44: 333–356

Coté GG, Crain RC (1994) Why do plants have phosphoinositides? BioEssays 16: 39–46

Deng X-W, Matsui M, Wei N, Wagner D, Chu AM, Feldmann KA, Quail PH (1992) *COP1*, an *Arabidopsis* regulatory gene, encodes a protein with both a zinc-binding motif and a Gβ homologous domain. Cell 71: 791–801

Dhallan RS, Yau K-W, Schrader KA, Reed RR (1990) Primary structure and functional expression of a cyclic nucleotide-activated channel from olfactory neurons. Nature 347: 184–187

Dillenschneider M, Hetherington A, Graziana A, Alibert G, Berta P, Haiech J, Ranjeva R (1986) The formation of inositol phosphate derivatives by isolated membranes from *Acer pseudoplatanus* is stimulated by guanine nucleotides. FEBS Lett 208: 413–417

Dolphin AC (1990) G protein modulation of calcium currents in neurons. Annu Rev Physiol 52: 243–255

Dolphin AC, Scott RH (1987) Calcium channel currents and their inhibition by (–)-baclofen in rat sensory neurones: modulation by guanine nucleotides. J Physiol 386: 1–17

Drøbak BK, Allan EF, Comerford JG, Roberts K, Dawson AP (1988) Presence of guanine nucleotide-binding proteins in a plant hypocotyl microsomal fraction. Biochem Biophys Res Commun 150: 899–903

Dunlap K, Holz GG, Rane SG (1987) G proteins as regulators of ion channel function. Trends Neurosci 10: 241–244

Fairley-Grenot KA, Assmann SM (1991) Evidence for G-protein regulation of inward K^+ channel current in guard cells of fava bean. Plant Cell 3: 1037–1044

Fairley-Grenot KA, Assmann SM (1992) Whole-cell K^+ current across the plasma membrane of guard cells from a grass, *Zea mays*. Planta 186: 282–293

Gautam N, Simon MI (1990) β Subunit of G proteins. In: Moss J, Vaughan M (eds) ADP-ribosylating toxins and G proteins. American Society for Microbiology, Washington, DC, pp 371–380

Gierschik P (1992) ADP-ribosylation of signal-transducing guanine nucleotide-binding proteins by pertussis toxin. Curr Top Microbiol Immunol 175: 69–96

Gierschik P, Jakobs KH (1990) Receptor-stimulated GTPase activity of G-proteins. In: Houslay MD, Milligan G (eds) G-proteins as mediators of cellular signalling processes. Wiley, Chichester, pp 67–82

Gilman AG (1987) G proteins: transducers of receptor-generated signals. Annu Rev Biochem 56: 615–649

Gilroy S, Fricker MD, Read ND, Trewavas AJ (1991) Role of calcium in signal transduction of *Commelina* guard cells. Plant Cell 3: 333–344

Grignon C, Sentenac H (1991) pH and ionic conditions in the apoplast. Annu Rev Plant Physiol Plant Mol Biol 42: 103–128

Gutowski S, Smrcka A, Nowak L, Wu D, Simon M, Sternweis PC (1991) Antibodies to the α_q subfamily of guanine nucleotide-binding regulatory protein α subunits attenuate activation of phosphatidylinositol 4,5-bisphosphate hydrolysis by hormones. J Biol Chem 266: 20519–20524

Hamill OP, Marty A, Neher E, Sakmann B, Sigworth FJ (1981) Improved patch-clamp techniques for high-resolution current recording from cells and cell-free membrane patches. Pflugers Arch 391: 85–100

Hamilton SL, Codina J, Hawkes MJ, Yatani A, Sawada T, Strickland FM, Froehner SC, Spiegel AM, Toro L, Stefani E, Birnbaumer L, Brown AM (1991) Evidence for direct interaction of $G_s\alpha$ with the Ca^{2+} channel of skeletal muscle. J Biol Chem 266: 19528–19535

Harper JF, Sussman MR, Schaller GE, Putnam-Evans C, Charbonneau H, Harmon AC (1991) A calcium-dependent protein kinase with a regulatory domain similar to calmodulin. Science 252: 951–954

Hasunuma K, Funadera K (1987) GTP-binding protein(s) in green plant, *Lemna paucicostata*. Biochem Biophys Res Commun 143: 908–912

Hasunuma K, Furukawa K, Funadera K, Kubota M, Watanabe M (1987a) Partial characterization and light-induced regulation of GTP-binding proteins in *Lemna paucicostata*. Photochem Photobiol 46: 531–535

Hasunuma K, Furukawa K, Tomita K, Mukai C, Nakamura T (1987b) GTP-binding proteins in etiolated epicotyls of *Pisum sativum* (Alaska) seedlings. Biochem Biophys Res Commun 148: 133–139

Hedrich RH, Busch H, Raschke K (1990) Ca^{2+} and nucleotide dependent regulation of voltage dependent anion channels in the plasma membrane of guard cells. EMBO J 9: 3889–3892

Heidbuchel H, Callewaert G, Vereecke J, Carmeliet E (1992) Membrane-bound nucleoside diphosphate kinase activity in atrial cells of frog, guinea pig, and human. Circ Res 71: 808–820

Hepler JR, Gilman AG (1992) G proteins. Trends Biol Sci 17: 383–387

Higashida H, Brown DA (1986) Two polyphosphatidylinositide metabolites control two K^+ currents in a neuronal cell. Nature 323: 333–335

Hille B (1992) Ionic channels of excitable membranes. Sinauer Associates, Sunderland, MA

Huang H, Weiss CA, Ma H (1994) Regulated expression of the *Arabidopsis* G protein α subunit gene *GPA1*. Int J Plant Sci 155: 3–14

Irving HR, Gehring CA, Parrish RW (1992) Changes in cytosolic pH and calcium of guard cells precede stomatal movements. Proc Natl Acad Sci USA 89: 1790–1794

Ishida S, Yohsuke T, Nagata T (1993) Isolation of cDNA of an auxin-regulated gene encoding a G protein β subunit-like protein from tobacco BY-2 cells. Proc Natl Acad Sci USA 90: 11152–11156

Jacobs M, Thelen MP, Farndale RW, Astle MC, Rubery PH (1988) Specific guanine nucleotide binding by membranes from *Cucurbita pepo* seedings. Biochem Biophys Res Commun 155: 1478–1484

Kaiser WM, Hartung W (1981) Uptake and release of abscisic acid by isolated photoautotrophic mesophyll cells, depending on pH gradients. Plant Physiol 68: 202–206

Katz A, Wu D, Simon MI (1992) Subunits βγ of heterotrimeric G protein activate β2 isoform of phospholipase C. Nature 360: 686–689

Kaufman LS (1994) GTP-binding signalling proteins in higher plants. J Photochem Photobiol B 22: 3–7

Kaziro Y, Itoh H, Kozasa T, Nakafuku M, Satoh T (1991) Structure and function of signal-transducing GTP-binding proteins. Annu Rev Biochem 60: 349–400

Kearns EV, Assmann SM (1993) The guard cell–environment connection. Plant Physiol 102: 711–715

Kelly WB, Esser JE, Schroeder JI (1995) Effects of cytosolic calcium and limited, possible dual, effects of G protein modulators on guard cell inward potassium channels. Plant J 8: 479–489

Kim D, Lewis DL, Graziadei L, Neer EJ, Bar-Sagi D, Clapham DE (1989) G-protein βγ-subunits activate the cardiac muscarinic K^+-channel via phospholipase A2. Nature 337: 527–536

Kincaid RL, Balaban CD, Billingsley ML (1987) Differential localization of calmodulin-dependent enzymes in rat brain: evidence for selective expression of cyclic nucleotide phosphodiesterase in specific neurons. Proc Natl Acad Sci USA 84: 1118–1122

Klee CB, Crouch TH, Krinks MH (1979) Calcineurin: a calcium- and calmodulin-binding protein of the nervous system. Proc Natl Acad Sci USA 76: 6270–6273

Klee CB, Draetta GF, Hubbard MJ (1987) Calcineurin. Methods Enzymol 139: 149–200

Kleuss C, Scherubl H, Hescheler J, Schultz G, Wittig B (1992) Different β-subunits determine G-protein interaction with transmembrane receptors. Nature 358: 424–426

Kozlowski RZ, Twist VW, Brown AM, Powell T (1991) Flash photolysis of intracellular caged GTPγS increases L-type Ca^{2+} currents in cardiac myocytes. Am J Physiol 30: H1665–H1670

Kurachi Y, Nakajima T, Sugimoto T (1986) On the mechanism of activation of muscarinic K^+ channels by adenosine in isolated atrial cells: involvement of GTP-binding proteins. Pflugers Arch 407: 264–274

Kurachi Y, Hiroyuki I, Sugimoto T, Shimizu T, Miki I, Ui M (1989) Arachidonic acid metabolites as intracellular modulators of the G protein-gated cardiac K^+ channel. Nature 337: 555–557

Lee HJ, Tucker EB, Crain RC, Lee Y (1993) Stomatal opening is induced in epidermal peels of *Commelina communis* L. by GTP analogs or pertussis toxin. Plant Physiol 102: 95–100

Legendre L, Heinstein PF, Low PS (1992) Evidence for participation of GTP-binding proteins in elicitation of the rapid oxidative burst in cultured soybean cells. J Biol Chem 267: 20140–20147

Lemtiri-Chlieh F, MacRobbie EAC (1994) Role of calcium in the modulation of *Vicia* guard cell potassium channels by abscisic acid: a patch-clamp study. J Membr Biol 137: 99–107

Leung J, Bouvier-Durand M, Morris P-C, Guerrier D, Chefdor F, Giraudat J (1994) *Arabidopsis* ABA response gene *ABI1*: features of a calcium-modulated protein phosphatase. Science 264: 1448–1452

Li W, Assmann SM (1993) Characterization of a G-protein-regulated outward K^+ current in mesophyll cells of *Vicia faba* L. Proc Natl Acad Sci USA 90: 262–266

Li W, Luan S, Schreiber SL, Assmann SM (1994a) Cyclic AMP stimulates K^+ channel activity in mesophyll cells of *Vicia faba* L. Plant Physiol 106: 957–961

Li W, Luan S, Schreiber SL, Assmann SM (1994b) Evidence for protein phosphatase 1 and 2A regulation of K^+ channels in two types of leaf cells. Plant Physiol 106: 963–970

Liu J, Farmer JD Jr, Lane WS, Friedman J, Weissman I, Schreiber SL (1991) Calcineurin is a common target of cyclophilin-cyclosporin A and FKBP-FK506 complexes. Cell 66: 807–815

Liu J, Albers MW, Wandless TJ, Luan S, Alberg DG, Belshaw PJ, Cohen P, MacKintosh C, Klee CB, Schreiber SL (1992) Inhibition of T cell signaling by immunophilin-ligand complexes correlates with loss of calcineurin phosphatase activity. Biochemistry 31: 3896–3901

Logothetis DE, Kurachi Y, Galper J, Neer EJ, Clapham DE (1987) The $\beta\gamma$ subunits of GTP-binding proteins activate the muscarinic K^+ channel in heart. Nature 325: 321–326

Luan S, Li W, Rusnak F, Assmann SM, Schreiber SL (1993) Immunosuppressants implicate protein phosphatase-regulation of K^+ channels in guard cells. Proc Natl Acad Sci USA 90: 2202–2206

Ma H (1994) GTP-binding proteins in plants: new members of an old family. Plant Mol Biol 26: 1611–1636

Ma H, Yanofsky MF, Meyerowitz EM (1990) Molecular cloning and characterization of *GPA1*, a G protein subunit gene from *Arabidopsis thaliana*. Proc Natl Acad Sci USA 87: 3821–3825

Ma H, Yanofsky MF, Huang H (1991) Isolation and sequence analysis of *TGA1* cDNAs encoding a tomato G protein α subunit. Gene 107: 189–195

McAinsh MR, Brownlee C, Hetherington AM (1990) Abscisic acid-induced elevation of guard cell cytosolic Ca^{2+} precedes stomatal closure. Nature 343: 186–188

McAinsh MR, Brownlee C, Hetherington AM (1992) Visualizing changes in cytosolic-free Ca^{2+} during the response of stomatal guard cells to abscisic acid. Plant Cell 4: 1113–1122

McCormick F (1993) Introduction. In: Marsh J, Goode J (eds) The GTPase superfamily. Wiley, Chichester, pp 1–5

McFadzean I, Brown DA (1992) Electrophysiological approaches to G-protein function. In: Milligan G (ed) Signal transduction. Oxford University Press, Oxford, pp 167–179

McLaughlin SK, McKinnon PJ, Margolskee RF (1992) Gustducin is a taste-cell-specific G protein closely related to the transducins. Nature 357: 563–569

MacRobbie EAC (1992) Calcium and ABA-induced stomatal closure. Philos Trans R Soc Lond Biol 338: 5–18

Marsh J, Goode J (eds) (1993) The GTPase superfamily. Wiley, Chichester

Meyer K, Leube MP, Grill E (1994) A protein phosphatase 2C involved in ABA signal transduction in *Arabidopsis thaliana*. Science 264: 1452–1455

Millner PA (1987) Are guanine nucleotide-binding proteins involved in regulation of thylakoid protein kinase activity? FEBS Lett 226: 155–160

Neer EJ, Clapham DE (1988) Roles of G protein subunits in transmembrane signalling. Nature 333: 129–134

Neher E, Sakmann B (1992) The patch clamp technique. Sci Am 266: 44–51

Neuhaus G, Bowler C, Kern R, Chua N-H (1993) Calcium/calmodulin-dependent and -independent phytochrome signal transduction pathways. Cell 73: 937–952

Parmar PN, Brearly CA (1993) Identification of 3- and 4-phosphorylated phosphoinositides and inositol phosphates in stomatal guard cells. Plant J 4: 255–263

Perdue DO, Lomax TL (1992) Characterization of GTP binding and hydrolysis in plasma membranes of zucchini. Plant Physiol Biochem 30: 163–172

Poffenroth M, Green DB, Tallman G (1992) Sugar concentrations in guard cells of *Vicia faba* illuminated with red or blue light: analysis by high performance liquid chromatography. Plant Physiol 98: 1460–1471

Pugh EN Jr, Miller WH (1987) Special topic: phototransduction in vertebrates. Annu Rev Physiol 49: 711–714

Reckmann U, Scheibe R, Raschke K (1990) Rubisco activity in guard cells compared with the solute requirement for stomatal opening. Plant Physiol 92: 246–253

Romero LC, Lam E (1993) Guanine nucleotide binding protein involvement in early steps of phytochrome-regulated gene expression. Proc Natl Acad Sci USA 90: 1465–1469

Romero LC, Biswal B, Song P-S (1991a) Protein phosphorylation in isolated nuclei from etiolated *Avena* seedlings: effects of red/far-red light and cholera toxin. FEBS Lett 282: 347–350

Romero LC, Sommer D, Gotor C, Song P-S (1991b) G-proteins in etiolated *Avena* seedlings: possible phytochrome regulation. FEBS Lett 282: 341–346

Schroeder JI (1988) K^+ transport properties of K^+ channels in the plasma membrane of *Vicia faba* guard cells. J Gen Physiol 92: 667–683

Schroeder JI, Hagiwara S (1989) Cytosolic calcium regulates ion channels in the plasma membrane of *Vicia faba* guard cells. Nature 338: 427–430

Schroeder JI, Hagiwara S (1990) Repetitive increases in cytosolic Ca^{2+} of guard cells by abscisic acid activation of nonselective Ca^{2+} permeable channels. Proc Natl Acad Sci USA 87: 9305–9309

Schroeder JI, Raschke K, Neher E (1987) Voltage dependence of K^+ channels in guard-cell protoplasts. Proc Natl Acad Sci USA 84: 4108–4112

Schroeder JI, Schmidt C, Sheaffer J (1993) Identification of high-affinity slow anion channel blockers and evidence for stomatal regulation by slow anion channels in guard cells. Plant Cell 12: 1831–1841

Schwartz A, Wu W, Tucker E, Assmann SM (1994) Inhibition of inward K^+ channels and stomatal response by abscisic acid: an intracellular locus of phytohormone action. Proc Natl Acad Sci USA 91: 4019–4023

Schwartz A, Ilan N, Schwarz M, Schaeffer J, Assmann SM, Schroeder JI (1995) Anion channel blockers inhibit S-type anion channels and abscisic acid responses in guard cells. Plant Physiol 109: 651–658

Serrano EE, Zeiger E, Hagiwara S (1988) Red light stimulates an electrogenic ion pump in *Vicia* guard cell protoplasts. Proc Natl Acad Sci USA 85: 436–440

Simon MI, Strathmann MP, Gautam N (1991) Diversity of G proteins in signal transduction. Science 252: 802–808

Smrcka AV, Hepler JR, Brown KO, Sternweis PC (1991) Regulation of polyphosphoinositide-specific phospholipase C activity purified by G_q. Science 251: 804–807

Sternweis PC, Smrcka AV (1993) G proteins in signal transduction: the regulation of phospholipase C. In: Marsh J, Goode J (eds) The GTPase superfamily. Wiley, Chichester, pp 96–111

Stewart AA, Ingebritsen TS, Manalan A, Klee CB, Cohen P (1982) Discovery of a Ca^{2+}- and calmodulin-dependent protein phosphatase. FEBS Lett 137: 80–84

Stryer L (1986) Cyclic GMP cascade of vision. Annu Rev Neurosci 9: 87–119

Sweeney MI, Dolphin AC (1992) 1,4-Dihydropyridines modulate GTP hydrolysis by G_o in neuronal membranes. FEBS Lett 310: 66–70

Tallman G, Zeiger E (1988) Light quality and osmoregulation in *Vicia* guard cells: evidence for involvement of three metabolic pathways. Plant Physiol 88: 887–895
Terryn N, Van Montagu M, Inze D (1993) GTP-binding proteins in plants. Plant Mol Biol 22: 143–152
Trewavas A, Gilroy S (1991) Signal transduction in plants. Trends Genet 7: 356–361
Ui M (1990) Pertussis toxin as a valuable probe for G-protein involvement in signal transduction. In: Moss J, Vaughan M (eds) ADP-ribosylating toxins and G proteins. American Society for Microbiology, Washington, DC, pp 45–77
Vera-Estrella R, Barkla BJ, Higgins VJ, Blumwald E (1994) Plant defense response to fungal pathogens. Plant Physiol 104: 209–215
Wang M, Sedee NJA, Heidekamp F, Snaar-Jagalska BE (1993) Detection of GTP-binding proteins in barley aleurone protoplasts. FEBS Lett 329: 245–248
Warpeha KMF, Hamm HE, Rasenick MM, Kaufman LS (1990) A blue-light-activated GTP-binding protein in the plasma membrane of etiolated peas. Proc Natl Acad Sci USA 88: 8925–8929
Weiss CA, Huang H, Ma H (1993) Immunolocalization of the G protein α subunit encoded by the *GPA1* gene in Arabidopsis. Plant Cell 5: 1513–1528
West RE Jr, Moss J, Vaughan M, Liu T, Lui T-Y (1985) Pertussis toxin-catalyzed ADP-ribosylation of transducin. J Biol Chem 260: 14428–14430
Wittinghofer A, Franken SM, Scheidig AJ, Rensland H, Lautwein A, Pai EF, Goody RS (1993) Three-dimensional structure and properties of wild-type and mutant H-*ras*-encoded p21. In: Marsh J, Goode J (eds) The GTPase superfamily. Wiley, Chichester, pp 6–27
Wu W, Assmann SM (1994) A membrane-delimited pathway of G-protein regulation of the guard cell inward K^+ channel. Proc Natl Acad Sci USA 91: 6310–6314
Yang Z (1996) Signal transducing proteins in plants: an overview. In: Verma DPS (ed) Signal transduction in plant growth and development. Springer Wien New York, pp 1–37 [Dennis ES et al (eds) Plant gene research. Basic knowledge and application]
Yatani A, Hamm HE, Codina J, Mazzoni MR, Birnbaumer L, Brown AM (1988) A monoclonal antibody to the α subunit of G_k blocks muscarinic activation of atrial K^+ channels. Science 241: 828–831
Zaina S, Reggiani R, Bertani A (1990) Preliminary evidence for involvement of GTP-binding protein(s) in auxin signal transduction in rice (*Oryza sativa* L.) coleoptile. J Plant Physiol 136: 653–658
Zeiger E (1983) The biology of stomatal guard cells. Annu Rev Plant Physiol 34: 441–475
Zeiger E, Farquhar GD, Cowan IR (eds) (1987) Stomatal function. Stanford University Press, Stanford

Metabolite Sensing and Regulatory Points of Carbon and Nitrogen Metabolic Pathways and Partitioning in Plants

Guo-Hua Miao

Contents

I. Introduction

Increased yield and quality improvement of harvested tissues or organs of economically important crop plants are two major goals of plant breeders. Crop yield is determined by multiple genetic and environmental factors, among these selective increase in partitioning of photoassimilates to harvested tissues or organs is of major importance (Gifford et al., 1984). Over the past few decades the yield of major crop plants has been improved substantially. However, further improvements can be made to achieve the genetic yield potential of major crop plants (Gifford et al., 1984; Thorne, 1985; Knight and Langston-Unkefer, 1988). Improvement in grain quality, composition and the relative distribution of storage compounds have also received increased attention over the last few years. Although conventional breeding practices will continue to play a vital role in crop improvement, plant metabolic engineering through biotechnology has the great potential of producing future crops with better quality and higher yield.

Carbohydrate, lipids, and proteins are the major storage compounds found in most oil seed and small grain crops; and their relative proportions of these components vary in storage and vegetative organs. The allocation of photoassimilates is mainly controlled by the regulated distribution of photoassimilates into different carbon and nitrogen metabolic pathways. Lipids and proteins are the major storage compounds in some plants such as peanuts and soybeans, whereas carbohydrate is the predominant form of storage compounds in cereals. In spite of its importance, the molecular mechanism for carbon and nitrogen partitioning among metabolic pathways are largely unknown in plants. Isolation of key genes affecting grain quality and composition, and development of ways of introducing these genes into major crop plants have created new approaches to study the molecular basis of regulation of metabolic pathways. Genes encoding regulatory enzymes and transporters controlling the flux of carbon and nitrogen through metabolic pathways have recently been isolated (Hu et al., 1992; Riesmeier et al., 1992; Schulz et al., 1993; Shorrosh et al., 1994; Gornicki et al., 1994). Manipulation of the activities of these enzymes and transporters in transgenic plants, along with genetic analysis, has identified the rate-limiting steps and regulatory control points of carbon and nitrogen metabolism (Von Shaewen et al., 1990; Sonnewald and Willmitzer, 1992; Stark et al., 1992; Kishore and Somerville, 1993; Riesmeier et al., 1993a; Voelker et al., 1993; Bate et al., 1994; Kavi Kishor et al., 1995). We are also beginning to understand the development and metabolic signals that coordinately regulate different metabolic pathways in plants. Many plant genes involved in photosynthesis and metabolic pathways are regulated by catabolite repression mechanism (Sheen, 1990; Krapp et al., 1993 Vincentz et al., 1993; Graham et al., 1992, 1994; Sano and Youssefian, 1994; Thomas and Rodrguez, 1994), which is a major regulatory strategy for controlling carbon and nitrogen metabolic pathways in prokaryotes (Fisher and Sonenshein, 1991; Saier, 1991) and in lower eukaryotes (Saier, 1991; Gancedo, 1992; Trumbly, 1992; Marzluf, 1993). Plant homologs of regulatory components of signal transduction networks involved in regulation of metabolic (Alderson et al., 1991; Sano and Youssefian, 1994) and other related (Chang et al., 1993; Kieber et al., 1993) pathways in bacteria and in yeast have been isolated, suggesting the existence of these signaling pathways in plants. Functional characterization of the signal transduction events and their role in regulating metabolic pathways in plants will be the task of future research in this area.

This chapter begins with a brief review of strategies controlling carbon and nitrogen metabolism in bacteria, yeast, and filamentous fungi, and their relevance to higher plants. The focus of the chapter is to highlight the recent advances in using molecular, genetic and transgenic approaches to unravel the regulatory control points of plant metabolism. These studies, along with extensive physiological and biochemical studies (for reviews, see Huppe and Turpin, 1994; S. Huber and Kaiser, 1996), have not only yielded new answers to the key regulatory events of metabolic pathways, but also have revealed the complexity of metabolic regulation in plants.

II. Global Regulatory Systems of Metabolic Pathways in Prokaryotes and Lower Eukaryotes

Catabolite repression is a general mechanism utilized by prokaryotes and lower eukaryotes to regulate carbon and nitrogen metabolism in response to changes in nutrient availability (Saier, 1991; Fisher and Sonenshein, 1991; Gancedo, 1992; Trumbly, 1992; Marzluf, 1993). In this mechanism, the expression of genes encoding alternative metabolic enzymes are repressed when preferred carbon and nitrogen sources are available. Although catabolite repression is involved in regulation of both carbon and nitrogen metabolism in diverse organisms, the molecular mechanisms in the different systems are not conserved. Below is a summary of the regulatory strategies of carbon and nitrogen repression in *Escherichia coli*, *Saccharomyces cerevisiae*, and *Neurospora crassa* to illustrate the relatedness and uniqueness of the regulatory aspects of different metabolism in these organisms.

A. Regulation of Carbon Metabolism

The best understood catabolite repression system is glucose repression in *E. coli* and in *S. cerevisiae*. The presence of glucose negatively regulates genes that encode enzymes required for metabolism of other carbon sources including sucrose, maltose, galactose and other non-fermentable carbon sources. In *E. coli,* glucose repression is controlled by CAMP-CAP regulon that is a global regulon consisting of operons regulated by the cyclic AMP and catabolite activator protein, CAP (Magasannik and Neidhardt, 1987). Interaction of CAP, in conjunction with high concentration of CAMP produced by adenylate cyclase, with CAP cis-regulatory elements of glucose repressible genes activates the operons. The mechanistic aspect of the interaction between CAP and its DNA binding element and activation of glucose repressible genes in *E. coli* have been studied in detail (Ren et al., 1988; Gaston et al., 1990; Joung et al., 1994). However, the CAMP-CAP mediated glucose repression mechanism does not operate in other bacteria species such as *B. subtilis* and *S. venezuelae* (Fisher and Sonenshein, 1991; Saier, 1991).

In yeast, carbon sensing and glucose repression is regulated by a CAMP independent pathway. Knowledge concerning the regulatory circuit of glucose repression is primarily obtained by analysis of glucose repression and derepression mutation of *SUC2*, which encodes an invertase in yeast. Many of the key components of the system have been identified and their epistatic relationships have been established (Gancedo, 1992; Trumbly, 1992). Among the known components in the pathway, the *SNF1* (sucrose non-fermenting) gene encodes a global regulator and plays a central role in the pathway (Celenza and Carlson, 1986). The *SNF1* mutant was initially isolated as one of the derepression mutants that does not express invertase (SUC2) in the absence of glucose. Therefore, *SNF1* is required for expression of glucose-repressible genes such as *SUC2*. Requirement of *SNF1* function for *SUC2* expression has been shown to be mediated at the transcriptional level (Celenza and Carlson, 1986). *SNF1* encodes a 72 kDa serine and threonine protein kinase (Celenza and Carlson, 1986), whose activity has been shown to be down regulated by the availability of glucose (Woods et al., 1994). The levels of the *SNF1* RNA and protein are not altered by glucose starvation (Celenza and Carlson, 1986). Thus the increase in SNF1 protein kinase activity under these conditions is likely to be mediated by a post-translational mechanism. Treatment of SNF1 with protein phosphatase 2A dramatically reduces SNF1 protein kinase activity. The inactivated SNF1 can be reactivated by mammalian AMP-activated protein kinase kinase (AMPKK; Mackinstosh et al. 1992; Woods, 1994). Since the AMP-activated protein kinase (AMPK) is a mammalian homolog of SNF1 protein kinase, the activation of SNF1 is likely to be regulated by an upstream kinase in vivo (Woods et al., 1994).

In addition to its global role in the derepression of glucose repressible genes, *SNF1* plays a role in the regulation of glycogen metabolism (Thompson-Jaeger et al., 1991) in yeast. The yeast strain with the allelic mutation of *SNF1* (*snf1*-172) exhibited significantly reduced glycogen accumulation under glucose and other nutrient starvation conditions (Thompson-Jaeger et al., 1991), suggesting that SNF1 is also required for glycogen biosynthesis. Failure to derepress gluconeogenic enzymes in the *snf1*-172 mutant may result in a shortage of glucose needed for glycogen synthesis under glucose derepression conditions. Recently, SNF1 protein kinase has also been shown to regulate lipid biosynthesis by controlling acetyl-CoA carboxylase (ACCase) activity (Woods et al., 1994; Mitchelhill et al., 1994;). In wild type yeast strains, the ACCase activity undergoes a time-dependent decrease under conditions of glucose derepression. However, such glucose mediated decrease of ACCase was not observed in *snf1* mutants (Woods et al., 1994), demonstrating that the decrease of ACCase activity under these conditions is dependent upon SNF1 protein kinase activity in vivo. The partially purified SNF1 protein kinase, like AMPK, also phosphorylates and inactivates yeast ACCase in vitro (Mitchelhill et al., 1994). These results provide convincing evidence suggesting that *SNF1* is involved in the regulation of fatty acid biosynthesis in addition to its role in carbon sensing and utilization. These studies are consistent with the earlier genetic evidence showing that *SNF1* is a global regulator of carbon metabolism (Celenza and Carlson, 1986). SNF1-like protein kinases have been found to be conserved in different organisms including higher plants (Alderson et al., 1990; Sano and Youssefian, 1994; Halford et al., 1994) and mammals (Carling et al., 1994). These plant and mammalian homologs of yeast SNF1 protein kinases also appear to be involved in regulation of carbon metabolism (see discussion in Sect. III. A, D).

The interactions between *SNF1* and other cellular components were identified by isolation and analysis of *snf1* suppression mutations that allow *snf1* mutants to utilize sucrose as carbon source under glucose derepression conditions (Carlson et al., 1984). *SNF4* has been shown to physically interact with SNF1 protein kinase in vivo and to functions as a positive effector of the SFN1 protein kinase (Celenza et al., 1989). Among the other known components in this pathway, hexokinase PII encoded by *HXK2* has been shown to act as a glucose sensor (Entian et al., 1985; Ma et al., 1989); several other components involved in signal relay from *HXK2* to *SNF1* have also been identified (Gancedo, 1992; Trumbly, 1992). However, the molecular and biochemical nature of these interactions have not yet been elucidated.

B. Regulation of Nitrogen Metabolism

Ammonia is the preferred nitrogen source for microorganisms. When bacteria are grown under conditions of limited ammonia, a number of functionally related genes (*Ntr* regulon) suppressed by the preferred nitrogen source are activated (Stock et al., 1989). The nitrogen derepression is regulated by a two component sensor and response regulator pair, *NtrB*/*NtrC*, in conjunction with other regulators (Albright et al., 1989). *NtrB* encodes a histidine kinase and *NtrC* encodes a transcriptional activator that is required for activation of glutamine synthetase (*glnA*) gene expression. The regulatory aspects of nitrogen sensing and repression are understood in this system. Ammonia deficiency, a high α-ketoglutarate to glutamine ratio, is monitored by nitrogen regulators encoded by *glnB* and *glnD*. *GlnB* and *glnD* in turn activate the autophosphorylation activity of NtrB histidine protein kinase, which subsequently transfers the phosphate to NtrC, leading to activation of the NtrC transcriptional activator. That the carbon and nitrogen ratio, rather than nitrogen concentration alone, is the metabolic signal for availability of nitrogen illustrates the coregulation of both carbon and nitrogen metabolism in bacteria. The two component regulatory systems are involved in signal transduction pathways that regulate a broad range of nutrient and environmental cues in bacteria (Stock et al., 1989; Parkinson and Kofoid et al., 1992; Alex and Simon, 1994). Recently, similar signal transduction modules have been identified in plants (Chang et al., 1993), yeast (Ota and Varshavsky, 1993; Maeda et al., 1994), and filamentous fungi (Alex and Simon, 1994), suggesting that the two component regulatory mechanisms have been conserved throughout evolution. In *Arabidopsis*, DNA sequences related to *ETR1*, encoding the two-component regulator of ethylene-response pathway, have been detected (Chang et al., 1993), indicating presence of multiple two-component sensor and regulator pairs in plants. However, the general utility of the system in plants and other eukaryotic organisms remains to be seen as more genes encoding eukaryotic two-component pairs are being isolated and characterized.

In *S. cerevisiae*, nitrogen sensing and response is primarily mediated by the RAS/adenylate cyclase pathway (Gilman, 1984; Malone, 1990), a cAMP dependent pathway. The presence of nutrients, primarily ammonia, activates adenylate cyclase (CDC35) through GTP-bound RAS and the GDP–GTP exchange factor (CDC25; Gilman, 1984; Broek et al., 1987). The increased cAMP level by adenylate cyclase activates the pathway through a cAMP-dependent protein kinase (Malone, 1990). Mutations in this pathway lead to cell cycle arrest in G_1 and hyperaccumulation of glycogen. In addition to the RAS/adenylate cyclase pathway, SLK1, a yeast homolog of mitogen-activated protein (MAP) kinase activators has been shown to play a role in nutrient sensing in a RAS/cAMP-independent pathway (Costigan and Snyder, 1994). Thus MAP kinase mediated signaling pathway also appears to be involved in control of nutrient sensing, in addition to regulating the pheromone response, cell wall formation and osmoregulation (for reviews, see Errede and Levin, 1993; Blumer and Johnson, 1994). The functional specificity of the structurally similar protein kinases appears to be achieved through the interacting proteins that specifically link multiple components within a signal pathway, thereby preventing cross-talk between different signaling pathways (Choi et al., 1994; Marcus et al., 1994). Multiple MAP kinases involved in different signal transduction pathways have been isolated in plants (Duerr et al., 1993; Kieber et al., 1993; Mizoguchi et al., 1994). Involvement of MAP kinase mediated signaling pathways has recently been established in the regulation of ethylene response process (Kieber et al., 1993). Studies are now being initiated to identify and characterize the different MAP kinase-mediated signal transduction pathways involved in the control of plant metabolism and development.

The global trans-activation mechanism of genes involved in nitrogen metabolism is illustrated by studies of the nitrogen regulatory circuit in *N. crassa* (Marzluf, 1993). In this organism, the genes required for alternative nitrogen assimilation pathways are also repressed by the presence of the preferred nitrogen sources. Activation of alternative nitrogen metabolism under low ammonia environment is mediated by a global positive-acting regulatory protein, NIT2 (Marzluf, 1993). NIT2 is a Cys2/Cys2-type zinc-finger global regulator that binds and activates many different co-regulated genes in the nitrogen regulatory circuit (Fu and Marzluf, 1990a, b), providing an efficient way to regulate unlinked genes in a common nitrogen pathway in eukaryotes. In addition to the global regulatory mechanism, many nitrogen suppressed genes are modulated by pathway-specific signals. Activation of nitrate reductase (*nit-3*) of *N. crassa* requires both a global signal (nitrogen starvation) and a pathway-specific signal (presence of nitrate; Fu and Marzluf, 1987). Similar trans-activation mechanism also exists in regulation of amino acid biosynthetic pathways, in which *GCN4* encodes a global regulator in yeast (Hope and Struhl, 1985).

The general outlines of the signal perception and transduction pathways as well as the mechanistic aspects of gene activation in transcriptional regulation of carbon and nitrogen metabolism have been established in prokaryotes and lower eukaryotes, although in many cases important gaps still remain. It is clear that catabolite repression is a general mechanism for both carbon and nitrogen metabolism in microorganisms. The diversity of regulatory strategies of catabolite repression has evolved in nature, however. Understanding the common and specific aspects of global regulation of metabolic pathways in these model experimental organisms will continue to provide new insights into regulation of metabolism.

III. Regulation of Carbon and Nitrogen Metabolism in Higher Plants

A. Control Points of Major Metabolic Pathways

One primary difference in metabolism between plant and animal cells is that plants are capable of fixing CO_2 to synthesize carbohydrate through photosynthesis. Therefore, the overall regulation of plant metabolism is balanced by

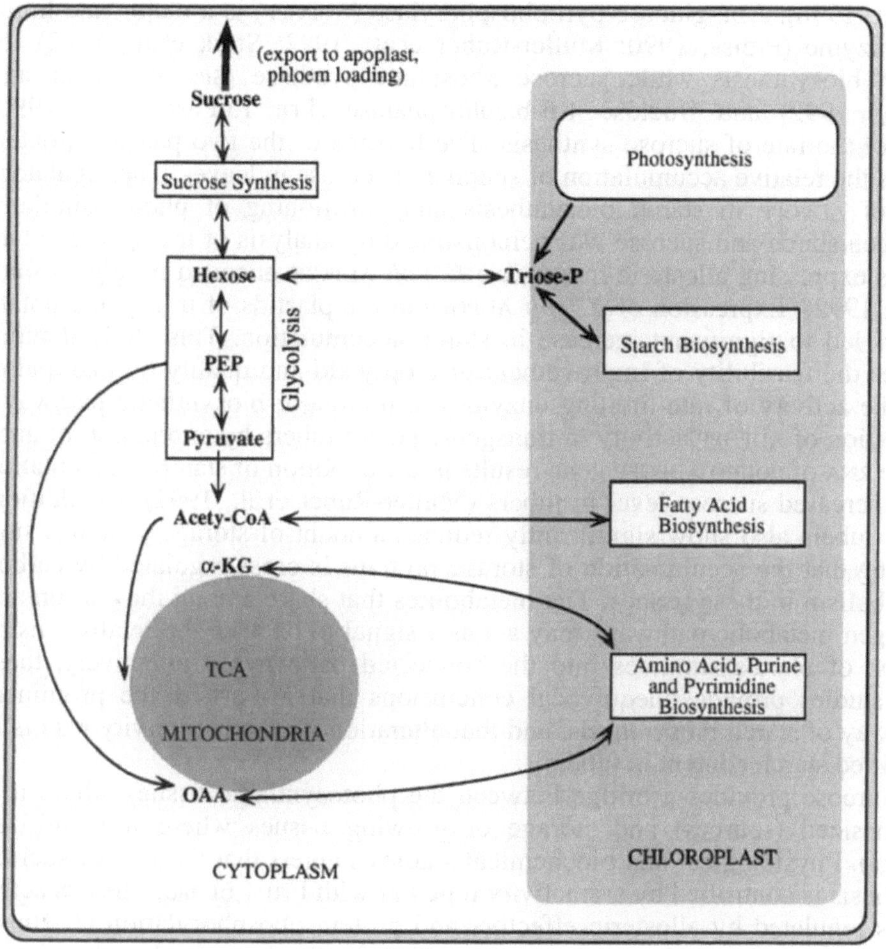

Fig.1. Interconnections and subcellular compartmentalization of major metabolic pathways in the photosynthetic leaves. Arrows point to the direction of the pathways; PEP, phosphoe*nol*pyruvate; OAA, oxaloacetate; α-KG, α-ketoglutarate; Triose-P, triose-phosphate; Acety-CoA, acetyl coenzyme A; TCA, the tricarboxylic acid or the citric acid cycle

catabolism, anabolism and photosynthesis. Figure 1 summarizes the major metabolic events and the primary regulatory points where these metabolic pathways are connected in leaves. These pathways merge at a few central metabolites such as acetyl-CoA, pyruvate, oxaloacetate and α-ketoglutarate in glycolysis and the citric acid cycle. Catabolic pathways are relatively convergent; biosynthetic pathways are, in contrast, highly branched. Therefore, the partitioning of photoassimilates into different biosynthetic pathways are expected to be regulated by global metabolic flux as well as pathway-specific regulatory mechanisms.

The photoassimilate distribution between starch and sucrose is best studied in photosynthetic leaves (Stitt, 1990; J. Huber and Huber, 1992; S. Huber and

Kaiser, 1996). ADP-glucose pyrophosphorylase (ADPGPP) is a major rate-limiting enzyme (Preiss, 1990; Muller-Ruber et al., 1992; Stark et al., 1992) for starch biosynthesis while sucrose phosphate synthase (SPS; J. Huber and Huber, 1992) and fructose 1,6-bisphosphatase (Fru 1,6Pase; Stitt, 1990) control the rate of sucrose synthesis. The balance of the two pathways determines the relative accumulation of starch and sucrose in leaves. The regulatory role of ADPGPP in starch biosynthesis and partitioning of photoassimilates between starch and sucrose was demonstrated by analysis of transgenic potato plants expressing allosteric insensitive *E. coli* ADPGPP encoded by *glgC* (Stark et al., 1992). Expression of *E. coli* ADPGPP in the plastids of transgenic potato tubers led to significant increase in starch accumulation. This study demonstrates the feasibility of improvement of crop yield and quality by manipulating the activity of rate-limiting enzymes in the major biosynthetic pathways. Inhibition of ADPGPP activity in transgenic potato tubers by expression of antisense RNA of potato ADPGPP gene results in the abolition of starch accumulation and increased sucrose level in tubers (Muller-Ruber et al., 1992). In addition, these tubers also show significantly reduced amount of storage proteins, suggesting that the acculumation of storage proteins is cross-regulated by carbon metabolism in these tissues. The metabolites that share among the carbon and nitrogen metabolic pathways may act as a signal to balance the relative distribution of photoassimilates into the connected pathways. Collectively, these two studies provide unequivocal conclusions that ADPGPP is the prominent pathway of starch biosynthesis, and that alteration of ADPGPP activity results in modified starch content in tubers.

Surcose provides a bridge between the photosynthetic tissues where it is synthesized (sources) and storage or growing tissues where it is utilized (sinks). Physiological and biochemical studies suggest that the rate of sucrose synthesis is controlled by SPS activity together with Fru 1,6Pase. The SPS activity is regulated by allosteric effectors and protein phosphorylation (J. Huber and Huber, 1992). Superimposed on the feedback and phosphorylation modification mechanism, SPS activity also appears to depend on levels of SPS transcripts (Worrell et al., 1991). Over-expression of a maize SPS in transgenic leaves of tomato plants results in up to six-fold increase of SPS enzyme activity, consequently causing a reduction of starch and an increase of sucrose in leaves (Worrell et al., 1991). These results further substantiate that SPS is a major regulator that controls sucrose biosynthesis, thereby altering carbon partitioning in leaves.

As the primary carbon photoassimilate, sucrose is distributed into both growing and storage sinks where it is cleaved by sucrose synthase into glucose and fructose to support various biosynthetic pathways. Therefore, sucrose synthase plays a pivotal role in source and sink regulation of photoassimilate distribution. This has now been demonstrated by the studies in transgenic plants (Von Schaewen et al., 1990; Sonnewald et al., 1991; Heneike et al., 1992). The introduction of the yeast invertase gene driven by a constitutive promoter, cauliflower mosaic virus 35S promoter, into transgenic potato plants, leads to an increase in the protein to starch ratio in the transgenic tubers (Heineke et al.,

1992). The level of hexoses as well as amino acids was found to be elevated significantly in the transgenic potato plants. These plants produce fewer tubers as compared to control plants. The markedly increased accumulation of protein contents is due to the increase of main storage protein (e.g., patatin). The higher invertase activity may direct the carbon flux to glycolysis and the citric acid cycle that supplies the carbon skeleton to amino acid biosynthetic pathways.

Another major branch point for carbon assimilation pathways is fatty acid biosynthesis. Fatty acids are the major components of cellular membranes. In some plants it is also the main storage compound in seeds. The biosynthesis of fatty acid can be considered as a branch pathway of carbon metabolism (Fig. 1). Acetyl-CoA is a precursor and a major branch point in the biosynthesis of fatty acids, amino acids, and carbohydrate. Therefore, the global signal sensing and regulatory strategies are expected at this branch point of the multiple pathways. In plants, ACCase catalyzing the first reaction of the fatty acid biosynthetic pathway has been implicated as a rate-limiting enzyme of the pathway (Harwood, 1988; Post-Beittenmiller et al., 1992), as in animal cells (Kim et al., 1989; Ha et al., 1994). However, the direct evidence is still missing. Genes encoding ACCase have now been isolated from a number of plants (Roesler et al., 1994; Gornicki et al., 1994; Shorrosh et al., 1994). Both cytosolic and plastidal form of ACCase have been found in pea (Konishi and Sasaki, 1994), suggesting the existence of a dual location of the pathway, although fatty acid biosynthesis takes place primarily in plastids. In animal cells, ACCase activity is rapidly regulated by reversible phosphorylation (Kim et al., 1989; Ha et al., 1994) in close cooperation with allosteric control mechanisms. AMPK phosphorylates serine residue at position −79 of ACCase (Ha et al., 1994) and thus inactivates the enzyme (Davies et al., 1992). In addition to ACCase, diacylglycerol acyltransferase (DGAT), catalyzing triacylglycerol synthesis in seeds, has also been suggested to be an important regulatory enzyme for oil biosynthesis in seeds (Wilson et al., 1992).

Identification of rate-limiting steps is the first step towards alteration of metabolic pathways. Probing of putative regulatory genes of metabolic pathways in transgenic plants offers new opportunity to identify the critical regulatory point(s) in the complex metabolic pathways. This is illustrated by work leading to determination of the rate-limiting step in proline biosynthetic pathway (Szoke et al., 1992; Kavi Kishor et al., 1995). Genes encoding all the enzymes in proline biosynthetic pathway have recently been isolated (Delauney and Verma, 1990; Hu et al., 1992; Delauney et al., 1993). Δ1-pyrroline-5-carboylate synthetase (P5CS; Hu et al., 1992) and Δ1-pyrroline-5-carboylate reductase (P5CR; Delauney and Verma, 1990) control the first and last steps of the proline biosynthetic pathway, respectively. Transgenic tobacco plants expressing up to 50-fold higher P5CR activity as compared to the control plants, did not synthesize more proline (Szoke et al., 1992). In contrast, overexpression of P5CS led to 10- to 18-fold increase in free proline levels in transgenic tobacco plants (Kavi Kishor et al., 1995). These results demonstrate that the rate-limiting step in proline biosynthesis pathway is controlled by P5CS and not by P5CR.

Carbon and nitrogen metabolism are interconnected (Fig. 1). Expression of photosynthetic genes in *Chlamydomonas reinhardtii* appears to be modulated by the availability of nitrogen (Plumley and Schmidt, 1989). Expression of the nitrate reductase gene in tobacco is antagonistically regulated by nitrogen and carbon metabolites (Vincentz et al., 1993). While glucose, sucrose and fructose up-regulate expression of nitrate reductase gene, glutamine represses expression of both nitrate and nitrite reductase genes. A pleiotropic tobacco mutation affecting both carbon and nitrogen metabolism has now been isolated (Faure et al., 1994). The biochemical and molecular basis of the interactions between carbon and nitrogen metabolism remains to be elucidated.

Glutamine synthetase (GS) plays a central role in nitrogen metabolism, as glutamine is the key metabolite that connects carbon and nitrogen metabolism through the citric acid cycle (Fig. 1) (also see Hirel et al., 1993). In general, GS genes are highly expressed in tissues where active nitrogen metabolism takes place (Hirel et al., 1993). In *Selenastrum*, higher glutamine to glutamate and α-ketoglutarate ratio determined by the GS/GOGAT activities appears to act as a metabolite signal to activate the phospho*enol*pyruvate carboxylase (PEPCase), a key enzyme of the anapleurotic pathway that replenishes depleted carbon skeleton in the citric acid cycle for amino acid biosynthesis. Recently, Huppe and Turpin (1994) further hypothesized that a similar metabolic signal may also activate PEPCase to coordinate carbon and nitrogen metabolism in plants. This is consistent with the fact that dicarboxylic acids generated by PEPCase-mediated reductive pathway are the primary carbon source for symbiotic nitrogen fixation and assimilation (Dilworth and Glenn, 1984; Vance and Heichel, 1991).

B. Cellular Compartmentalization and Metabolite Transport

Metabolic pathways are compartmentalized in cells, separating distinct metabolic processes from each other. The intercellular and intracellular metabolite pools are established and maintained by metabolite transporter systems (Bush, 1993) and metabolic multienzyme complexes (metabolons; Srere 1987; Hrazdina and Jensen, 1992) formed by protein–protein interactions of enzymatic components of the same pathways. Specific sugar and amino acid transporter systems regulate the metabolite pools in subcellular compartments (Bush, 1993). Therefore, they play a central role in regulation of diverse metabolic pathways and of overall flux of photoassimilates. In plants and other eukaryotic cells, separation of metabolic pathways is further reinforced by subcellular compartmentation, a simple regulatory mechanism in which the enzymes for distinct pathways are physically separated by membrane-bound organelles. Catabolic and biosynthetic pathways are two different processes that generally take place in different subcellular compartments. The primary catabolic pathways, glycolysis and the citric acid cycle, are located in cytoplasm and mitochondria respectively. In plants, however, glycolysis also occurs in the plastids to support photosynthesis and other biosynthetic pathways (Hrazdina and Jensen, 1992). Most of the biosynthetic pathways including starch, lipid, and amino acid biosynthesis are compartmentalized in plastids in plants while these pathways are generally located in the cytoplasm in other organisms.

Taking advantage of the large numbers of well defined sugar and amino acid transporter mutants in yeast and of the T-DNA tagged *Arabidopsis* metabolite transporter mutants, plant genes encoding sugar, amino acid, and other nutrient transporters have now been isolated (Riesmeier et al., 1992; Schachtman et al., 1992; Frommer et al., 1993; Tsay et al., 1993; Ninnemann et al., 1994; Schachtman and Schroeder, 1994). The activity of transporters for cellular metabolites plays an important role in partitioning of photoassimilates. The potato sucrose transporter gene, *StSUT1*, is highly expressed in mature leaves, whereas stem and sink tissues, such as developing leaves, show only low expression. The expression of *StSUT1* is specifically localized in the phloem. Transgenic potato plants with reduced sucrose transporter transcripts achieved by expression of antisense RNA dramatically increased leaf carbohydrate content (Riesmeier et al., 1993b), providing direct molecular evidence for the proposed apoplastic transport of sucrose in photosynthetic leaves, a mechanism for phloem loading (Giaquinta, 1983). Triose phosphate (triose-P) represents the net product of the Calvin cycle of photosynthesis. The export of triose-P into the cytosol is of central importance with respect to photoassimilate partitioning between sources and sinks as well as allocation between various pathways such as sucrose and starch synthesis. The chloroplast triose-P translocator (TPT) mediates transport of triose-P and 3-phosphoglycerate (3-PGA) out of the chloroplast in a strict counter exchange for P_i. Expression of tobacco TPT gene is repressed by increased sucrose levels (Knight and Gray, 1994). Antisense RNA repression of TPT gene expression in potato plants results in a 20–30% reduction of P_i import (Riesmeier et al., 1993a). The reduced P_i import to chloroplasts leads to a 40–60% reduction in photosynthetic rate and an increase of starch accumulation in leaves at the expense of sucrose and amino acids. Therefore, the TPT is a major rate-limiting factor that regulates the flow of photoassimilates into different biosynthetic pathways. Molecular characterization and targeted manipulation of metabolite transporter systems in plants will continue to help in our understanding of how metabolite pools coordinately affect carbon partitioning at the whole plant levels.

C. Carbon Metabolite Repression in Plants

Although catabolite repression is well known in regulation of carbon and nitrogen metabolism in microorganisms, repression of plant genes involved in metabolic pathways was discovered recently. Sheen (1990) reported that presence of elevated sucrose and glucose levels can specifically repress the expression of seven maize photosynthetic genes in maize mesophyll protoplasts, thus demonstrating that genes involved in metabolic pathways are subject to regulation by the fluctuation of cellular metabolites in plants, a mechanism similar to catabolite repression in prokaryotes (Fisher and Sonenshein, 1991; Saier, 1991) and lower eukaryotes (Gancedo, 1992; Trumbly, 1992; Marzluf, 1993).

The metabolic repression provides a mechanism for plant cells to efficiently respond to the fluctuation of cellular metabolites, and to maintain a balance of different metabolic pathways. The glucose repression of ribulosebisphosphate carboxylase small subunit (*rbcS*) gene has also been demonstrated in a whole plant system (Krapp et al., 1993). Expression of invertase in the cell wall of tobacco leaves resulted in accumulation of carbohydrate and inhibition of photosynthesis (Von Schaewen et al., 1990). This was accompanied by significant decrease of *rbcS* transcript levels in the transgenic plants (Krapp et al., 1993), demonstrating that repression of photosynthetic genes by glucose and sucrose is of physiological significance. That photosynthetic genes such as *rbcS* are inhibited by carbohydrate, most probably by metabolic factors related to the metabolism of carbohydrate (Krapp et al., 1993), suggests a possible control mechanism for the sink regulation of photosynthesis (Stitt, 1990).

Sugar levels also modulate differential expression of two maize sucrose synthase genes, *Sh1* and *Sus1*, which are expressed in different tissues and cells of developing maize (Koch et al., 1992). The levels of *Sh1* transcripts increase in excised root tips under glucose deprivation conditions; addition of glucose (4%) can repress the accumulation of *Sh1* transcripts (Koch et al., 1992). The repression of *Sh1* transcript levels by sugar is controlled at the transcriptional level (Maas et al., 1990). In contrast, *Sus1* transcript levels are increased when 4% of glucose was included in the culture media. Although *Sh1* and *Sus1* are expressed in different tissues and regulated in an inverse mode by sugars, these two genes are functionally redundant (Chourey and Taliercio, 1994). Normally, only the *Sus1* encoded SS2 is detected in embryos. In *sus* mutant embryos, however, *Sh1* encoded protein SS1 is detected. Furthermore, there is no phenotypic change in the *sus* mutant embryos, demonstrating the functional compensation of *Sh1* to *Sus1* in the embryos. Ectopic expression of *Sh1* encoded SS1 in embryos where otherwise only *Sus1* is expressed suggests that metabolic regulatory control overrides the developmental regulation of the two genes. The molecular mechanism underlying this phenomenon remains unsolved. Nevertheless, these results illustrate the complexity of regulation of these genes by metabolites as well as plant developmental status.

Expression of genes involved in carbon catabolism is also regulated by carbohydrate starvation. Sucrose repression of rice α-amylase gene expression has been demonstrated in both suspension-cultured cells (Yu et al., 1991) and in germinating seeds (Karrer and Rodriguez, 1992). The α-amylase catalyzes the hydrolysis of α-1,4 glucan bonds of amylose and amylopectin in the starchy endosperm of cereals. It plays an important role in mobilizing storage starch into various biosynthetic pathways in germinating seedlings. Activation of α-amylase by sugar starvation allows cells to keep the demand of active biosynthetic pathways in the germinating seedlings. Phytohormone gibberellic acid (GA) also induces the expression of α-amylase genes in cereals. Therefore, both carbon metabolite signals and GA signal are involved in control of carbon metabolism in the germinating cereal seedling (for a review, see Thomas and Rodriguez, 1994).

The glyoxalate cycle converts fatty acids to carbohydrate and other metabolites, a unique metabolic pathway of plants and microorganisms. In oilseed plants like castor beans, storage fatty acids are metabolized to carbohydrate through the pathway to support postgerminative growth. Isocitrate lyase (ICL) and malate synthase (MS) are the two key enzymes regulating the pathway (Kornberg and Beevers, 1957). Sugar starvation of cucumber cell culture results in the significant induction of the expression of both genes, and this effect is reversible by addition of sucrose, glucose and fructose to the culture media (Graham et al., 1992, 1994). Induction of ICL and MS genes under carbohydrate deficient conditions activates the glyoxylate cycle that plays an anapleurotic role in replenishing the metabolites of the citric acid cycle. Collectively, these studies demonstrate that sugar repression, as in microorganisms, is a regulatory mechanism controlling expression of a variety of genes involved in different metabolic pathways (Table 1), and that such a regulatory mechanism is of physiological importance in plants.

The changes in gene expression by carbon metabolites are controlled at the transcriptional level (Maas et al., 1990; Sheen, 1990). Promoter analysis of maize pyruvate phosphodikinase (PPDK) and chlorophyll *a/b*-binding protein genes reveals that repression is mediated by the positive upstream regulatory elements. The analysis of seven different photosynthetic gene promoters of maize did not reveal any consensus sequences among the upstream regulatory elements (Sheen, 1990). Acetate also repressed the expression of the same set of genes. Moreover, the repression of most of these sugar repressible genes by acetate was ten times stronger than by sugar, indicating that two distinctive pathways may be involved in sugar and acetate repression of photosynthetic genes (Sheen, 1990). Acetate also repressed photosynthetic genes and activated genes in the glyoxylate pathway of green algae, in which acetate is the preferred carbon source (Monroy and Schwartzbach, 1984; Kindle, 1987).

Table 1. List of selected plant genes repressed by carbon metabolites

Enzymes encoded by the repressible genes	References
Pyruvate phosphodikinase	Sheen, 1990
Phospho*enol*pyruvate carboxylase	Sheen, 1990
Malic enzyme	Sheen, 1990
Chlorophyll *a/b*-binding protein	Sheen, 1990; Criqui et al., 1992
Ribulosebisphosphate carboxylase small subunit	Sheen, 1990; Criqui et al., 1992; Krapp et al., 1993
Sucrose synthase (Sh1)	Mass et al., 1990; Koch et al., 1992
α-Amylase	Yu et al., 1991; Karrer and Rodriguez, 1992
Isocitrate lyase	Graham et al., 1994
Malate synthase	Graham et al., 1992, 1994
Triose-phosphate translocator	Knight and Gray, 1994
SNF1-like protein kinase	Sano and Youssefian, 1994

Although it is now generally accepted that carbon metabolites repress genes of metabolic pathways in microorganisms and in plants (Table 1) as well, the regulatory strategies are likely to be diverse among these organisms. In bacteria and yeast, carbon sources in the form of sugars can be supplied exogenously in the surrounding environments. Presence of the preferred carbon source, glucose, leads to repression of utilization of the other metabolizable sugars such as sucrose. Higher plants are capable of synthesizing reduced carbon including glucose indigenously through photosynthesis in chloroplasts. The reduced carbon in plastids can then be translocated in the form of sucrose into various sinks to support biosynthetic pathways (Giaquinta, 1983). Carbon metabolite repression of plant genes by both glucose and sucrose as opposed to by glucose alone in microorganisms may reflect the regulatory adaptation of plants to the fundamental difference in carbon source utilization and metabolism. Furthermore, expression of sugar repressible genes in plants is also subject to regulation by plant development status and other environmental cues. However, metabolic repression may override the developmental and light regulation of these genes (Sheen, 1990; Cheng et al., 1993; Chourey and Taliercio, 1994). Understanding the signal transduction pathways leading to the regulated expression of these genes by developmental status and different metabolites will provide significant insight as to how metabolite pools in planta affect plant development and metabolism.

D. Metabolite Sensing and Nutrient-mediated Signal Transduction Pathways in Plants

The metabolite signals that give rise to repression of ICL and MS genes of the cucumber glyoxylate cycle appear to be phosphorylated hexoses or the flux of hexoses into glycolysis (Graham et al., 1994). 2-Deoxyglucose, which is phosphorylated by hexokinase but not further metabolized in cells, effectively represses both MS and ICL gene expression. In contrast, the non-phosphorylated glucose analog, 3-methyglucose, does not cause repression of either ICL or MS. Therefore, the level of phosphorylated hexose produced by hexokinase appears to be the metabolic signal molecule that leads to repression of ICL and MS. This is reminiscent of the sugar sensor mechanism of glucose repression in bacteria (Saier, 1991; Angell et al., 1994) and yeast (Gancedo, 1992; Trumbly, 1992). In yeast, hexokinase PII (*HXK2*) has been implicated as the initial sensor for glucose levels. Mutations in *hxk2* result in derepression of glucose repressible genes (Entian et al., 1985). Furthermore, the phosphorylation activity of HXK2 is directly correlated with its glucose repression capability (Ma et al., 1989). Repression of photosynthetic genes also requires further metabolized glucose rather than the absolute concentration and transport of glucose (Krapp et al., 1993). Treatment of maize mesophyll protoplasts with the specific inhibitors to HXK2 blocks the glucose repression of photosynthetic genes, further confirming that glucose phosphorylated by HXK2 is required for glucose repression of these genes (Jang and Sheen, 1994). These recent studies indicate a common role of HXK2 in sugar metabolite repression in plants (Krapp et al., 1993; Graham et al., 1994; Jang and Sheen, 1994) and in microorganisms (Saier, 1991; Gancedo, 1992; Trumbly, 1992; Angell et al., 1994). However, it is still not known how hexokinase senses glucose levels and transmits the signals to the immediate targets in both systems.

Many diverse developmental and metabolic responses are mediated through a protein phosphorylation and dephosphorylation hierarchical regulatory systems (Ranjeva and Boudet, 1987; Jiao and Chollet, 1989; Allen, 1992; Champigny and Foyer, 1992; Chang et al., 1993; Kieber et al., 1993; Miao et al., 1993; Sheen, 1993; Halford et al., 1994; S. Huber, 1996). Reversible protein phosphorylation is a major means of metabolic regulation in plants (Budde and Chollet, 1988). One such example is the inverse regulation of SPS and PEPCase activity by reversible phosphorylation (Jiao and Chollet, 1988; J. Huber et al., 1989; Bakrim et al., 1992; Champigny and Foyer, 1992; S. Huber and Huber, 1992). SPS plays a major role in regulating the flux of carbon into sucrose (S. Huber and Huber, 1992) and PEPCase is a key enzyme of the anapleurotic pathway. PEPCase is activated by phosphorylation of the protein by a serine protein kinase (Jiao and Chollet, 1990). In contrast, SPS activity is inhibited by protein phosphorylation, and dephosphorylation of the protein activates SPS activity (S. Huber and Huber, 1992). It has been hypothesized that nitrate is the common metabolite signal that coordinately regulates the activity of SPS and PEPCase, possibly by activating the cytosolic protein kinases (Champigny and Foyer, 1992; Van Quy and Champigny, 1992). Several other enzymes involved in carbon and nitrogen metabolism and photosynthetic proteins have been shown to be subject to regulation by protein kinase and protein phosphatase-mediated reversible phosphorylation (J. Huber et al., 1992; Sheen, 1993; LaBrie and Crawford, 1994).

It is now clear that phosphorylation and dephosphorylation is a major event regulating the key enzymes of metabolic pathways. Identification of protein kinases and phosphatases responsible for the modification of the enzymes should allow a better understanding of the regulatory cascades. Molecular characterization of mutations that fail to phosphorylate or dephosphorylate the respective enzymes will create an entry to elucidation of the metabolite-mediated phosphorylation and dephosphorylation systems regulating metabolic pathways.

Plant homolog of the yeast SNF1 protein kinase gene have been isolated (Alderson et al., 1991, Halford et al., 1992, Le Guen et al, 1992; Muranake et al., 1994; Sano and Youssefian, 1994). Rye *SNF1*-like cDNA sequence, c*RKIN1*, is capable of rescuing *snf1* mutation in yeast, suggesting *RKIN1* may also be involved in regulation of carbon metabolism in plants. Expression of *RKIN1* is primarily in endosperm of rye (Alderson et al., 1991), where storage starch is synthesized and deposited. However, *SNF1*-like genes from other plants appear to be constitutively expressed (Le Guen et al., 1992; Muranake et al., 1994). Expression of wheat *SNF1* homolog, *Wpk4*, is regulated by nutrient status (Sano and Youssefian, 1994). Transfer of seedlings from a nutrient-rich MS culture medium to water results in a dramatic increase in *Wpk4* transcripts in leaves. Addition of glucose down regulates expression of *Wpk4*. Genomic Southern analysis indicated that *SNF1*-like protein kinase genes are encoded by a gene family in plants. These related sequences may encode isozymes that play similar roles in different tissues and at developmental stages. Alternatively, these sequences may encode SFN1-related protein kinases that regulate different metabolic pathways or downstream targets.

To establish the role of SNF1-like protein kinases in carbon metabolism in plants, effort has been focused on manipulation of SNF1-like protein kinase activity in the tuber of potato plants by antisense RNA approach. Transgenic potato plants that have reduced SNF1 protein kinase activity (up to 90% inhibition compared to control plants) specifically in potato tubers have been recovered (N. G. Halford, pers. comm.). No apparent morphological differences in the aerial parts of the transgenic potato plants were observed. However, preliminary results also indicated that constitutive expression of the antisense potato *SNF1*-like sequence may be lethal to plants. In yeast, SNF1 protein kinase activity is regulated by phosphorylation and dephosphorylation (Woods et al., 1994). The *SNF1* transcript and protein level are not altered under both glucose repression and derepression conditions (Celenza and Carlson, 1986). On the other hand, the steady levels of transcripts for wheat *Wpk4* is significantly increased under glucose derepression conditions (Sano and Youssefian, 1994). Therefore, regulation of plant *SNF1*-like genes may differ from that of *SNF1* in yeast.

AMPK regulates mammalian lipid metabolism by phosphorylating ACCase, thereby inactivating the committing enzyme in fatty acid biosynthesis (Davies et al., 1992; Ha et al., 1994). Recently a cDNA encoding AMPK has been isolated from rat liver (Carling et al., 1994). DNA sequence analysis reveals that it shares a significant degree of sequence homology to yeast *SNF1* and plant *SNF1*-like sequences (Carling et al 1994). Thus AMPK appears to be a member of SNF1 protein kinase family. *SNF1* plays a major role in regulation of carbon metabolism in yeast. The structural similarity of the protein kinases in animal, plant and yeast suggests possible involvement of SNF1 protein kinase in the regulation of lipid biosynthetic pathway in addition to that of sugar metabolism. This possibility has now been experimentally confirmed in vitro (Mitchelhill et al., 1994) and in vivo (Woods et al., 1994). In wild type yeast the ACCase activity undergoes a time-dependent decrease in glucose derepression condition. However, such glucose mediated decrease of ACCase activity was not observed in *snf1* mutants (Woods et al., 1994), suggesting that regulation of ACCase by glucose is specifically mediated by SNF1 protein kinase activity. The glucose deprivation activated SNF1 protein kinase appears to phosphorylate ACCcase, thus inactivating the enzyme in vivo. The *SNF1*-like sequences have also been isolated from oilseed plants (Miao and Heppard, unpubl. results). With the cloned plant genes for SNF1-related kinases and plant ACCase (Gornicki et al., 1994; Roesler et al., 1994; Shorrosh et al., 1994), the potential regulation of ACCase by SNF1-like protein kinases can be tested directly.

VI. Future Prospects: Challenges for Metabolic Engineering of Crop Plants

Significant progress has been made towards understanding the regulatory events of metabolic pathways and partitioning over the past decade. Recent developments in gene transfer technology in major crop plants have created new opportunities to enhance the yield and quality of crops by manipulating the major metabolic pathways. Several successful experiments have recently demonstrated the feasibility and potential benefits of metabolic engineering through biotechnology to future agriculture (Voelker et al., 1992; Stark et al., 1992; Riesmeier et al., 1993; Tarczynski et al., 1993; Kavi Kishor et al., 1995). These pioneering researches have laid a foundation for the emerging science: plant metabolic engineering. However, plant metabolic engineering is still in its infancy. A fundamental understanding of the molecular and mechanistic aspects of regulation of metabolic pathways and photoassimilate partitioning is of central importance for the rapid realization of the promised benefits of this technology to agricultural industry.

Information from understanding strategies of metabolic regulation in microorganisms has provided some very useful insights into the relatedness between these organisms and higher plants. On the other hand, many fundamental differences in metabolism exist between multicellular higher plants and microorganisms. Therefore, in addition to recognizing the similarity of regulatory components in different organisms, we should continue to anticipate novel and unique metabolic regulatory strategies operating in plants. Metabolic control architecture are nonlinear and interconnected at multiple key branch points. Alteration of a single rate-limiting step of the pathways may not necessarily lead to overproduction of many metabolites that requires significant redirection of flux distributions in primary metabolism, because the network rigidity has evolved to resist metabolic flux changes at branch points (Stephanopoulos and Vallino, 1991). Existence of alternative routes of many plant metabolic pathways (Knight and Langston-Unkefer, 1988; Delauney and Verma, 1993) that are functionally redundant adds another layer of complexity to the regulation of metabolic pathways. Furthermore, regulation of metabolic pathways is coupled to plant development. Accumulation of photoassimilates in storage organs may not only be determined by relative metabolic flux controlled by global and pathway-specific regulatory mechanisms but also influenced by developmental status, e.g., cell numbers and structure of storage organs, determined by cell division and pattern formation.

Despite the fact that several regulatory points of some metabolic pathways have been identified, our knowledge of metabolite-mediated signal perception and transduction events regulating these key branching points is still very limited and fragmented. What are the different developmental and metabolic regulatory strategies operating in the oil- and starch-producing seeds? What is the molecular nature of metabolic signals that are involved in coordinating the interactions between carbon and nitrogen metabolic pathways? How are these signals perceived and transduced to affect plant metabolism as well as development? Answers to these questions are pertinent to the future success of rational design of plant metabolism through biotechnology.

Acknowledgments

I thank N. G. Halford for generously providing preprints and unpublished data, and R. M. Broglie, T. Klein, Y. S. Yamamoto, and S. Luan for their comments on the manuscript.

V. References

Alex LA, Simon MI (1994) Protein histidine kinases and signal transduction in prokaryotes and eukaryotes. Trends Genet 10: 133–138

Albright LM, Huala E, Ausubel FM (1989) Prokaryotic signal transduction mediated by sensor and regulator protein pairs. Annu Rev Genet 23: 311–336

Alderson A, Sabelli PA, Dickinson R, Cole D, Richardson M, Kreis M, Shewary PR, Halford NG (1991) Complementation of *snf1*, a mutation affecting global regulation of carbon metabolism in yeast, by a plant protein kinase CDNA. Proc Natl Acad Sci USA 88: 8602–8605

Allen FJ (1992) How does protein phosphorylation regulate photosynthesis? Trends Biochem Sci 17: 12–17

Angell S, Lewis CG, Buttner MJ, Bibb MJ (1994) Glucose repression in *Streptomyces coelicolor* A3(2): a likely regulatory role for glucose kinase. Mol Gen Genet 244: 135–143

Bakrim N, Echevarria C, Cretin C, Arrio-Dupont M, Pierre N, Vidal J, Chollet R, Gadal P (1992) Regulatory phosphorylation of *Sorghum* leaf phospho*enol*pyruvate carboxylase: identification of the protein-serine kinase and some elements of the signal-transducing cascade. Eur J Biochem 204: 821–830

Bate NJ, Orr J, Ni W, Meromi A, Nadler-Hassar T, Doerner PW, Dixon RA, Lamb CJ, Elkind Y (1994) Quantitive relationship between phenylalanine ammonialyse levels and phenylpropanoid accumulation in transgenic tobacco identifies a rate-determining step in natural product synthesis. Proc Natl Acad Sci USA 91: 7608–7612

Blumer KJ, Johnson GL (1994) Diversity in function and regulation of MAP kinase pathway. Trends Biochem Sci 19: 236–240

Broek D, Toda T, Michaeli T, Levin L, Birchmeier C, Zoller M, Powers S, Wigler M (1987) The *S. cerevisiae* CDC25 gene product regulates the RAS/adenylate cyclase pathway. Cell 48: 789–799

Budde RJA, Chollet R (1988) Regulation of enzyme activity in plants by reversible phosphorylation. Physiol Plant 72: 435–439

Bush DR (1993) Proton-coupled sugar and amino acid transporters in plants. Annu Rev Plant Physiol Plant Mol Biol 44: 513–542

Carling D, Aguan K, Woods A, Verhoeven AJM, Beri RK, Brennan CH, Sidebottom C, Davison MD, Scott J (1994) Mammalian AMP-activated protein kinase is homologous to yeast and plant protein kinases involved in the regulation of carbon metabolism. J Biol Chem 269: 11442–11448

Carlson M, Osmond BC, Neigeborn L, Botstein D (1984) A suppressor of *snf1* mutations causes constitutive high-level invertase synthesis in yeast. Genetics 107: 19–32

Celenza JL, Carlson M (1986) A yeast gene that is essential for release from glucose encodes a protein kinase. Science 233: 1175–1180

Celenza JL, Eng FJ, Carlson M (1989) Molecular analysis of the *SNF4* gene of *Saccharomyces cerevisiae*: evidence for physical association of the SNF4 protein with the SNF1 protein kinase. Mol Cell Biol 9: 5045–5045

Champigny M-L, Foyer C (1992) Nitrate activation of cytosolic protein kinases diverts photosynthetic carbon from sucrose to amino acid biosynthesis. Plant Physiol 100: 7–12

Chang C, Kwok SF, Bleecker AB, Meyerowitz EM (1993) *Arabidopsis* ethylene-reseponse gene *ETR1*: similarity of product to two-component regulators. Science 262: 539–544

Cheng C-L, Acedo GN, Christin M, Conkling MA (1992) Sucrose mimics the light induction of *Arabidopsis* nitrate reductase gene transcription. Proc Natl Acad Sci USA 89: 1861–1864

Choi K-Y, Satterberg B, Lyons DM, Elion EA (1994) Ste5 tethers multiple protein kinases in the MAP kinase cascade required for mating in *S. cerevisiae*. Cell 78: 499–512

Chourey RS, Taliercio EW (1994) Epistatic interaction and functional compensation between the two tissue- and cell-specific sucrose synthase genes in maize. Proc Natl Acad Sci USA 91: 7917–7921

Costigan C, Snydern M (1994) SLK1, a yeast homolog of MAP kinase activators, has a RAS/cAMP-independent role in nutrient sensing. Mol Gen Genet 243: 286–296

Criqui MC, Durr A, Parmentier Y, Marrach J, Fleck J, Jamet E (1992) How are photosynthetic genes repressed in freshly isolated mesophyll protoplasts of *Nicotiana sylvestris*? Plant Physiol Biochem 30: 597–601

D'Amico L, Valsasina B, Daminati G, Fabbrini S, Nitti G, Bollini R, Ceriotti A, Vitale A (1992) Bean homologs of the mammalian glucose-regulated proteins: induction by tunicamycin and interaction with newly synthesized seed storage proteins in the endoplasmic reticulum. Plant J 2: 443: 455

Davies SP, Carling D, Mundy MR, Hardie DG (1992) Diurnal rhythm of phosphorylation of rat liver acetyl-CoA carboxylase by the AMP-activated protein kinase, demonstrated using freeze-clamping. Eur J Biochem 203: 615–623

Delauney AJ, Verma DPS (1990) A soybean gene encoding Δ1-pyrroline-5-carboxylate reductase was isolated by functional complementation in *Escherichia coli* and is found to be osmoregulated. Mol Gen Genet 221: 299–305

Delauney AJ, Verma DPS (1993) Proline biosynthesis and osmoregulation in plants. Plant J 4: 215–223

Delauney AJ, Hu C-A, Kavi Kishor PB, Verma DPS (1993) Ornithine amintransferase and proline biosynthesis in *Vigna aconitifolia*. J Biol Chem 268: 18673–18678

Dilworth M, Glenn A (1984) How does a legume nodule work? Trends Biochem Sci 9: 519–523

Duerr B, Gawienowski M, Ropp T, Jacob T (1993) MsERK1: a mitogen-activated protein kinase from a flowering plant. Plant Cell 5: 87–96

Entian KD, Hilberg F, Opitz H, Mecke D (1985) Cloning of hexokinase structural genes from *Saccharomyces cerevisiae* mutants with regulatory mutations responsible for glucose repression. Mol Cell Biol 5: 3035–3040

Errede B, Levin DE (1993) A conserved kinase cascade for MAP kinase activation in yeast. Curr Opin Cell Biol 5: 254–260

Faure J-D, Jullien M, Caboche M (1994) Zea3: a pleiotropic mutation affecting cotyledon development, cytokinin resistance and carbon-nitrogen metabolism. Plant J 5: 481–491

Fisher SH, Sonenshein AL (1991) Control of carbon and nitrogen metabolism in *Bacillus subtilis*. Annu Rev Microbiol 45: 107–135

Frommer WB, Hummel S, Riesmeier JW (1993) Expression cloning in yeast of a cDNA encoding a broad specificity amino acid permease from *Arabidopsis thaliana*. Proc Natl Acad Sci USA 90: 5944–5948

Fu YH, Marzluf GA (1987) Molecular cloning and analysis of the regulation of *nit-3*, the structural gene for nitrate reductase in *Neurospora crassa*. Proc Natl Acad Sci USA 84: 8243–8247

Fu YH, Marzluf GA (1990a) *Nit-2*, the major positive acting nitrogen regulatory gene of *Neurospora crassa*, encodes a sequence-specific DNA-binding protein. Proc Natl Acad Sci USA 87: 5331–5335

Fu YH, Marzluf GA (1990b) *Nit-2*, the major nitrogen regulatory gene of *Neurospora crassa*, encodes a protein with a putative zinc finger DNA-binding domain. Mol Cell Biol 10: 1056–1065

Gancedo JM (1992) Carbon catabolite repression in yeast. Eur J Biochem 206: 297–313

Gaston K, Bell A, Kolb A, Buc H, Busby S (1990) Stingent spacing requirements for transcription activation by CRP. Cell 62: 733–743

Giaquinta RT (1983) Phloem loading of sucrose. Annu Rev Plant Physiol 34: 347–387

Gifford RM, Thorne JH, Hitz WD, Giaquinta RT (1984) Crop productivity and photoassimilate partitioning. Science 225: 801–808

Gilman AG (1984) Proteins and dual control of adenylate cyclase. Cell 36: 577–579.

Gornicki P, Podkowinski L, Scappino LA, DiMaio J, Ward E, Haselkorn R (1994) Wheat acetyl-coenzyme A carboxylase: cDNA and protein structure. Proc Natl Acad Sci USA 91: 6860–6864

Graham IA, Leaver CJ, Smith SM (1992) Induction of malate synthase gene expression in senescent and detached organs of cucumber. Plant Cell 4: 349–357

Graham IA, Denby KJ, Leaver CJ (1994) Carbon catabolite repression regulates glyoxylate cycle gene expression in cucumber. Plant Cell 6: 761–772

Ha J, Daniel S, Broyles SS, Kim K-H (1994) Critical phosphorylation sites for acetyl-CoA carboxylase activity. J Biol Chem 269: 22162–22168

Halford NG, Man AL, Barker JHA, Monger W, Shewry PR, Smith A, Purcell PC (1994) Investigating the role of plant SNF1-related protein kinases. Biochem Soc Trans 22: 953–957

Halford NG, Vicente-Carbajosa J, Sabelli PA, Shewry PR, Hannappel U, Kreis M (1992) Molecular analyses of a barley multigene family homologous to the yeast protein kinase gene *SNF1*. Plant J 2: 791–797

Harwood JL (1988) Fatty acid metabolism. Annu Rev Plant Physiol Plant Mol Biol 39: 101–138

Heineke DU, Sonnewald D, Bussis G, Gunter K, Leidreiter I, Wilke K, Raschke L, Willmitzer, Heldt H W (1992) Apoplastic expression of yeast-derived invertase in potato: effects on photosynthesis, leaf solute composition, water relations, and tuber composition. Plant Physiol 100: 301–308

Hirel B, Miao G-H, Verma DPS (1993) Metabolic and developmental control of glutamine synthetase genes in legume and non-legume plants. In: Verma DPS (ed) Control of plant gene expression. CRC Press, Boca Raton, pp 443–458

Hope IA, Struhl K (1985) GCN4 protein, synthesized in vitro, binds *HIS3* regulatory sequences: implication for general control of amino acid biosynthetic genes in yeast. Cell 43: 177–188

Hrazdina G, Jensen RA (1992) Spatial organization of enzymes in plant metabolic pathways. Annu Rev Plant Physiol Plant Mol Biol 43: 241–267

Hu C-A, Delauney AJ, Verma DPS (1992) A bifunctional enzyme (Δ1-pyrroline-5-carboxylate reductase) catalyzes the first two steps in proline biosynthesis in plants. Proc Natl Acad Sci USA 89: 9354–9358

Huber JLA, Huber SC, Nielsen TH (1989) Protein phosphorylation as a mechanism for regulation of spinach leaf sucrose-phosphate synthase activity. Arch Biochem Biophys 270: 681–690

Huber JL, Huber SC, Campbell WH, Redinbaugh MG (1992) Reversible light/dark modulation of spinach leaf nitrate reductase activity involves protein phosphorylation. Arch Biochem Biophys 296: 58–65

Huber SC, Huber JL (1992) Role of sucrose-phosphate synthase in sucrose metabolism in leaves. Plant Physiol 99: 1275–1278

Huber SC, Kaiser WM (1996) Regulation of C/N interaction in higher plants by protein phosphorylation. In: Verma DPS (ed) Signal transduction in plant growth and development. Springer, Wien New York, pp 87–112 [Dennis ES et al (eds) Plant gene research. Basic knowledge and application]

Huppe HC, Turpin DH (1994) Interaction of carbon and nitrogen metabolism in plant and algal cells. Annu Rev Plant Physiol Plant Mol Biol 45: 577–607

Jang J-C, Sheen J (1994) Sugar sensing and signal transduction in higher plants. In: 4th International Congress of Plant Molecular Biology, June 19–24, 1994, abstract 1329

Jiao J-A, Chollet R (1989) Regulation seryl-phosphorylation of C_4 phospho*enol*pyruvate carboxylase by a soluble protein kinase from maize leaves. Arch Biochem Biophys 269: 526–535

Joung JK, Koepp DM, Hochschild A (1994) Synergistic activation of transcription by bacteriophage-λ cl protein and *E. coli cAMP* receptor protein. Science 265: 1863–1866

Karrer EE, Rodriguez RL (1992) Metabolic regulation of rice α-amylase and sucrose synthase gene in planta. Plant J 2: 517–523

Kavi Kishor PB, Hong Z, Miao G-H, Hu C-A, Verma DPS (1995) Overexpression Δ1-pyrroline-5-carboxylate reductase increases proline production and helps maintain osmotic potential in transgenic plants during water stress. Plant Physiol 108: 1387–1394

Kieber JJ, Rothenberg M, Roman G, Feldmann KA, Ecker JR (1993) *CTR1*, a negative regulator of the ethylene response pathway in *Arabidopsis*, encodes a member of the Raf family of protein kinases. Cell 72: 427–441

Kim K-H, Lopez-Casillas F, Bai D-H (1989) Role of reversible phosphorylation of acetyl-CoA carboxylase in long-chain fatty acid synthesis. FASEB J 3: 2250–2256

Kindle KL (1987) Expression of a gene for a light-harvesting chlorophyll *a/b*-binding protein in *Chlamydomonas reinhardtii*: effect of light and acetate. Plant Mol Biol 9: 547–563

Kishore GM, Somerville CR (1993) Genetic engineering of commercially useful biosynthetic pathways in transgenic plants. Curr Opin Biotechnol 4: 152–158

Knight JS, Gray JC (1994) Expression of genes encoding the tobacco chloroplast phosphate translocator is not light-regulated and is repressed by sucrose. Mol Gen Genet 242: 586–594

Knight TJ, Langston-Unkefer PJ (1988) Enhancement of symbiotic dinitrogen fixation by a toxin-releasing plant pathogen. Science 241: 951–954

Koch KE, Nolte KD, Duke ER, McCarty DR, Avigne WT (1992) Sugar levels modulate differential expression of maize sucrose synthase genes. Plant Cell 4: 59–69

Kornberg HL, Beevers H (1957) The glyoxylate cycle as a stage in the conversion of fat to carbohydrate in castor beans. Biochim Biophys Acta 26: 531–537

Konishi T, Sasaki Y (1994) Compartmentalization of two forms of acetyl-CoA carboxylase in plants and the origin of their tolerance toward herbicides. Proc Natl Acad Sci USA 91: 3598–3601

Krapp A, Hofmann B, Schafer C, Stitt M (1993) Regulation of the expression of *rbcS* and other photosynthetic genes by carbohydrates: a mechanism for the sink regulation of photosynthesis? Plant J 3: 817–828

LaBrie ST, Crawford NM (1994) A glycine to aspartic acid change in the MoCo domain of nitrate reductase reduces both activity and phosphorylation levels in *Arabidopsis*. J Biol Chem 269: 14497–14501

Le Guen L, Thomas M, Bianchi M, Halford NG, Kreis M (1992) Structure and expression of a gene from *Arabidopsis thaliana* encoding a protein related to SNF1 protein kinase. Gene 120: 249–254

Ma H, Bloom LM, Walsh CT, Botstein D (1989) The residual enzymatic phosphorylation activity of hexokinase II mutants is correlated with glucose repression in *Saccharomyces cerevisiae*. Mol Cell Biol 9: 5643–5649

Maas C, Schaal S, Werr W (1990) A feedback control element near the transcription start site of the maize *Shrunken* gene determines promoter activity. EMBO J 9: 3447–3452

MacKintosh RW, Davies SP, Clarke PR, Weekes J, Gillespie JG, Gibb BJ, Hardie DG (1992) Evidence for a protein kinase cascade in higher plants. Eur J Biochem 209: 923–931

Maeda T, Wurgler-Murphy SM, Saito H (1994) A two-component system that regulates an osmosensing MAP kinase cascade in yeast. Nature 369: 242–245

Magasannik B, Neidhardt FC (1987) Regulation of carbon and nitrogen utilization. In: Ingraham JL, Law KB, Magasanik B, Schaechter M, Umbarger HE (eds) *Escherichia coli* and *Salmonella typhimurium*: cellular and molecular biology, vol 2. American Society for Microbiology, Washington, DC, pp 1318–1325

Malone RE (1990) Dual regulation of meiosis in yeast. Cell 61: 375–378

Marcus S, Polverino A, Barr M, Wigler M (1994) Complexes between STE5 and components of the pheromone-responsive mitogen-activated protein kinase module. Proc Natl Acad Sci USA 91: 7762–7766

Marzluf GA (1993) Regulation of sulfur and nitrogen metabolism in filamentous fungi. Annu Rev Microbiol 47: 31–55

Miao G-H, Hong Z, Verma DPS (1993) Two functional soybean genes encoding $p34^{cdc2}$ protein kinases are regulated by different plant developmental pathways. Proc Natl Sci USA 90: 943–947

Mitchelhill KL, Stapleton D, Gao G, House C, Michell B, Katsis F, Witters LA, Kemp BE (1994) Mammalian AMP-activated protein kinase shares structural and functional homology with the catalytic domain of yeast SNF1 protein kinase. J Biol Chem 269: 2361–2364

Mizoguchi T, Gotoh Y, Nishida E, Yamaguchi-Schinozaki K, Hayashida N, Iwasaki T, Kamada H, Shinozaki K (1994) Characterization of two cDNAs that encode MAP kinase homologues in *Arabidopsis thaliana* and analysis of the possible role of auxin in activating such kinase activities in cultured cells. Plant J 5: 111–122

Monroy AF, Schwartzbach SD (1984) Catabolite repression of chloroplast development in *Euglena*. Proc Natl Acad Sci USA 81: 2786–2790

Muller-Ruber B, Sonnewald U, Willmitzer L (1992) Inhibition of the ADP-glucose pyrophosphorylase in transgenic potatoes leads to sugar-sorting tubers and influences tuber formation and expression of tuber storage protein genes. EMBL J 11: 1229–1238

Muranaka T, Banno H, Machida Y (1994) Characterization of tobacco protein kinase NPK5, a homolog of *Saccharomyces cerevisiae* SNF1 that constitutively activates expression of the glucose-repressible *SUC2* gene for a secreted invertase of *S. cerrevisiae*. Mol Cell Biol 14: 2958–2965

Ninnemann O, Jauniaux J-C, Frommer WB (1994) Identification of a high affinity NH^+ transporter from plants. EMBO J 13: 3464–3471

Ota I, Varshavsky A (1993) A yeast protein similar to bacterial two-component regulators. Science 262: 566–569

Parkinson JS, Kofoid EC (1992) Communication modules in bacterial signaling proteins. Annu Rev Genet 26: 71–112

Plumley FG, Schmidt GW (1989) Nitrogen-dependent regulation of photosynthetic gene expression. Proc Natl Acad Sci USA 86: 2678–2682

Post-Beittenmiller D, Roughan PG, Ohlrogge JB (1992) Regulation of plant fatty acid biosynthesis: analysis of actyl-CoA and acyl-ACP substrate pools in spanish and pea chloroplasts. Plant Physiol 100: 923–930

Preiss J (1991) Biology and molecular biology of starch synthesis and its regulation. Oxford Surv Plant Mol Cell Biol 7: 59–114

Ranjeva R Boudet AM (1987) Phosphorylation of proteins in plants: regulatory effects and potential involvement in stimulus/response coupling. Annu Rev Plant Physiol 38: 73–93

Ren YL, Garges S, Adhya S, Krakow JS (1988) Cooperative DNA binding of heterologous proteins: evidence for contact between the cyclic AMP receptor protein and RNA polymerase. Proc Natl Acad Sci USA 85: 4138–4142

Riesmeier JW, Willmitzer L, Frommer WB (1992) Isolation and characterization of a sucrose carrier CDNA from spinach by functional expression in yeast. EMBO J 11: 4705–4713

Riesmeier JW, Flugge UI, Schulz B, Heineke D, Heldt HW, Willmitzer L, Frommer WB (1993a) Antisense repression of the chloroplast triose phosphate translocator affects carbon partitioning in transgenic potato plants. Proc Natl Acad Sci USA 90: 6160–6164

Riesmeier JW, Hirner B, Frommer WB (1993b) Potato sucrose transporter expression in minor veins indicates a role in phloem loading. Plant Cell 5: 1591–1598

Roesler KR, Shorrosh BS, Ohlrogge JB (1994) Structure and expression of an *Arabidopsis* acetyl-coenzyme A carboxylase gene. Plant Physiol 105: 611–617

Saier MH (1991) A multiplicity of potential carbon catabolite repression mechanisms in prokaryotic and eukaryotic microorganisms. New Biol 3: 1137–1147

Sano H, Youssefian S (1994) Light and nutritional regulation of transcripts encoding a wheat protein kinase homolog is mediated by cytokinins. Proc Natl Acad Sci USA 91: 2582–2586

Schachtman DP, Schroeder JI (1994) Structure and transport mechanism of a high-affinity potassium uptake transporter from higher plants. Nature 370: 655–658

Schachtman DP, Schroeder JI, Lucas WJ, Anderson JA, Gaber RF (1992) Expression of an inward-rectifying potassium channel by the *Arabidopsis KAT 1* cDNA. Science 258: 1654–1658

Schulz B, Frommer WB, Flugge UI, Hummel S, Fisher K, Willmitzer L (1993) Expression of the triose phosphate translocator gene from potato is light dependent and restricted to green tissues. Mol Gen Genet 238: 357–361

Sheen J (1990) Metabolic repression of transcription in higher plants. Plant Cell 2: 2107–1038

Sheen J (1993) Protein phosphatase activity is required for light-inducible gene expression in maize. EMBO J 12: 3497–3505

Shintani DK, Ohlrogg JB (1993) Feedback regulation of fatty acid synthesis in tobacco cell suspension cultures. Plant Physiol 102: S-11

Shorrosh BS, Dixon RA, Ohlrogge JB (1994) Molecular cloning, characterization, and elicitation of acetyl-CoA carboxylase from alfalfa. Proc Natl Acad Sci USA 91: 4323–4327

Sonnewald U, Willmitzer L (1992) Molecular approach to sink–source interactions. Plant Physiol 99: 1267–1270

Sonnewald U, Brauer M, Von Schaewen A, Stitt M, Willmitzer L (1991) Transgenic tobacco plants expressing yeast-derived invertase either in the cytosol, vacuole or apoplast: a powerful tool for studying sucrose metabolism and sink/source interactions. Plant J 1: 95–106

Srere PA (1987) Complexes of sequential metabolic enzymes. Annu Rev Biochem 56: 21–56

Stark DM, Timmerman KP, Barry GF, Preiss J, Kishore GM (1992) Regulation of the amount of starch in plant tissues by ADP-glucose pyrophosphorylase. Science 258: 287–292

Stephanopoulos G, Vallino JJ (1991) Network rigidity and metabolic engineering in metabolite overproduction. Science 252: 1675–1681

Stitt M (1990) Fructose-2,6-bisphosphate as a regulatory metabolite in plants. Annu Rev Plant Physiol Plant Mol Biol 42: 153–185

Stock JB, Ninfa AJ, Stock AM (1989) Protein phosphorylation and regulation of adaptive responses in bacteria. Microbial Rev 53: 450–490

Szoke A, Miao G-H, Hong Z, Verma DPSV (1992) Subcellular localization of Δ1-pyrroline-5-carboxylate reductase in root / nodule and leaf of soybean. Plant Physiol 99: 1642–1649

Tarczynski MC, Jensen RG, Bohnert HJ (1993) Stress protection of transgenic tobacco by production of the osmolyte mannitol. Science 259: 508–510

Thomas BR, Rodriguez (1994) Metabolite signals regulate gene expression and source/sink relations in cereal seedlings. Plant Physiol 106: 1235–1239

Thompson-Jaeger S, Francis J, Gaughran JP, Tatchell K (1991) Deletion of SNF1 affects the nutrient response of yeast and resembles mutations which activate the adenylate cyclase pathway. Genetics 129: 697–706

Thorne JH (1985) Phloem unloading of C and N assimilates in developing seeds. Annu Rev Plant Physiol 36: 317–343

Tsay Y-F, Schroeder JI, Feldmann KA, Crawford NM (1993) The herbicide sensitivity gene *CHL1* of *Arabidopsis* encodes a nitrate-inducible nitrate transporter. Cell 72: 705–713

Trumbly RJ (1992) Glucose repression in the yeast *Saccharomyces cerevisiae*. Mol Microbiol 6: 15–21

Yu S-M, Kuo Y-H, Sheu G, Liu L-F (1991) Metabolic depression of α-amylase gene expression in suspension-cultured cells of rice. J Biol Chem 266: 21131–21137

Vance CP, Heichel GH (1991) Carbon in N_2 fixation: limitation or exquisite adaptation. Annu Rev Plant Physiol Plant Mole Biol 42: 373–392

Van Quy L, Foyer C, Champigny ML (1991) Effect of light and nitrate on wheat leaf phospho-*enol*pyruvate carboxylase and sucrose synthase proteins in wheat leaves. Plant Physiol 99: 344–347

Vincentz M, Moureaux T, Leydecker M-T, Vauchere H, Caboche (1993) Regulation of nitrate and nitrite reductase expression in *Nicotiana plumbaginfalia* leaves by nitrogen and carbon metabolites. Plant J 3: 315–324

Voelker TA, Worrell AC, Anderson L, Bleibaum J, Fan C, Hawkins DJ, Radke SE, Davies HM (1992) Fatty acid biosynthesis redirected to medium chains in transgenic oilseed plants. Science 257: 72–74

Von Schaewen A, Stitt M, Schmidt R, Sonnewald U, Willmitzer L (1990) Expression of yeast-derived invertase in the cell wall of tocacco and *Arabidopsis* plants leads to inhibition of sucrose export, accumulation of carbohydrate, and inhibition of photosynthesis, and strongly influences the growth and habitus of transgenic tobacco. EMBO J 9: 3033–3044

Willmitzer L, Topfer R (1992) Manipulation of oil, starch and protein composition. Curr Opin Biotechnol 3: 176–180

Wilson RF, Kwanyuen P, Dewey RE, Settlage SB (1992) Recent developments in the molecular biochemistry and genetics of diacylglycerol acyltransferase from soybean. In: Mackenzie SL, Taylor DC (eds) Seed oils for the future. American Oil Chemists' Society, Champaign, IL, pp 116–135

Woods A, Munday MR, Scott J, Yang X, Carlson M, Carling D (1994) Yeast SNF1 is functionally related to mammalian AMP-activated protein kinase and regulates acetyl-CoA carboxylase in vivo. J Biol Chem 269: 19509–19515

Worrell AC, Bruneau J-M, Summerfelt K, Boersig M, Voelker TA (1991) Expression of a maize sucrose phosphate synthase in tomato alters leaf carbohydrate partitioning. Plant Cell 3: 1121–1130

Regulation of C/N Interactions in Higher Plants by Protein Phosphorylation

Steven C. Huber and Werner M. Kaiser

Contents

I. Introduction

A. Role of Leaves in C/N Metabolism

In the light, leaves are the primary site of photosynthetic CO_2 fixation for synthesis of carbohydrates and also for the reduction of nitrate to ammonium for biosynthesis of amino acids. In many species, sucrose is the major soluble carbohydrate synthesized and also serves as the primary transport form of reduced carbon. Many of the amino acids, synthesized in mesophyll cells are also exported from the leaf, along with sucrose, via the phloem (Riens et al., 1991). Based on measurements of the concentrations of metabolites in the phloem sap versus the mesophyll cytoplasm, Riens et al. (1991) concluded that amino-acid uptake into the phloem is passive in nature and therefore translocation occurs in response to active sucrose uptake and movement from the leaf. Both carbohydrates and amino acids are also stored in the leaf during the photoperiod such that export of reduced-C and -N can continue throughout the night period. In spinach (*Spinacia oleracea* L.) and barley (*Hordeum vulgare* L.), the rate of export from leaves at night was found to be about 40% of the rate during the light period (Riens et al., 1994).

Carbohydrate biosynthesis and N-assimilation (into amino acids) are potentially competitive processes as both involve inputs of reduced carbon (derived from photosynthesis) and energy. It is clear that competition per se does not occur, as both processes take place simultaneously and balance between C- and N-metabolism in the long term is maintained. This is achieved, in part, because their metabolism is highly interrelated. It should also be noted that while nitrate assimilation in leaves uses photosynthetically generated reductant, the process of nitrite reduction (which occurs in the chloroplast stroma) does not compete with CO_2 fixation for electrons (de la Torre et al., 1991; Robinson and Baysdorfer, 1985); apparently, there is sufficient capacity of the thylakoid electron transport system to provide reduced ferredoxin for both nitrite reductase and NADPH formation (Robinson, 1988).

From a standpoint of function, a leaf must be engaged in both processes simultaneously. Hence, effective regulation exists to control: (*i*) the flux of carbon into amino acids versus sucrose; and (*ii*) the partitioning of fixed-carbon between the chloroplast (for starch biosynthesis) and the cytosol (for sucrose biosynthesis). Control of both aspects involves regulation of cytoplasmic metabolism, which will be the focus of this review. Specifically, we will consider sucrose biosynthesis and nitrate reduction/assimilation as key component processes of plant C/N balance. One of the recent developments in this area is the realization that at least three cytoplasmic enzymes involved in C/N metabolism are regulated by reversible protein phosphorylation; viz. NADH : nitrate reductase (NR); phospho*enol*pyruvate (PEP) carboxylase (PEPC); and sucrose-phosphate synthase (SPS). This review will summarize the current understanding of the role that phosphorylation plays in the regulation of these key target enzymes.

B. Role of Roots in C/N Interactions

In roots, like in leaves, the diversion of C- and N-fluxes into structural or storage carbohydrates, nitrogenous compounds, or export-metabolites has to be under mutual control. While in this respect, sucrose synthesis is a major metabolic branchpoint in leaf metabolism, this does not hold true for roots. In contrast to leaves, the root is inevitably always a sink for carbohydrates. Further, it is a common experience that roots have to cope with the contradictory task to grow more in length or biomass when nutrient (especially N) availability is low, in order to explore new soil space.

The contribution of roots to whole plant N-assimilation may vary with the N-form and concentration in the soil solution surrounding the roots. As a general but oversimplified rule, it may be stated that at very low nitrate supply, most of the nitrate is reduced in the root. Also, nitrogen from ammonium is usually processed only in the root (Lewis, 1986). In both cases, energy and C-skeletons for amino-acids biosynthesis are provided via the phloem, and part of that carbon is returned to the shoot via the xylem in the form of amino acids or amides (Pate, 1983). Contrary to leaves, excessive production of potentially toxic primary products of nitrate reduction (nitrite or ammonium) may not be as dangerous for the roots because they can be excreted into the soil solution. However, as root N-assimilation is most important at low N-availability, loss of reduced nitrogen must be avoided.

Accordingly, control of N-assimilation in the root and adaptation to carbohydrate availability is as essential as in the shoot. While the anaplerotic reaction of PEP carboxylation for organic-acid production is required both in shoots and roots, NADH supply for nitrate reduction in the root is probably provided by glycolysis and mitochondrial activity (Emes and Bowsher, 1991). Accordingly, oxygen availability is probably much more important as an environmental factor for root metabolism as compared to leaves. The basic enzymatic steps of nitrate reduction and ammonium assimilation in roots and shoots are similar. Evidence is increasing that regulatory mechanisms are also similar, and that protein phosphorylation/dephosphorylation is just as important for modulating cytosolic enzymes, such as NR, as in shoots (Glaab and Kaiser, 1993).

C. Pathways Involved in C- and N-Metabolism

Carbon and nitrogen metabolism are not competitive processes, in part because of coordinated modulation of the respective metabolic pathways. This coordination can involve several levels of regulation, including molecular genetic, metabolic, and posttranslation modification. The primary focus of this review will be on covalent modification of key target enzymes via protein phosphorylation, and the possible implications for C/N interactions. As this is a new and rapidly expanding area of research, much of what will be presented reflects the current working models.

The pathways involved in sucrose biosynthesis and nitrate reduction/assimilation are generally established (see Fig. 1). Triose-P, derived from CO_2 assimilated within the chloroplast stroma by the C_3-reductive pentose phosphate pathway, are exported to the cytoplasm via the phosphate translocator (for review, see Flügge and Heldt, 1991). In the cytosol, the triose-P are used both for sucrose biosynthesis as well as to provide glycolytic PEP for the PEPC-catalyzed formation of oxaloacetate (OAA). The OAA is taken up by mitochondria by a high-affinity translocator (Ebbighausen et al., 1985), and a major product of OAA metabolism in the mitochondria is citrate (Hanning and Heldt, 1993). Formation of citrate would require acetyl-CoA, which can be generated from pyruvate within the mitochondria by the pyruvate dehydrogenase complex (PDC). The requisite pyruvate could be derived from PEP in the cytosol by pyruvate kinase, or it could be generated within the mitochondrial matrix by NAD-malic enzyme (not shown in Fig. 1). Regardless of the source of pyruvate, the in vitro studies of Hanning and Heldt (1993) provided strong experimental support for the earlier suggestion of Chen and Gadal (1990) that mitochondria might export citrate rather than 2-oxoglutarate (2-OG). The

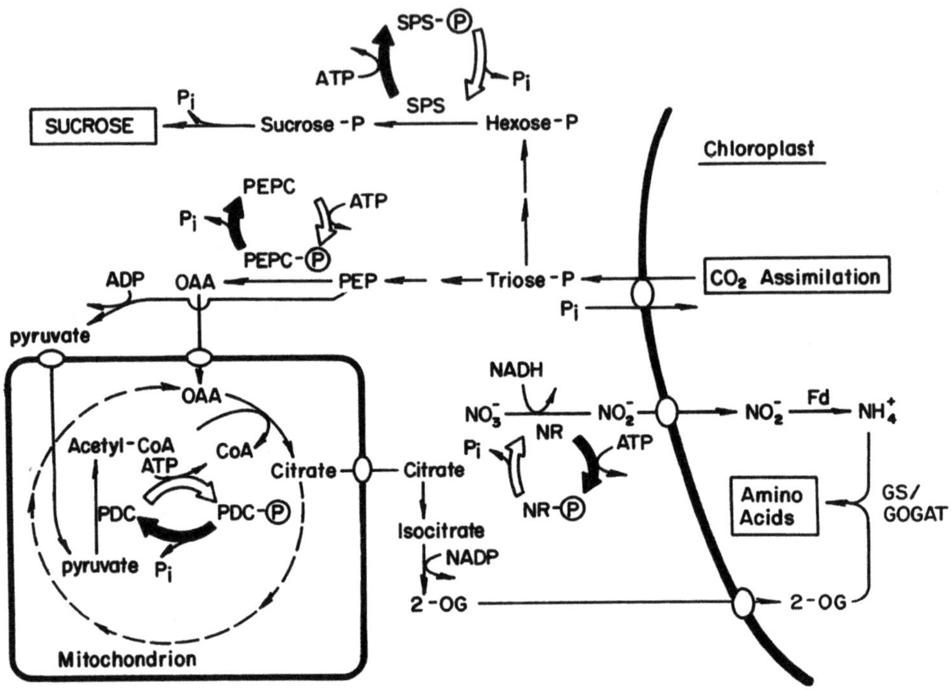

Fig. 1. Simplified scheme showing the pathways of C-flow into sucrose and amino acids in a C_3-leaf mesophyll cell. The modulation of four key target enzymes—cytoplasmic SPS, NR and PEPC, and mtPDC—by reversible protein phosphorylation in response to light/dark signals is shown. The interconversion reactions promoted by light are shown with open arrows, and those promoted in the dark with closed arrows. Note that phosphorylation activates PEPC but inactivates SPS, NR, and mtPDC. Thus, the three cytosolic enzymes are activated in the light whereas mtPDC is at least partially inactivated in the light

citrate could then be converted by cytosolic aconitase and NADP-isocitrate dehydrogenase to 2-OG for incorporation of ammonium via the chloroplastic Gln synthetase/Glu synthase (GS/GOGAT) system (Hanning and Heldt, 1993) (see Fig. 1).

In the simplified scheme depicted in Fig. 1, there are four key reactions catalyzed by enzymes that are potentially regulated by protein phosphorylation: cytosolic SPS, PEPC and NR, and mitochondrial PDC (mtPDC). With respect to PEPC, regulation by reversible protein phosphorylation is well established in C_4, and crassulacean acid metabolism (CAM) plants (Jiao and Chollet, 1991; Nimmo, 1992; see Vidal et al, 1996), but there are also indications for control by phosphorylation in C_3 species (Van Quy et al., 1991). SPS, NR, and mtPDC are inactivated by phosphorylation, whereas PEPC is activated (for reviews, see Budde and Randall, 1990; Jiao and Chollet, 1991; S. Huber et al., 1994). Light/dark signals (Fig. 1) are major external factors controlling the phosphorylation status, and hence enzymatic activity, of these target enzymes (for review, see Budde and Randall, 1990). As illustrated in Fig. 1, SPS, PEPC, and

NR are light activated, whereas mtPDC is light inactivated. However, even though mtPDC has been inactivated to about 20% of its maximum activity (Budde and Randall, 1990), the activity present in illuminated leaves appears adequate to provide C-skeletons for nitrate assimilation (D.D. Randall, pers. comm., 1994). Presumably, there is a much greater demand for mtPDC activity in the dark when the entire TCA cycle is fully operative. In the light, manipulation of CO_2 (or the CO_2/O_2 ratio) will affect the balance between photosynthetic CO_2 fixation and photorespiratory metabolism. As photosynthetic rate is increased and photorespiration is decreased (by whatever means), the reduced flux of Gly and its subsequent oxidation in the mitochondria will result in increased mtPDC activity (Gemmel and Randall, 1992). At the same time, elevated levels of photosynthetic intermediates (P-esters) will activate NR (discussed below). Thus, an overall balance between NO_3^- reduction and provision of 2-OG in the cytosol will be maintained.

II. Regulation of Key Cytosolic Enzymes by Reversible Phosphorylation

In order to understand the role of protein phosphorylation in the coordination of C- and N-metabolism, it is essential that regulation of the interconverting enzymes (i.e., protein kinases and protein phosphatases) be understood. A priori, regulation of the interconverting enzymes could occur at two distinct levels: (*i*) "coarse" control, involving changes in maximum extractable activity (e.g., covalent modification or changes in enzyme protein level); or (*ii*) metabolic "fine" control. For many of the plant enzymes regulated by phosphorylation and subject to metabolic fine control, it is not known whether metabolites interact with the interconverting enzyme(s) or with the target protein substrate. It is thought that there is at least one example of each type (discussed below). However, regardless of the site of action, regulation by metabolites seems to be important. There are also several examples of apparent coarse control of interconverting enzymes, illustrating the diversity of possible regulatory options.

A. *Phosphoenolpyruvate Carboxylase*

In C_4 and CAM plants, PEPC plays an important role as the primary carboxylating enzyme of atmospheric CO_2. It is well established that reversible phosphorylation of C_4- and CAM-leaf PEPC increases enzyme activity (particularly under suboptimal assay conditions) and perhaps more importantly, reduces inhibition by L-malate and thereby activates the enzyme when assayed in the presence of its negative allosteric effector. In particular, the regulation of C_4-leaf PEPC by reversible seryl phosphorylation has been both extensively studied and recently reviewed (Jiao and Chollet, 1991; Nimmo, 1992; S. Huber et al., 1994; Rajagopalan et al., 1994; Vidal et al., 1996). However, briefly stated, PEPC is phosphorylated on a single seryl residue near the N-terminus (Ser-15 in maize or Ser-8 in sorghum; Jiao and Chollet, 1991). The phosphorylation state of C_4-leaf PEPC is apparently controlled primarily by the light activation of

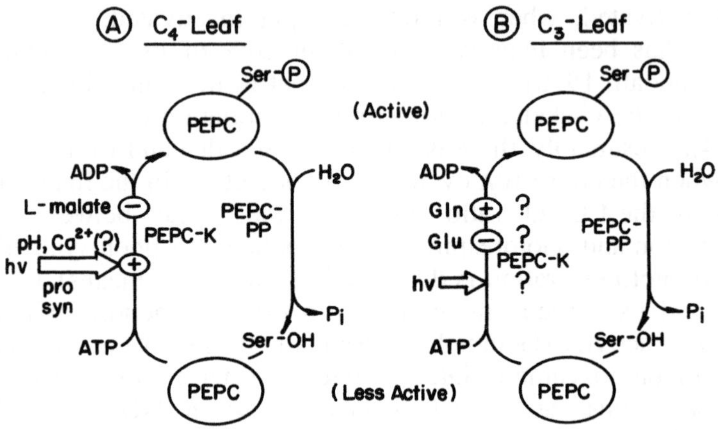

Fig. 2. Schematic representation of the regulation of PEPC by reversible seryl-phosphorylation in A a C_4 leaf (Ser-8, sorghum; Ser-15, maize) and B a C_3 leaf. The C_4-leaf PEPC-kinase has been shown to be activated itself by light, in a process involving the illuminated chloroplast, cytosolic protein synthesis, and perhaps Ca^{2+} and alkalization of the cytoplasm (Jiao and Chollet, 1991, 1992; Jiao et al., 1991; Pierre et al., 1992; Nimmo, 1993; Bakrim et al., 1992; Vidal et al., 1996). This is denoted by the open arrow in A. It is not clear whether similar regulation occurs in the C_3-leaf. However, the C_3-leaf PEPC-kinase is, perhaps, regulated by Gln and Glu (Manh et al., 1993). Pro syn, protein synthesis; hv, light

PEPC-kinase—a process that involves, in some fashion, the illuminated chloroplast and cytoplasmic protein synthesis (Jiao et al., 1991; Jiao and Chollet, 1992; Li and Chollet, 1993; McNaughton et al., 1991; Bakrim et al., 1992). In addition, cytoplasmic Ca^{2+} and alkalization of cytosolic pH may be required for the light activation of C_4-leaf PEPC-kinase (Pierre et al., 1992). The phosphorylation/activation reaction is inhibited by L-malate, and it is thought that malate affects the reaction primarily by interaction with the target protein per se, not the protein kinase (Wang and Chollet, 1993b) (see Fig. 2).

There are potentially two protein kinases (both Ca^{2+}-independent, with subunit M_r about 37 and 30 kDa) that phosphorylate the regulatory site of maize-leaf PEPC and are themselves light activated (Li and Chollet, 1993; Wang and Chollet, 1993b). In addition, there is a Ca^{2+}-dependent protein kinase (M_r ~57 kDa) that phosphorylates PEPC but is not regulated by light as is the regulatory protein kinase (Li and Chollet, 1993), and presumably does not play a role in the regulation of PEPC in vivo.

Phospho-PEPC is dephosphorylated by a type 2A protein phosphatase (PP2A), but little is known about the holoenzyme(s) that occurs in vivo and its possible modes of regulation (McNaughton et al., 1991; Nimmo, 1993). It could well be that primary control of C_4-leaf PEPC phosphorylation status, and hence enzymatic activity, resides with modulation of PEPC-kinase activity.

In CAM plants, PEPC is also regulated by phosphorylation, but the enzyme is phosphorylated and more active at night than during the day (Nimmo, 1992).

As in C_4 plants, the phosphorylation state of CAM-leaf PEPC is controlled by changes in PEPC-kinase activity, that is itself controlled by an endogenous rhythm, rather than by light/dark signals, and involves cytoplasmic protein synthesis (Carter et al., 1991).

In leaves of C_3 plants, PEPC activity is much lower compared to leaves of C_4 and CAM species, but serves several important functions. First, it feeds carbon into the TCA cycle to provide precursors for various biosynthetic processes, including amino-acid biosynthesis (see Fig. 1). Second, PEPC also plays an important role by producing protons to balance the hydroxyl ions generated during nitrate reduction (Raven and Smith, 1976). Consequently, C_3-leaf PEPC is at a pivotal point to play a central role in the integration and coordination of C- and N-metabolism.

Recent evidence suggests that C_3-leaf PEPC may also be regulated by phosphorylation. Some light activation of the C_3-leaf enzyme has been reported in a range of species, but the extent of activation was less and occurred more rapidly than in C_4 species (Rajagopalan et al., 1993). Like C_4- and CAM-leaf PEPC, the C_3-leaf enzyme has the regulatory phosphorylation motif Acidic-Basic-X-X-*Ser*-Ile near the N-terminus (Rajagopalan et al., 1994). Moreover, phosphorylation of wheat-leaf PEPC in vivo following $[^{32}P]P_i$ feeding has been reported (Van Quy et al., 1991), and purified tobacco-leaf PEPC has been phosphorylated in vitro using a Ca^{2+}-independent PEPC-kinase partially purified from either tobacco or maize leaves (Wang and Chollet, 1993a). Recent preliminary results also suggest that phosphorylation modulates C_3-leaf PEPC activity and regulatory properties (e.g., L-malate inhibition) in vitro (Duff and Chollet, 1994). However, it is not clear whether C_3-leaf PEPC-kinase is itself light-activated as in C_4-leaves. Preliminary results suggest that C_3-leaf PEPC-kinase may be regulated by certain amino acids: Gln has been reported to activate and Glu to inhibit wheat-leaf PEPC-kinase activity in vitro (Manh et al., 1993) (see Fig. 2). In this experimental system, which consisted of immunopurified wheat-leaf PEPC and a partially purified wheat-leaf kinase preparation (from chromatography on blue-dextran agarose), Gln and Glu effects on phosphorylation were inferred from changes in PEPC enzymatic activity, but neither malate sensitivity nor phosphorylation state were examined. The potential regulation of C_3-leaf PEPC-kinase by amino acids must be confirmed directly using $[\gamma\text{-}^{32}P]$ATP, and should be extended to test for effects of other amino acids. However, these results may provide a partial explanation, at least, for activation of PEPC activity in response to nitrate feeding (see below).

The reciprocal effects of Gln and Glu on apparent wheat-leaf PEPC-kinase activity (Manh et al. 1993) are analogous to the direct effects of these amino acids on enzymatic activity of PEPC extracted from the unicellular green alga *Selenastrum minutum* (Vanlerberghe et al., 1990). Two molecular forms of PEPC were chromatographically resolved from N-sufficient *Selenastrum* cells; the minor form ($PEPC_1$, about 30% of total activity) was regulated by Gln (activated) and Glu (inhibited), whereas the major form ($PEPC_2$) was not (Schuller et al., 1990). However, recent preliminary evidence suggests that $PEPC_1$ may be derived from $PEPC_2$ by proteolytic modification in vitro (Dunford et al.,

1994). $PEPC_2$, which may be the only molecular form in vivo, appears not sufficiently sensitive to Gln and Glu to explain the dramatic activation of PEPC activity that occurs in response to nitrate or ammonium feeding. Consequently, the regulation of PEPC in *Selenastrum* remains to be resolved; conceivably, phosphorylation may play a role, but this remains to be established.

B. Nitrate Reductase

Reversible phosphorylation of NR in vivo has been demonstrated in leaves of spinach (J. Huber et al., 1992), *Arabidopsis* (LaBrie and Crawford, 1994), and maize (J. Huber et al., 1994). With all three species, phosphorylation occurred exclusively on seryl residues, and peptide mapping suggested the potential for multiple phosphorylation sites. As none of the sites have been identified thus far, the occurrence of multisite phosphorylation remains to be proven.

The regulation of NR activity by reversible phosphorylation has been recently reviewed (S. Huber et al., 1994; Kaiser and Huber, 1994; Kaiser et al.,

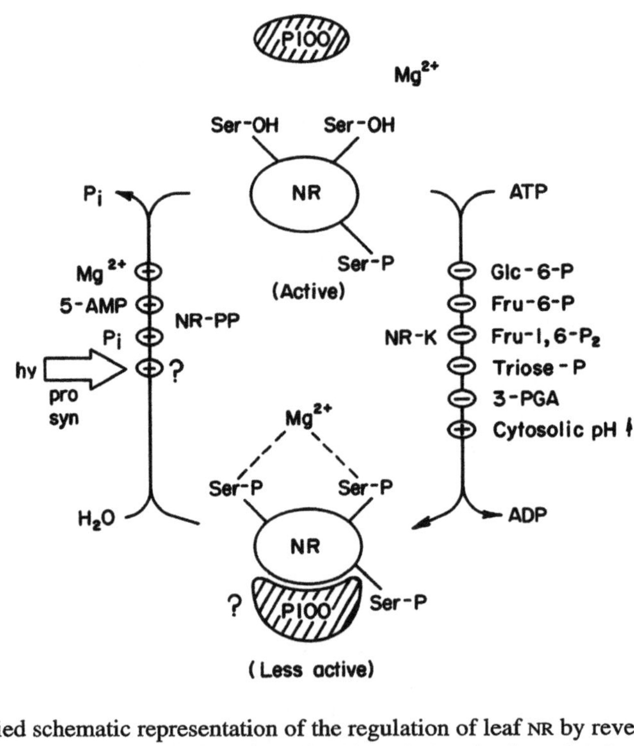

Fig. 3. Simplified schematic representation of the regulation of leaf NR by reversible seryl-phosphorylation. Multiple phosphorylation sites, thought to occur in the spinach enzyme (J. Huber et al 1992), are indicated. It is postulated that inactivation of phospho-NR involves binding of Mg^{2+}, and another protein factor designated as "P100" (Spill and Kaiser, 1994; Glaab and Kaiser, 1994). Positive (+) and negative (–) effectors of the reactions are indicated. The possible light activation of NR-PP (S. Huber et al., 1992) is indicated by the open arrow. Pro syn, protein synthesis; hv, light

1993). Briefly stated, in vitro studies have suggested that phosphorylation of NR modulates enzymatic activity by affecting sensitivity to Mg^{2+} inhibition: dephospho-NR is unaffected by Mg^{2+} whereas phospho-NR is strongly inhibited (Kaiser and Brendle-Behnisch, 1991; Kaiser and Spill, 1991; J. Huber et al., 1992; MacKintosh, 1992). As shown schematically in Fig. 3, the putative Mg^{2+} binding site postulated to exist in phospho-NR may involve the negatively charged P-seryl groups, although there is no direct evidence for this. Another point requiring further study concerns whether or not another protein factor is required for Mg^{2+}-inhibition of phospo-NR. Spill and Kaiser (1994) partially purified and chromatographically resolved two proteins with molecular masses of about 100,000 and 67,000 (P100 and P67, respectively). With highly purified NR, both protein factors were required for the time- and ATP-dependent inactivation of NR in vitro. As expected, inhibition of NR activity was only observed when assays contained free Mg^{2+}. Recent experiments have shown that P67 alone can phosphorylate NR in vitro with [γ-^{32}P]ATP in a time-dependent fashion, but without affecting NR activity; however, subsequent addition of P100 resulted in the *immediate* inactivation of NR (only when assayed with Mg^{2+}) (Glaab and Kaiser, 1994). This had led to the recent proposal that P100 may be a phospho-NR inhibitor protein.

Regulation of NR phosphorylation state in vivo may involve: (*i*) light-activation of NR-PP by a mechanism somehow involving cytosolic protein synthesis (S. Huber et al., 1992); (*ii*) metabolite regulation of NR-PP in vivo (Kaiser and Huber, 1994); and (*iii*) metabolite regulation of NR-kinase in vivo (Bachmann et al., 1994) (see Fig. 3). Studies of metabolite effectors that modulate NR activation/dephosphorylation have been largely done with desalted crude leaf extracts, whereas NR-kinase has been studied both in crude extracts as well as after partial purification. Of the potential regulatory metabolites shown in Fig. 3, it is not known which, if any, play a role in vivo. However, it has been speculated that 5′-AMP activation of NR dephosphorylation may play a role in the activation of NR during anaerobiosis (Kaiser and Huber, 1994). Inhibition of NR-kinase by P-esters may play an important role in light/dark modulation of NR and in the "hyperactivation" of NR (S. Huber et al., 1992) observed when assimilates accumulate in leaves (see below). It is important to note that while many P-esters inhibit the ATP-dependent inactivation of NR a number of compounds have no effect, including P_i, GlcN-6-P, 5′-AMP, free sugars, nitrate, ammonium, and amino acids (Bachmann et al. 1994). It is also not clear whether the metabolites are interacting with the protein kinase per se or with the target substrate protein.

Another factor that may impact on the modulation of NR is cytosolic pH. Manipulation of the intracellular pH of spinach leaf discs by acid- or base-loading in the dark resulted in activation or inactivation of NR, respectively (Kaiser and Brendle-Behnisch, 1994). The effects of *intracellular pH* were reversible and were comparable in magnitude to those obtained with light/dark transitions of intact leaves. Thus, acidification of the cytosol promotes dephosphorylation/activation whereas alkalization promotes phosphorylation/inactivation. How these effects of cytosolic pH are mediated is not clear, but we speculate that cytosolic pH may primarily influence the activity of NR-kinase in vivo. This is based on the observations that NR-kinase activity in vitro has an alkaline pH optimum, and is essentially inactive at pH 6.5 (Kaiser and Brendle-Behnisch, 1994, unpubl.; Bachmann et al., 1994). In contrast, NR-PP activity in vitro has a relatively broad pH optimum (Kaiser and Huber, 1994). These results are significant because it provides the first actual experimental evidence in support of a long postulated role of NR as part of the "biochemical pH-stat" of plant cells (for review, see Raven 1985; Kurkdjian and Guern, 1989). It is not clear, however, under what conditions (if any) the regulation by pH might play a role in vivo. For example, it probably does not contribute to light-activation of NR because light results in only a small alkalization of the cytosol in C_3 leaves (Raghavendra et al., 1993). However, a decrease in pH with anaerobiosis or other forms of stress might play a role in activation of NR under these conditions.

C. *Sucrose-Phosphate Synthase*

Phosphorylation of leaf SPS in vivo has only been directly demonstrated in two species—spinach (J. Huber et al., 1992) and maize (S. Huber et al., 1995). However, based on modulation of SPS activity in relation to light/dark signals and other experimental treatments, it probably occurs in leaves of a variety of

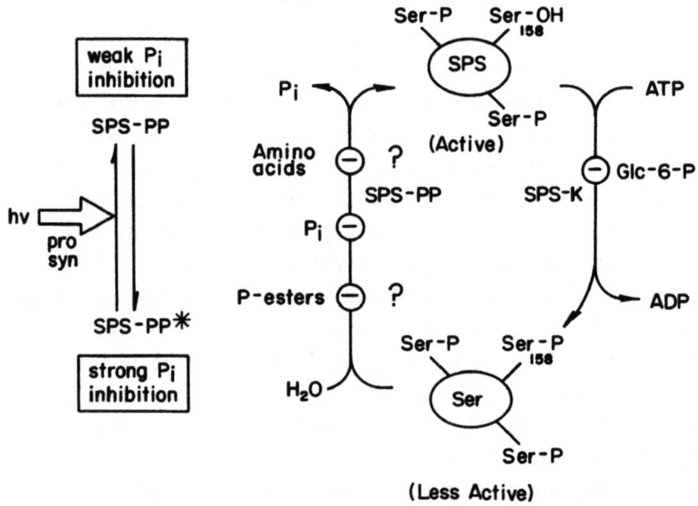

Fig. 4. Schematic representation of the regulation of spinach leaf SPS by reversible seryl-phosphorylation. Multisite phosphorylation (J. Huber and Huber, 1992) and the identification of the major regulatory site as Ser-158 are indicated. SPS-kinase is specifically inhibited by Glc-6-P, and SPS-PP is potentially inhibited by P_i, various P-esters (Weiner et al., 1993) and amino acids (S. Huber et al. 1993; Shannon et al. 1994). The effect of light on the interconversion of presumed molecular forms of SPS-PP differing in sensitivity to P_i inhibition (SPS-PP, SPS-PP*) is shown (Weiner et al., 1993). Pro syn, protein synthesis; hv, light

species (S. Huber et al., 1989) as well as certain non-photosynthetic tissues such as potato tubers (Reimholz et al., 1994).

With both spinach and maize, evidence for multisite seryl-phosphorylation was obtained. Only one or two of the P-seryl groups are apparently involved in the light/dark modulation of enzyme activity (J. Huber and Huber, 1992); the major regulatory phosphorylation site in spinach SPS has recently been identified as Ser-158 (McMichael et al., 1993).

The control of SPS phosphorylation has been recently reviewed (J. Huber and Huber, 1992; S. Huber et al. 1993; S. Huber et al. 1994b). Briefly stated, control of the phosphorylation state of SPS could involve: (*i*) "coarse control" of SPS-PP, somehow involving cytosolic protein synthesis (Weiner et al., 1992); and (*ii*) metabolite "fine control" of SPS-PP and/or SPS-kinase in vivo (S. Huber et al. 1994b; Weiner et al., 1992) (see Fig. 4). Both levels of control seem to play a role in vivo, because preventing the light-activation of SPS-PP with cycloheximide, a cytoplasmic protein synthesis inhibitor, dramatically reduced the rate of SPS activation/dephosphorylation in situ (Weiner et al., 1992). Similarly, metabolites clearly play a role as well, because feeding P_i to excised leaves can reduce SPS activation whereas feeding mannose, a phosphate-sequestering agent (Lewis and Herold, 1977), results in SPS activation in the dark (Stitt et al., 1988).

Recent results suggest that the "light-activation" of SPS-PP may actually involve an interconversion between two forms of the enzyme (in some manner dependent upon protein synthesis) that differ in sensitivity to inhibition by P_i and several P-esters including 3-phosphoglycerate (3-PGA) and dihydroxy-acetone (DHAP or triose-P) (Weiner et al., 1993): the SPS-PP extracted (and partially purified) from dark spinach leaves was strongly inhibited by these metabolities whereas SPS-PP from light leaves was markedly less sensitive. The interconversion of SPS-PP between the metabolite-sensitive and -insensitive forms could also be induced in the dark by feeding mannose to leaves, suggesting that light per se was not required. Rather, changes in leaf phosphate status might be critical (Weiner et al., 1993). Because of uncertainties in the details of SPS-PP interconversion and regulation, Fig. 4 simply indicates that P_i and a variety of P-esters are potential inhibitors of SPS-PP and that different forms of the enzyme may exist that vary in sensitivity to metabolite regulation. Another important class of compounds that might serve to regulate SPS phosphorylation state are amino acids, which can inhibit SPS-PP (extracted from dark leaves) in vitro (S. Huber et al., 1993). It is not known whether different molecular forms of SPS-PP vary in sensitivity to amino-acid inhibition; however, the possible role of amino acids as regulators of SPS will be considered further below.

The kinetic effect of phosphorylation of SPS varies somewhat depending upon the source of enzyme, but in general, affinities for substrates (Fru-6-P and/or UDP-glucose) and effectors (Glc-6-P and P_i) can be affected (Reimholz et al., 1994; for review, see S. Huber et al., 1994). Consequently, to observe effects of phosphorylation on SPS activity in vitro, assays are often conducted using "selective" conditions, i.e., rate-limiting substrates and a mixture of

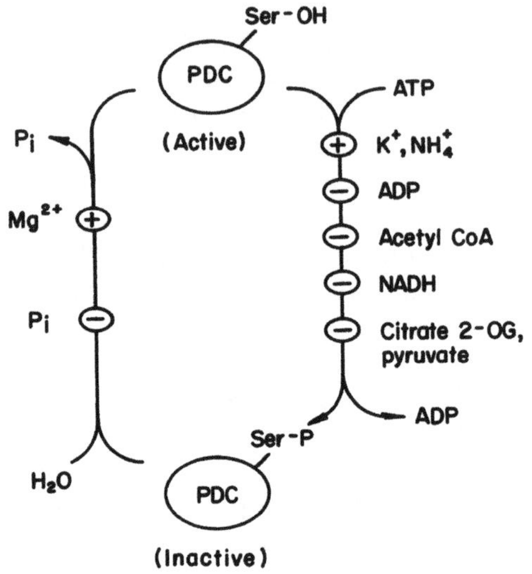

Fig. 5. Schematic representation of the regulation of mtPDC activity by reversible seryl phosphorylation. Positive (+) and negative (–) effectors are shown

allosteric effectors which approximate concentrations in the cytoplasm of leaf cells.

D. Mitochondrial Pyruvate Dehydrogenase Complex

The regulation of mtPDC has been recently reviewed (Budde and Randall et al., 1990a, b; S. Huber et al., 1994b), and metabolites and ions that influence the phosphorylation state of the enzyme will be discussed only briefly. The mtPDC is inactivated by seryl-phosphorylation, and it is thought that control of the kinase, which is associated with the PDC, will be the primary factor controlling PDC activity. The PDC-kinase is activated by several cations (millimolar K^+; micromolar NH_4^+) and inhibited by metabolites including ADP, NADH, acetyl-CoA, and certain organic acids (see Fig. 5). Because PDC functions to produce acetyl-CoA, and thereby introduce carbon into the TCA cycle, the inhibition of PDC-kinase by cycle intermediates (citrate and 2-OG) seems unusual. If the only function of mtPDC was to provide cycle intermediates, then citrate and 2-OG might be expected to stimulate the kinase (and thereby reduce additional C-flow into the cycle). The unusual regulation observed may indicate that citrate is also required outside of the mitochondrion, e.g., for biosynthesis of 2-OG and assimilation of ammonium (see Fig. 1).

The light-dependent inactivation of mtPDC is clearly linked to photosynthesis, and more specifically to photorespiration (Budde and Randall, 1990; Gemmel and Randall, 1992). It has been suggested that production of Gly during photorespiration is part of the link between light (photorespiration) and the inactivation of mtPDC, because conversion of Gly to Ser within the matrix (by glycine decarboxylase) would produce both NADH and ATP (for PDC-kinase) and also generate NH_4^+, a potent activator of PDC kinase. Both factors could contribute to the observed light-inactivation of mtPDC.

Regulation of mtPDC in the light in relation to other aspects of cytosolic metabolism (see Fig. 1) is not well understood, but conceivably, organic acids and NH_4^+ could play a role in coordinating provision of 2-OG with nitrate reduction. In the remainder of this review, we will restrict our attention to the cytosolic enzymes PEPC, NR, and SPS and consider how the phosphorylation of these target enzymes may contribute to control of C/N balance under different conditions.

III. Control of C/N Interactions

Having considered the factors that might regulate the phosphorylation status of several key cytosolic and mitochondrial enzymes, we can now address how some of these regulatory properties might relate to the metabolic coordination that occurs in vivo. The interrelationships between photosynthetic C-metabolism and nitrate assimilation have been explored using several experimental systems. In broad terms, these include: (*i*) varying CO_2 or nitrate supply to plants or excised leaves; (*ii*) feeding carbohydrates or amino acids (end products); and (*iii*) the use of mutant and transgenic plants with altered capacities for photosynthetic CO_2 fixation, starch metabolism, or nitrate reduction (for reviews, see Stitt and Schulze, 1994; Hoff et al., 1994). Results from these different approaches all tend to support the notion that photosynthesis and nitrate assimilation are closely coupled in leaves. It is becoming clear that several steps of reciprocal control are involved in the "cross-talk," or coregulation, of pathways involved in C- and N-metabolism.

A. *Impact of C on N-Metabolism*

1. Light, CO_2, and Photosynthesis

It has long been recognized that restricting photosynthesis by decreasing atmospheric CO_2, or imposing a rapid water stress (which closes stomata), inhibits nitrate assimilation in leaves (Aslam et al., 1979; Pace et al., 1990; Kaiser and Förster, 1989). Another important observation is that light and CO_2, or conditions that promote high carbohydrate levels, increase NR activity (Kaiser and Brendle-Behnisch, 1991) and nitrate assimilation in leaves (Aslam et al., 1979; Beevers and Hageman, 1980; Pace et al., 1990). The stimulation of nitrate assimilation by carbohydrates has often been interpreted as a reflection of the demand of nitrate reduction for NADH (Klepper et al., 1971). However, nitrate assimilation in leaves in the light utilizes electrons derived directly from photosynthetic electron transport (de Cires et al., 1993). It is now clear that the carbon dependence of nitrate reduction/assimilation also involves a regulatory component, and several specific aspects of this regulatory network are being elucidated. One important observation is that sucrose (or more likely some metabolite(s) derived from sucrose metabolism) can promote NR gene expression (Cheng et al., 1992). Another aspect, and the central focus of this article, is the control of NR activity (and other cytosolic enzymes) by reversible protein phosphorylation.

There is increasing evidence that protein phosphorylation contributes to the coordination between CO_2 fixation, sucrose biosynthesis, and nitrate reduction. The light activation of SPS, NR, and PEPC requires CO_2; hence, the light effect is indirectly mediated via induction of photosynthesis. As expected, various inhibitors of photosynthesis (e.g., DCMU) can also prevent the light activation of these enzymes. Light activation is not an "on–off" switch, but rather reflects a mechanism to fine tune the steady-state activities of these target enzymes in response to changes in CO_2 assimilation rate; e.g., as irradiance is varied. This has been demonstrated for spinach SPS (Battistelli et al., 1991), barley NR (de Cires et al., 1993), and wheat-leaf PEPC (Van Quy et al., 1991). How is this coordination among cytosolic enzymes brought about in response to photosynthesis? Unfortunately, not enough is known about the control of C_3-leaf PEPC by phosphorylation (see Sect. II.A) in order to speculate about the signals that mediate the activation of this non-photosynthetic isoform in the light. However, there is some suggestive evidence that nitrate, or a metabolite derived from nitrate reduction (e.g., glutamine), modulates the relative kinase/phosphatase ratio to promote phosphorylation and activation of C_3-leaf PEPC (discussed further below) (Manh et al., 1993, and references therein).

In contrast, it appears that cytosolic phosphate status is a critical component underlying the activation of SPS and NR. As photosynthetic rate increases, e.g., with irradiance, the concentration of many metabolic intermediates remains relatively constant, or even decreases in the case of UDP-glucose (for review, see Stitt et al., 1987). Thus, the flux of carbon into sucrose can be increased during a dark-to-light transition without large change in metabolite pools; this is accomplished by regulation of enzyme activities with regulatory metabolites (such as fructose-2,6-P_2) and protein phosphorylation. In contrast to other metabolites, it has been confirmed by subcellular fractionation of spinach leaves that the cytosolic concentration of triose-P is extremely low in the dark and increases rapidly with photosynthetic rate (Gerhardt et al., 1987). It already has been discussed in detail how an increase in cytosolic triose-P, necessarily coupled with a corresponding decrease in cytosolic P_i, will result in the coordinate control of cytosolic fructose-1,6-bisphosphatase (FBPase), by lowered [Fru-2,6-P_2], and SPS, by allosteric regulation (for review, see Stitt et al., 1987). We can now rationalize that reciprocal changes in the levels of cytosolic triose-P and P_i will also result in the activation of SPS by dephosphorylation, as cytosolic phosphate status affects SPS-PP activity (see above). An increase in cytosolic triose-P will also inhibit NR-kinase, and thereby promote NR dephosphorylation/activation. Although P_i activates NR-dephosphorylation (S. Huber et al., 1994), the concentrations required are quite high and it is unlikely that the relatively small changes in [P_i] that might occur in vivo would be of physiological importance. Changes in C_3-leaf PEPC activity in response to photosynthesis are relatively small (i.e., the enzyme must be active in the dark as well as in the light), and the metabolic signals that might be involved in promoting phosphorylation in the light are not known.

2. Photoassimilate Accumulation

Accumulation of photoassimilates in leaves can occur in vivo, as a result of a source/sink imbalance, or experimentally when phloem transport is either reduced (e.g., cold block) or stopped completely by petiole excision. When spinach leaves are excised and illuminated, photosynthesis continues but carbohydrate partitioning changes quickly: the flux of C into sucrose slows while starch biosynthesis increases (for review, see Stitt et al., 1987). In marked contrast, reduction of nitrate (from endogenous leaf reserves which are substantial in spinach) continues at a nearly constant rate for several hours (Kaiser and Förster, 1989; S. Huber et al., 1992). Changes in the phosphorylation state of cytosolic enzymes appear to be at least one component responsible for the shift in C-flow within the cytosol (sucrose versus amino acids) and also across the chloroplast envelope (sucrose versus starch).

In an excised, illuminated spinach leaf, as photoassimilates accumulate, SPS is rapidly inactivated/phosphorylated while NR is "hyperactivated" (probably by dephosphorylation) (S. Huber et al., 1992). The inactivation of SPS by protein phosphorylation is thought to be a cardinal event directly responsible for slowing the flux of C into sucrose, and indirectly responsible for the increased starch synthesis within the chloroplast (Stitt et al., 1987). If so, the critical question is what factors are responsible for the inactivation of NR? A priori, either SPS-kinase has been activated and/or SPS-PP inhibited such that the kinase/phosphatase ratio is increased. It has been assumed that accumulation of sucrose itself was somehow involved in the inactivation of SPS. Unfortunately, no effects of sucrose (or sucrose-P) on either of the interconverting enzymes have been observed in vitro (S. Huber et al., 1993). Preliminary results suggest that accumulation of amino acids may play a role. Amino acids are synthesized in leaves and the majority are exported in the phloem; they are taken up via a passive process that is driven by the active uptake and transport of sucrose (Winter et al., 1992). Consequently, amino acids will accumulate in excised leaves, when all export is blocked, and in attached leaves when sucrose export is restricted via a source/sink imbalance. Thus, amino acids could play an important role as "signal metabolites," with an accumulation indicating decreased plant demand for sucrose. Indeed, there are several lines of indirect evidence consistent with this postulate. First, feeding Gln to excised wheat leaves via the transpiration stream decreased the rate of photosynthetic sucrose formation in a concentration-dependent manner; feeding only 5 mM Gln reduced sucrose synthesis by approx. 40% (Manh et al., 1993). Second, in analogous experiments with spinach, we have observed that feeding amino acids (in particular Gln or Gly) to excised leaves resulted in inactivation of SPS relative to the control leaves (S. Huber et al., 1993; Shannon et al., 1994). Amino acids had little effect on net photosynthetic rate, and also had no effect on SPS-kinase activity in vitro, but inhibition of SPS-PP (extracted from dark leaves) was observed (see Fig. 4). These results certainly provide a possible explanation for the experimental observation noted above. A third line of indirect evidence concerns differences among species in accumulation of amino acids relative to partitioning of carbon. Attached, N-sufficient leaves of spinach and barley differ considerably in the accumulation of photoassimilates during the photoperiod (Riens et al., 1994). Spinach leaves accumulate more starch than sucrose, particularly towards the end of the photoperiod, whereas barley leaves accumulate high levels of sucrose, in a relatively linear fashion with time, but accumulate little starch. In addition to storing more sucrose, barley leaves also have a slightly higher rate

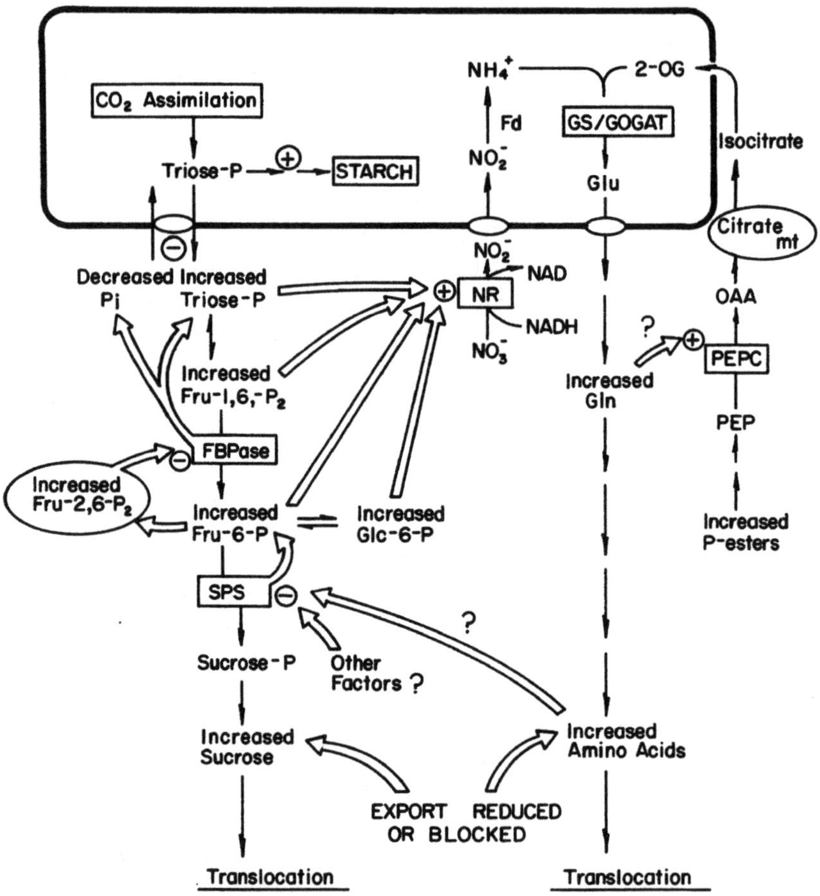

Fig. 6. Possible role of reversible protein phosphorylation in the response of C_3-leaf metabolism to an accumulation of photoassimilates ("feedback regulation"). It is postulated that a build-up of amino acids (especially Gln and Gly) contributes, perhaps along with other unidentified factors, to the inactivation of SPS by inhibition of SPS-PP (see Fig. 4). This would result in the accumulation of sucrose-pathway intermediates, as well as increased levels of the regulatory metabolite Fru-2,6-P_2 (Stitt et al. 1987). The cumulative build-up of phosphorylated metabolites would activate NR, by inhibition of NR-kinase (see Fig. 3). Thus, end-product accumulation can result in inactivation of SPS (and diversion of C-flux into starch) but activation of NR (S. Huber et al. 1992). Consult the preceding figures and the text, for specific details

of assimilate translocation during the light period. In terms of reduced-N, barley leaves accumulate about 60% less amino acids (on a N-basis) than do spinach leaves (Riens et al., 1994). It is simply worth noting that the higher sustained rate of sucrose formation in barley leaves was associated with reduced accumulation of amino acids. This is consistent with, but by no means proves, that amino acids may somehow modulate sucrose formation.

A simplified and highly speculative working model for "feedback regulation" of C- and N-metabolism involving protein phosphorylation is shown in Fig. 6. As discussed above, a decrease in export (e.g., by root cooling) will result in an accumulation of both sucrose and amino acids in source leaves. We postulate that an increase in certain amino acids (especially Gln and Gly) could inhibit SPS-PP and thereby promote phosphorylation and inactivation of SPS. Inhibition of SPS would result in increased concentrations of its substrates, in particular Fru-6-P. The rise in Fru-6-P will result in an increased concentration of the regulatory metabolite Fru-2,6-P_2, which will inhibit cytosolic FBPase and thereby increase cytosolic triose-P levels. The increase in P-esters will necessarily result in a corresponding decrease in P_i. These changes, and the basis for increased starch synthesis within the chloroplast, are discussed in detail elsewhere (Stitt et al., 1987). The general rise in cytosolic P-esters, such as triose-P and hexose-P, will inhibit NR-kinase and thereby promote dephosphorylation/activation of NR (Bachmann et al., 1994). While SPS-kinase is inhibited by Glc-6-P, the inhibition is specific for this P-ester and SPS-kinase is much less sensitive to Glc-6-P compared with NR-kinase (McMichael et al., 1994). Thus, inhibition of SPS-kinase should occur more slowly and to a lesser extent compared with NR-kinase. As a result, nitrate reduction will continue, and to the extent that amino acids accumulate in the cytosol, will tend to amplify the regulatory cycle postulated above. It should be noted that NR activity will ultimately be inhibited as a result of Gln repression of NR enzyme protein biosynthesis (Shiraishi et al., 1992; for review, see Hoff et al., 1922). If Gln, or other amino acids, acted to prevent synthesis of NR (either at the transcriptional or translational levels), then NR activity would be a function of how rapidly the protein is degraded. At least one of the major factors controlling NR protein degradation is thought to be the availability of nitrate (Li and Oaks, 1993). Thus, in the short-term (minutes to hours), nitrate reduction would continue although in the long-term (days), it would be restricted along with sucrose biosynthesis.

B. Impact of N on C-Metabolism

Adequate N-nutrition is necessary for all phases of plant growth and development, including biosynthesis of the photosynthetic apparatus. Thus, N-availability can influence critical factors such as gene expression and protein synthesis, and thus have a profound impact on C-metabolism in the long-term. For example, in maize, Gln is required for expression of several photosynthetic genes, including the C_4 PEPC gene (Sugiharto et al., 1992). Short-term effects are also apparent, and are indicative of the underlying mechanisms that serve to coordinate C- and N-metabolism; some of these effects are considered further below.

1. General Role for Cytoplasmic Protein Synthesis

At least in some species, certain of the interconverting enzymes that act on PEPC, SPS, and NR are regulated by short-term turnover of the interconverting enzyme itself or some other essential protein component (e.g., regulatory subunit). This has been documented for C_4-leaf PEPC-kinase (Jiao et al., 1991; Bakrim et al., 1992; Li and Chollet, 1993), and suggested to be the case for spinach leaf SPS-PP (Weiner et al., 1992, 1993), and spinach leaf NR-PP (S. Huber et al., 1992). Thus, when leaves are pretreated in the dark with cycloheximide (CHX), an inhibitor of cytoplasmic protein synthesis, the subsequent light activation of the target enzymes is severely reduced. When maize leaves were pretreated with CHX, the light activation of PEPC-kinase (and hence, PEPC activity) was prevented and maize leaf photosynthetic CO_2 assimilation was strongly inhibited (Jiao et al., 1991; Bakrim et al., 1993), indicating the critical importance of PEPC activation by protein phosphorylation. It has recently been demonstrated that two distinct protein kinases (catalytic subunits ~30 and ~37 kDa) are light-activated in situ and both are Ca^{2+}-independent and exclusively phosphorylate the regulatory serine residue (Li and Chollet, 1993). The light activation of both C_4 PEPC-kinase isoforms was prevented by CHX. Interestingly, in maize leaves, pretreatment with CHX had no effect on the light activation of SPS (Jiao et al., 1991).

In contrast, pretreatment of spinach leaves in the dark with CHX can both prevent and reverse the light activation of SPS (Weiner et al., 1992), suggesting that differences may exist among species. Feeding CHX to leaves also strongly reduced the mannose-dependent activation of SPS in the dark (Weiner et al., 1993). As discussed above (see Sect. II.C), illumination of leaves (or treatment with mannose in the dark) results in decreased sensitivity of SPS-PP to P_i and presumably several other P-ester inhibitors. The shift in sensitivity to inhibitors was also prevented by CHX, suggesting that cytosolic protein synthesis is somehow involved in the interconversion of the different kinetic forms of SPS-PP (Weiner et al., 1993). Similarly, CHX can reverse the light activation of NR in spinach leaves (S. Huber et al., 1992), suggesting that NR-PP has been restricted in some fashion; however, this has not been proven directly.

In all of these cases, it is presumed that some protein component is being synthesized either continuously or in response to the light signal. It is not known whether synthesis of the protein component utilizes existing amino-acid pools in the plant cell or whether newly synthesized amino acids are preferentially incorporated. Nonetheless, the effects of CHX described above suggest that in vivo, if nitrate and/or endogenous amino acids are limiting, the activation of C_4-leaf PEPC and C_3-leaf SPS and NR might be decreased accordingly. Note that in the case of PEPC, SPS, and NR, the interconverting enzyme that is regulated in this fashion is the one that *activates* the target protein. This could be an extremely important regulatory component underlying C/N balance, as a short-term restriction in N-availability would directly restrict C-metabolism.

2. Nitrate and Ammonium as Regulators

As mentioned above, the concentration and form of nitrate in the soil solution is a decisive factor controlling the distribution of N-assimilation between the root and shoot, and the corresponding carbohydrate and amino acid fluxes. It is to be expected that N-compounds (nitrate, ammonium, amino acids) might be effectors of enzymes of carbohydrate metabolism.

Short-term or transient effects of N on C-metabolism have been studied extensively by Champigny and co-workers. The system studied involves supply of nitrate (or other nitrogenous compounds) to detached, nitrate-depleted wheat leaves. The basic observation is that nitrate re-supply results in inhibition of sucrose synthesis (Van Quy et al., 1991). The effect of NO_3^- was blocked by pretreatment of seedlings with tungstate, which effectively prevents nitrate reduction (Heimer et al., 1969), indicating that NO_3^- per se was not an inhibitor of sucrose synthesis. Rather, it was suggested that NO_3^- reduction and incorporation of ammonium into amino acids was competitive with sucrose formation for either C-skeletons and/or photogenerated reductant (Van Quy et al., 1991). However, supply of malate (as a source of reductant) and isocitrate (as a precursor of 2-oxoglutarate) increased NO_3^- assimilation but further inhibited sucrose synthesis (Van Quy et al., 1991), which suggests that direct competition for energy or C-skeletons is not involved.

It was subsequently demonstrated that the short-term (i.e., within 1 h) effect of NO_3^- was unaffected by CHX and, therefore, independent of cytoplasmic protein synthesis. The short-term effects included activation of PEPC and inactivation of SPS—both presumably by enhanced phosphorylation (Van Quy et al., 1991). The effect of NO_3^- on the activity of the two enzymes was transient (Van Quy et al., 1991; Champigny and Foyer, 1992). Recent results suggest that PEPC and SPS are modulated by NO_3^- in a similar fashion in the C_4 plant *Zea mays* (Foyer et al., 1994). It has been suggested that the effect of NO_3^- in wheat (Manh et al., 1993) and maize (Foyer et al., 1994) is mediated by increased levels of Gln in response to NO_3^- supply. The inactivation of SPS in response to Gln is certainly consistent with the notion that Gln (and certain other amino acids) can inhibit SPS-PP (see above and Fig. 4). The basis for effects of Gln and Glu on C_3-leaf PEPC activity has been suggested to involve modulation of PEPC-kinase activity. Preliminary results suggest that Gln activates, while Glu inhibits, C_3-leaf PEPC-kinase activity in vitro (Manh et al., 1993). A rise in Gln would signal availability of reduced-N and a need for additional C-skeletons for amino-acid biosynthesis; activation of PEPC-kinase would increase the phosphorylation state and, hence, enzymatic activity of PEPC to stimulate C-flow into 2-OG (see Fig. 1). Conversely, Glu might signal decreased N-availability, and thus, a diminished need for C-skeletons. In many systems, the Gln/Glu ratio reflects the availability of N and thus could be an important aspect of PEPC regulation. The effects of Gln and Glu, singly and in combination, on C_3 PEPC-kinase activity and SPS-PP activity in vitro needs more detailed study.

The source of nitrogen also has pronounced effects on C-metabolism at the cellular level. This is due, in part, to the requirement for maintaining ionic charge- and pH-balance. When nitrate is the only N-source, it is taken up and transported in its anionic form, together with a cation (e.g., K^+). Nitrate reduction results in hydroxyl production, thus replacing KNO_3 by a strong base KOH. In contrast, at the pH values existing in soil solutions and plant cell walls and cells, ammonia is taken up and transported as the cation NH_4^+ (pKa 9.37). Ammonium assimilation may result in H^+ formation. Thus, different N-sources will have opposite effects not only on soil pH, but intracellular pH as well (Raven and Smith, 1976). Plants may respond by producing or consuming organic acids. In fact, it has long been recognized that plants grown exclusively on NO_3^--N are high in malate or other organic acids. Upon transfer of hydroponically-grown pea plants to NH_4^+-N, malate levels in shoots (but less so in roots) fell rapidly to very low concentrations (Speer and Kaiser, 1991), while free sugars increased (Lang and Kaiser, 1994). On the other hand, short-term addition of NH_4^+ to plants often results in increased amino-acid production (Müller et al., 1990). In all of these cases, one of the most probable control points of C/N-interaction would be expected to be at PEPC, but enzymes involved in amino-acid biosynthesis and/or organic-acid decarboxylation must also be considered. As mentioned above, PEPC is modulated at various levels. Accordingly, the absence of NO_3^- (as in the case of plants grown on NH_4^+) might be expected to decrease the activity of PEPC, which could explain the lower organic-acid content of NH_4^+-grown plants (Schweizer and Erismann, 1985). However, the situation is probably much more complex. There is some evidence that C_3-leaf PEPC activity in NH_4^+-grown plants is actually increased, but without much change in the amino-acid pools (Leport et al., unpubl.). Thus, either decarboxylation has been increased, or the OAA formed was diverted into other pathways such as polyamine biosynthesis. Alternatively, the assay conditions in vitro may not have detected down-regulation of PEPC by protein dephosphorylation that could determine enzymatic activity in vivo. The influence of N-nutrition of C_3-leaf PEPC activity needs to be reevaluated in light of recent developments.

IV. Conclusions

It is clear that C- and N-metabolism are highly coordinated as a result of reciprocal control mechanisms that operate at levels ranging from gene expression to posttranslational modification. In this review, we have tried to summarize our current understanding of the regulation of key cytosolic enzymes by reversible protein phosphorylation, and the role that regulatory phosphorylation may play in C/N interactions. Several generalizations emerge. First, C-metabolism (photosynthesis) is required for activation of NR, SPS, and PEPC. This appears to be a type of "feedforward" regulation whereby the availability of C-skeletons modulates the activities of enzymes that utilize metabolic intermediates. This provides coordination which prevents competition between processes (sucrose biosynthesis and nitrate reduction/amino-acid biosynthesis) that utilize common intermediates. The leaf cell "phosphate status" appears to mediate the activation of NR and SPS (and possibly PEPC?) in relation to photosynthetic activity.

N-metabolism (and nitrogenous compounds) also modulates the phosphorylation state of certain enzymes of C-metabolism (PEPC and SPS). Gln (indicative of active NO_3^- reduction) may activate C_3-leaf PEPC (perhaps by activation of PEPC-kinase?) whereas "excess" accumulation of Gln may inactivate SPS by inhibition of SPS-PP. Thus, leaf "N-status" may inversely affect PEPC and SPS activation.

In the future, it is quite likely that other leaf enzymes involved in specific aspects of C- and N-metabolism will be identified that are regulated by protein phosphorylation. However, the target enzymes identified to date occupy key positions that influence the partitioning of carbon between amino acids and sucrose in the cytosol, and between sucrose (in the cytosol) and starch (in the chloroplast). However, much more work needs to be done to confirm and extend the preliminary observations concerning the molecular mechanisms that may underlie this coordination.

Acknowledgements

Cooperative investigations of the U.S. Department of Agriculture, Agricultural Research Service, and the N.C. Agricultural Research Service, Raleigh, NC. Research in the authors' laboratories is also supported in part by grants from the U.S. Department of Energy (Grant 93-37305-9231 to SCH) and Deutsche Forschungsgemeinschaft (WMK). Mention of a trademark or proprietary product does not constitute a guarantee or warranty of the product by the N.C. Agricultural Research Service or the U.S. Department of Agriculture and does not imply its approval to the exclusion of other products that may also be suitable. The authors thank Drs. J. Huber, R. Chollet, and D. Randall for many discussions of the topic and for critical reading of the manuscript.

V. References

Aslam M, Huffaker RC, Rains DW, Rao KP (1979) Influence of ambient carbon dioxide concentration on nitrate assimilation by intact barley seedlings. Plant Physiol 63: 1205–1209

Bachmann M, McMichael RW Jr, Huber SC (1994) Nitrate reductase protein kinase in spinach leaves. Plant Physiol 104 [Suppl]: 91

Bakrim N, Echeverria C, Cretin C, Arrio-Dupont M, Pierre JN, Vidal J, Chollet R, Gadal P (1992) Regulatory phosphorylation of *Sorghum* leaf phospho*enol*pyruvate carboxylase. Identification of the protein-serine kinase and some elements of the signal-transduction cascade. Eur J Biochem 204: 821–830

Bakrim N, Prioul J-L, Deleens E, Rocher J -P, Arrio-Dupont M, Vidal J, Gadal P, Chollet R (1993) Regulatory phosphorylation of C_4 phospho*enol*pyruvate carboxylase: a cardinal event influencing the photosynthesis rate in C_4 plants. Plant Physiol 101: 891–897

Battistelli A, Adcock MD, Leegood RD (1991) The relationship between the activation state of sucrose phosphate synthase and the rate of CO_2 assimilation in spinach leaves. Planta 183: 620–622

Beevers L, Hageman RH (1980) Nitrate and nitrite reduction. In: Miflin BJ (ed) The biochemistry of plants, vol 5. Academic Press, New York, pp 115–168

Bialojan C, Takai A (1988) Inhibitory effect of a marine-sponge toxin, okadaic acid, on protein phosphatases. Biochem J 256: 283–290

Budde RJA, Randall, DD (1990a) Light as a signal influencing the phosphorylation status of plant proteins. Plant Physiol 94: 1501–1504

Budde RJA, Randall DD (1990b) Pea leaf mitochondrial pyruvate dehydrogenase complex is inactivated in vivo in a light-dependent manner. Proc Natl Acad Sci USA 87: 673–676

Carter PJ, Nimmo HG, Fewson CA, Wilkins MB (1991) Circadian rhythms in the activity of a plant protein kinase. EMBO J 10: 2063–2068

Champigny M-L, Foyer C (1992) Nitrate activation of cytosolic protein kinases diverts photosynthetic carbon from sucrose to amino acid biosynthesis. Plant Physiol 100: 7–12

Chen RD, Gadal P (1990) Do mitochondria provide the 2-oxoglutarate needed for glutamate synthesis in higher plant chloroplasts? Plant Physiol Biochem 28: 141–145

Cheng CL, Acedo GN, Cristinsin M, Conkling MA (1992) Sucrose mimics the light induction of *Arabidopsis* nitrate reductase gene transcription. Proc Natl Acad Sci USA 89: 1861–1864

De Cires A, de la Torre A, Delgado B, Lara C (1993) Role of light and CO_2 fixation in the control of nitrate-reductase activity in barley leaves. Planta 190: 277–283

De la Torre A, Delgado B, Lara C (1991) Nitrate-dependent O_2 evolution in intact leaves. Plant Physiol 96: 898–901

Duff SMG, Chollet R (1994) Regulation of wheat-leaf phospho*enol*pyruvate carboxylase by reversible phosphorylation. Plant Physiol 105 [Suppl]: 40

Dunford R, Plaxton WC, Turpin DH (1994) Kinetic properties of phospho*enol*pyruvate carboxylase are altered on ammonium resupply to N-limited cells of the green algae *Selenestrum minutum*. Plant Physiol 105 [Suppl]: 60

Ebbighausen H, Chen J, Heldt HW (1985) Oxaloacetate translocator in plant mitochondria. Biochim Biophys Acta 810: 184–199

Emes MJ, Bowsher CG (1991) Integration and compartmentation of carbon and nitrogen metabolism in roots. In: Emes MJ (ed) Compartmentation of plant metabolism in non-photosynthetic tissues. Cambridge University Press, Cambridge, pp 147–165

Flügge U-I, Heldt HW (1991) Metabolite translocators of the chloroplast envelope. Annu Rev Plant Physiol Plant Mol Biol 42: 129–144

Foyer C, Noctor G, Lelandais M, Lescure JC, Valadier MH, Boutin JP, Horton P (1994) Short-term effects of nitrate, nitrite and ammonium assimilation on photosynthesis, carbon partitioning and protein phosphorylation in maize. Planta 192: 211–220

Gemel J, Randall DD (1992) Light regulation of leaf mitochondrial pyruvate dehydrogenase complex. Role of photorespiratory carbon metabolism. Plant Physiol 100: 908–914

Glaab J, Kaiser WM (1993) Rapid modulation of nitrate reductase in pea roots. Planta 191: 173–179

Glaab J, Kaiser WM (1995) Inactivation of nitrate reductase is a two step mechanism, involving NR-protein phosphorylation and subsequent 'binding' of an inhibitor protein. Planta 195: 514–518

Hanning I, Heldt HW (1993) On the function of mitochondrial metabolism during photosynthesis in spinach (*Spinacia oleracea* L.) leaves. Partitioning between respiration and export of redox equivalents and precursors for nitrate assimilation products. Plant Physiol 103: 1147–1154

Hardie DG (1993) Use of protein phosphatase inhibitors in intact cells. In: Hardie DG (ed) Protein phosphorylation, practical approach. Oxford University Press, New York, pp 109–120

Hoff T, Stummann BM, Henningsen KW (1992) Structure, function and regulation of nitrate reductase in higher plants. Physiol Plant 84: 616–624

Hoff T, Trung H-N, Caboche M (1994) The use of mutants and transgenic plants to study nitrate assimilation. Plant Cell Environ 17: 489–506

Huber JL, Huber SC (1992) Site-specific serine phosphorylation of spinach leaf sucrose-phosphate synthase. Biochem J 283: 877–882.

Huber JL, Huber SC, Nielsen TH (1989) Protein phosphorylation as a mechanism for regulation of spinach leaf sucrose-phosphate synthase activity. Arch Biochem Biophys 270: 681–690

Huber JL, Huber, SC, Campbell WH, Redinbaugh MG (1992) Reversible light/dark modulation of spinach leaf nitrate reductase activity involves protein phosphorylation. Arch Biochem Biophys 296: 58–65

Huber, JL, Huber SC, McMichael RW Jr, Redinbaugh MG, Campbell WH (1993) Modulation of nitrate reductase by protein phosphorylation. Curr Top Plant Biochem Physiol 12: 7–8

Huber JL, Redinbaugh MG, Huber SC, Campbell WH (1994) Regulation of maize leaf nitrate reductase activity involves both gene expression and protein phosphorylation. Plant Physiol 106: 1667–1674

Huber, SC, Huber JL (1990a) Regulation of maize leaf sucrose-phosphate synthase by protein phosphorylation. Plant Cell Physiol 32: 319–326

Huber SC, Huber HL (1990b) In vitro phosphorylation and inactivation of spinach leaf sucrose-phosphate synthase by an endogenous protein kinase. Biochim Biophys Acta 1091: 393–400

Huber SC, Huber JL, (1990c) Regulation of spinach leaf sucrose-phosphate synthase by multi-site phosphorylation. Curr Top Plant Biochem Physiol 9: 329–343

Huber SC, Huber JL, McMichael RW Jr, Campbell WH Redinbaugh MG (1992) Regulation of cytoplasmic C- and N- metabolism by protein phosphorylation. In: Murata N (ed) Research in photosynthesis, vol III. Kluwer, Dordrecht, pp 675–682

Huber SC, Huber JL McMichael RW Jr, Campbell WH, Redinbaugh MG (1993) Sucrose-phosphate synthase: regulation by protein phosphorylation and possible role in C/N interactions. Curr Top Plant Biochem Physiol 12: 9–10

Huber SC, Huber JL, Kaiser WM (1994a) Differential response of nitrate reductase and sucrose-phosphate synthase-activation to inorganic and organic salts, in vitro and in situ. Physiol Plant 92: 302–310

Huber SC, Huber JL, McMichael RW Jr (1994b) Control of plant enzyme activity by reversible protein phosphorylation. Int Rev Cytol 149: 47–98

Huber SC, McMichael RW Jr, Huber JL, Bachmann M, Yamamoto YT, Conkling MA (1995) Light regulation of sucrose synthesis: role of protein phosphorylation and possible involvement of cytosolic [Ca^{2+}]. In: Madore MA, Lucas WJ (eds) Carbon partitioning and source–sink interactions. American Society of Plant Physiologists, Rockville, MD, pp 35–44

Huppe HC, Turpin DH (1994). Integration of carbon and nitrogen metabolism in plant and algal cells. Annu Rev Plant Physiol Plant Mol Biol 45: 577–607

Jiao J-A, Chollet R (1991) Posttranslational regulation of phospho*enol*pyruvate carboxylase in C_4 and crassulacean acid metabolism plants. Plant Physiol 95: 981–985

Jiao J-A, Chollet R (1992) Light activation of maize phospho*enol*pyruvate carboxylase protein-serine kinase activity is inhibited by mesophyll and bundle sheath-directed photosynthesis inhibitors. Plant Physiol 98: 152–156

Jiao J-A, Echeverria C, Vidal J, Chollet R (1991) Protein turnover as a component in the light/dark regulation of phospho*enol*pyruvate carboxylase protein-serine kinase activity in C_4 plants. Proc Natl Acad Sci USA 88: 2712–2715

Kaiser WM, Brendle-Behnisch E (1991) Rapid modulation of spinach leaf nitrate reductase activity by photosynthesis. I. Modulation in vivo by CO_2 availability. Plant Physiol 96: 363–367
Kaiser WM, Förster J (1989) Low CO_2 prevents nitrate reduction in leaves. Plant Physiol 91: 970–974
Kaiser WM, Huber SC (1994) Modulation of nitrate reductase in vivo and in vitro: effects of phosphoprotein phosphatase inhibitors, free Mg^{2+} and 5-AMP. Planta 93: 358–364
Kaiser WM, Spill D (1991) Rapid modulation of spinach leaf nitrate reductase activity by photosynthesis. II. In vitro modulation by ATP and AMP. Plant Physiol 96: 368–375
Kaiser WM, Spill D, Brendle-Behnisch E (1992) Adenine nucleotides are apparently involved in the light-dark modulation of spinach leaf nitrate reductase. Planta 186: 236–240
Kaiser WM, Spill D, Glaab J (1993) Rapid modulation of nitrate reductase in leaves and roots: indirect evidence for involvement of protein phosphorylation/dephosphorylation. Physiol Plant 89: 557–562
LaBrie ST, Crawford NM (1994) Glycine to aspartic acid change in the MoCo domain of nitrate reductase reduces both activity and phosphorylation levels in *Arabidopsis*. J Biol Chem 269: 14497–14501
Lewis OAM (1986) Plants and nitrogen. Edward Arnold, London.
Li B, Chollet R (1993) Resolution and identification of C_4 phospho*enol*pyruvate carboxylase protein-kinase polypeptides and their reversible light activation in maize leaves. Arch Biochem Biophys 307: 416–419
Li X-Z, Oaks A (1993) Induction and turnover of nitrate reductase in *Zea mays*. Influence of NO_3^-. Plant Physiol 102: 1251–1257
Lillo C (1994) Light regulation of nitrate reductase in green leaves of higher plants. Physiol Plant 90: 616–620
MacKintosh C (1992) Regulation of spinach-leaf nitrate reductase by protein phosphorylation. Biochim Biophys Acta 1137: 121–126
McMichael RW Jr, Klein RR, Salvucci ME, Huber SC (1993) Identification of the major regulatory phosphorylation site in sucrose-phosphate synthase. Arch Biochem Biophys 307: 248–252
McMichael RW Jr, Kochansky J, Klein R, Huber SC (1994) Characterization of sucrose-phosphate synthase protein kinase. Plant Physiol 104 [Suppl]: 91.
McNaughton GAL, MacKintosh C, Fewson CA, Wilkins MB, Nimmo HG (1991) Illumination increases the phosphorylation state of maize leaf phospho*enol*pyruvate carboxylase by causing an increase in the activity of a protein kinase. Biochim Biophys Acta 1093: 189–195
Manh T, Bismuth E, Boutin J-P, Provot M, Champigny M-L (1993) Metabolite effectors for short-term nitrogen-dependent enhancement of phospho*enol*ypyruvate carboxylase activity and decrease of net sucrose synthesis in wheat leaves. Physiol Plant 89: 460–466
Müller R, Steigner W, Gimmler H, Kaiser WM (1990) Effect of ammonium on dark-CO_2-fixation and on cytosolic and vacuolar pH values in *Eremosphaera viridis* de Bary (Chlorococcales). J Exp Bot 41: 441–448
Müller R, Baier M, Kaiser WM (1991) Differential stimulation of PEP-carboxylation in guard cells and mesophyll cells by ammonium or fusicoccin. J Exp Bot 42: 215–220
Neuhaus HE, Quick WP, Siegl G, Stitt M (1990) Control of photosynthate partitioning in spinach leaves. Analysis of the interaction between feedforward and feedback regulation of sucrose synthesis. Planta 181: 583–592
Nimmo HG (1992) The regulation of phospho*enol*pyruvate carboxylase by reversible phosphorylation. In: Battey NH, Dickinson HG, Hetherington AM (eds) Post-translational modification in plants. Cambridge University Press, Cambridge, pp 161–170. (Society for Experimental Biology seminar series 53)
Pace GM, Volk RJ, Jackson WA (1990) Nitrate reduction in response to CO_2-limited photosyn-

thesis. Relationship to carbohydrate supply and nitrate reductase activity in maize seedlings. Plant Physiol 92: 286–292

Pate JS (1983) Patterns of nitrogen metabolism in higher plants and their ecological significance. In: Lee JA, McNeill S, Rorison IH (eds) Nitrogen as an ecological factor. Blackwell, Oxford, pp 225–255

Pierre J-N, Pacquit V, Vidal J, Gadal P (1992) Regulatory phosphorylation of phospho*enol*pyruvate carboxylase in protoplasts from *Sorghum* mesophyll cells: pH and Ca^{2+} as possible components of the light transduction pathway. Eur J Biochem 210: 531–537

Raghavendra AS, Yin Z-H, Heber U (1993) Light-dependent pH changes in leaves of C_4 plants. Comparison of the pH response to carbon dioxide and oxygen with that of C_3 plants. Planta 189: 278–287

Rajagopalan AV, Tirumala Devi M, Raghavendra AS (1993) Patterns of phospho*enol*pyruvate carboxylase activity and cytosolic pH during light activation and dark deactivation in C_3 and C_4 plants. Photosynth Res 38: 51–60

Rajagopalan AV, Tirumala Devi M, Raghavendra AS (1994) Molecular biology of C_4 phospho*enol*pyruvate carboxylase: structure, regulation and genetic engineering. Photosynth Res 39: 115–135

Raven JA, Smith FA (1976) Nitrogen assimilation and transport in vascular land plants in relation to intracellular pH regulation. New Phytol 76: 415–431

Reggiani R, Mattana M, Aurisano N, Bertani A (1993) Utilization of stored nitrate during the anaerobic germination of rice seeds. Plant Cell Physiol 34: 379–383

Reimholz R, Geigenberger P, Stitt M (1994) Sucrose-phosphate synthase is regulated via metabolites and protein phosphorylation in potato tubers, in a manner analogous to the enzyme in leaves. Planta 192: 480–488

Riens B, Heldt HW (1992) Decrease of nitrate reductase activity in spinach leaves during a light-dark transition. Plant Physiol 98: 573–577

Riens B, Lohaus G, Heineke D, Heldt HW (1991) Amino acid and sucrose content determined in the cytosolic, chloroplastic, and vacuolar compartments and in the phloem sap of spinach leaves. Plant Physiol 97: 227–233

Riens B, Lohaus G, Winter H, Heldt HW (1994) Production and diurnal utilization of assimilates in leaves of spinach (*Spinacia oleracea* L.) and barley (*Hordeum vulgare* L.). Planta 192: 497–501

Robinson JM (1988) Spinach leaf chloroplast CO_2 and NO_2^- photoassimilations do not compete for photogenerated reductant. Plant Physiol 88: 1373–1380

Robinson JM, Baysdorfer C (1985) Interrelationships between carbon and nitrogen metabolism in mature soybean leaves and isolated leaf mesophyll cells. In: Heath RL, Preiss J (eds) Regulation of carbon partitioning in photosynthetic tissues. American Society of Plant Physiologists, Rockville, MD, pp 333–357

Schuller KA, Plaxton WC, Turpin DH (1990) Regulation of phospho*enol*pyruvate carboxylase from the green alga *Selenastrum minutum*. Plant Physiol 93: 1303–1311

Schweizer P, Erismann KH (1985) Effect of nitrate and ammonium nutrition on non-nodulated *Phaseolus vulgaris* L. on phospho*enol*pyruvate carboxylase and pyruvate kinase activity. Plant Physiol 78: 455–458

Shannon JC, Paul MJ, Huber SC (1994) Amino acids as effectors of nitrate reductase and sucrose-phosphate synthase activities in spinach leaves. Plant Physiol 104 [Suppl]: 91

Shiraishi N, Sato T, Ogura N, Nakagawa H (1992) Control by glutamine of the synthesis of nitrate reductase in cultured spinach cells. Plant Cell Physiol 33: 727–731

Siegl G, MacKintosh C, Stitt M (1990) Sucrose-phosphate synthase is dephosphorylated by protein phosphatase 2A in spinach leaves. Evidence from the effects of okadaic acid and microcystin. FEBS Lett 270: 198–202

Speer M, Kaiser WM (1991) Ion relations of symplastic and apoplastic space in leaves from *Spinacia oleracea* L. and *Pisum sativum* L. under salinity. Plant Physiol 97: 990–997

Spill D, Kaiser WM (1994) Partial purification of two proteins (100 kDa and 67 kDa), cooperating in the ATP-dependent inactivation of spinach leaf nitrate reductase. Planta 192: 183–188

Stitt M, Schulze D (1994) Does Rubisco control the rate of photosynthesis and plant growth? An excercise in molecular ecophysiology. Plant Cell Environ 17: 465–487

Stitt M, Huber SC, Kerr P (1987) Control of photosynthetic sucrose synthesis. In: Hatch MD, Boardman NR (eds) The biochemistry of plants, vol 10. Academic Press, New York, pp 328–429

Stitt M, Wilke I, Feil R, Heldt HW (1988) Coarse control of sucrose-phosphate synthase in leaves: alterations of the kinetic properties in response to the rate of photosynthesis and the accumulation of sucrose. Planta 174: 217–230

Sugiharto B, Suzuki L, Burnell JN, Sugiyama T (1992) Glutamine induces the N-dependent accumulation of mRNAs encoding phospho*enol*pyruvate carboxylase and carbonic anhydrase and detached maize leaf tissue. Plant Physiol 100: 2066–2070

Van Quy L, Champigny M-L (1992) NO_3^- enhances the kinase activity for phosphorylation of phospho*enol*pyruvate carboxylase and sucrose phosphate synthase proteins in wheat leaves. Plant Physiol 99: 344–347

Van Quy L, Foyer C, Champigny M-L (1991) Effect of light and nitrate on wheat leaf phospho*enol*pyruvate carboxylase activity. Evidence for covalent modulation of the C_3 enzyme. Plant Physiol 97: 1476–1482

Vanleberghe GC, Schuller KA, Smith RG, Feil R, Plaxton WC, Turpin DH (1990) Relationship between NH_4^+ assimilation rate and in vivo phospho*enol*pyruvate carboxylase activity. Regulation of anaplerotic carbon flow in the green alga *Selenastrum minutum*. Plant Physiol 94: 284–290

Vidal J, Pierre J-N, Echevarria C (1996) The regulatory phosphorylation of C_4 phospho*enol*pyruvate carboxylase: a cardinal event in C_4 photosynthesis. In: Verma DPS (ed) Signal transduction in plant growth and development. Springer, Wien New York, pp 141–166 [Dennis ES et al (eds) Plant gene research. Basic knowledge and application]

Wang Y-H, Chollet R (1993a) In vitro phosphorylation of purified tobacco-leaf phospho*enol*pyruvate carboxylase. FEBS Lett 328: 215–218

Wang Y-H, Chollet R (1993b) Partial purification and characterization of phospho*enol*pyruvate carboxylase protein-serine kinase from illuminated maize leaves. Arch Biochem Biophys 304: 496–502

Weiner H, McMichael RW, Huber SC (1992) Identification of factors regulating the phosphorylation status of sucrose-phosphate synthase in vivo. Plant Physiol 99: 1435–1442

Weiner H, Weiner H, Stitt M (1993) Sucrose-phosphate synthase phosphatase, a type 2A protein phosphatase, changes its sensitivity towards inhibition by inorganic phosphate in spinach leaves. FEBS Lett 333: 159–164

Winter H, Lohaus G, Heldt HW (1992) Phloem transport of amino acids in relation to their cytosolic levels in barley leaves. Plant Physiol 99: 996–1004

Phosphorylation and the Cytoskeleton

Siân M. Ritchie and Nicholas H. Battey

Contents

I. Introduction

The cytoskeleton is a dynamic network of microtubules, microfilaments and intermediate filaments. The special properties of these proteins enable them to participate in a wide range of functions in plant cells (see Lloyd, 1991). Their fundamental characteristic is reversible polymerisation. This assembly and disassembly has been likened to the transition between gas and liquid, because the elements are constantly fluxing between the soluble and insoluble state (Oosawa and Kasai, 1962). This behaviour means that the proteins are highly dynamic, and so must also be closely regulated. Phosphorylation regulates the dynamics of intermediate filaments directly, by effects on the polymerisation capability of the subunit proteins. In contrast, microtubule and microfilament dynamics are regulated largely by their associated proteins, whose activity is further controlled by phosphorylation.

In this review we first describe the features of cytoskeletal protein polymerisation which are central for the dynamic behaviour of these proteins in the cell. Then we illustrate the critical role of phosphorylation in the regulation of the cytoskeleton during interphase and cell division. Our aim is to highlight the potential importance of phosphorylation in the cytoskeleton as a mechanism for linking external signals (light, temperature, ligand binding) to cellular reorganisation in plants. Further, emerging evidence indicates the use of phosphorylation as a transduction mechanism for internal signals such as those that co-ordinately regulate the cellular rearrangements required during cell division.

II. Cytoskeletal Dynamics and Phosphorylation of the Polymer Proteins

A fundamental property of the proteins which constitute the cytoskeleton is the ability to polymerise. Equally important is the characteristic of random reversibility. Conditions necessary to induce polymerisation and depolymerisation of actin, tubulin, and intermediate filament proteins in vitro have been defined and are summarised in Table 1. However, protein polymerisation is not simply a matter of using up the free subunits until they are all incorporated into polymer, nor is it a stable equilibrium between subunit incorporation and release, as originally proposed for microtubules by Oosawa and Kasai (1962). In the following section we discuss current understanding of the behaviour of the polymer proteins.

A. Microtubules

Allen and Borisy (1974), Dentler et al. (1974), and Bergen and Borisy (1980), showed that the rate of subunit incorporation into microtubules differs at the two ends of the polymer (Fig. 1 a). Margolis and Wilson (1978) proposed that

Table 1. Standard conditions for cytoskeletal protein extraction (methods developed for animal tissue)

	Intermediate filament protein	Microfilaments	Microtubules
Extraction conditions	10–50 mM Tris, pH 7.4, 5 mM Mg^{2+}	low ionic strength, pH 7.5, 0.1 mM Ca^{2+} (plus subsequent chromatography) (acetone powder preparation suitable for sources of high actin concentration)	50 mM Pipes, pH 6.9, 1 mM Mg^{2+}
State of protein when extracted	insoluble (i.e. polymerised)	monomers	dimers
Conditions used to alter assembly state of polymer protein	8 M urea or 5–6 M guanidinium hydrochloride	0.1 M KCl, 1–2 mM $MgCl_2$	raised temperature
References	Renner et al., 1981	Pardee and Spudich 1982; Gordon et al., 1976	Shelanski et al., 1973

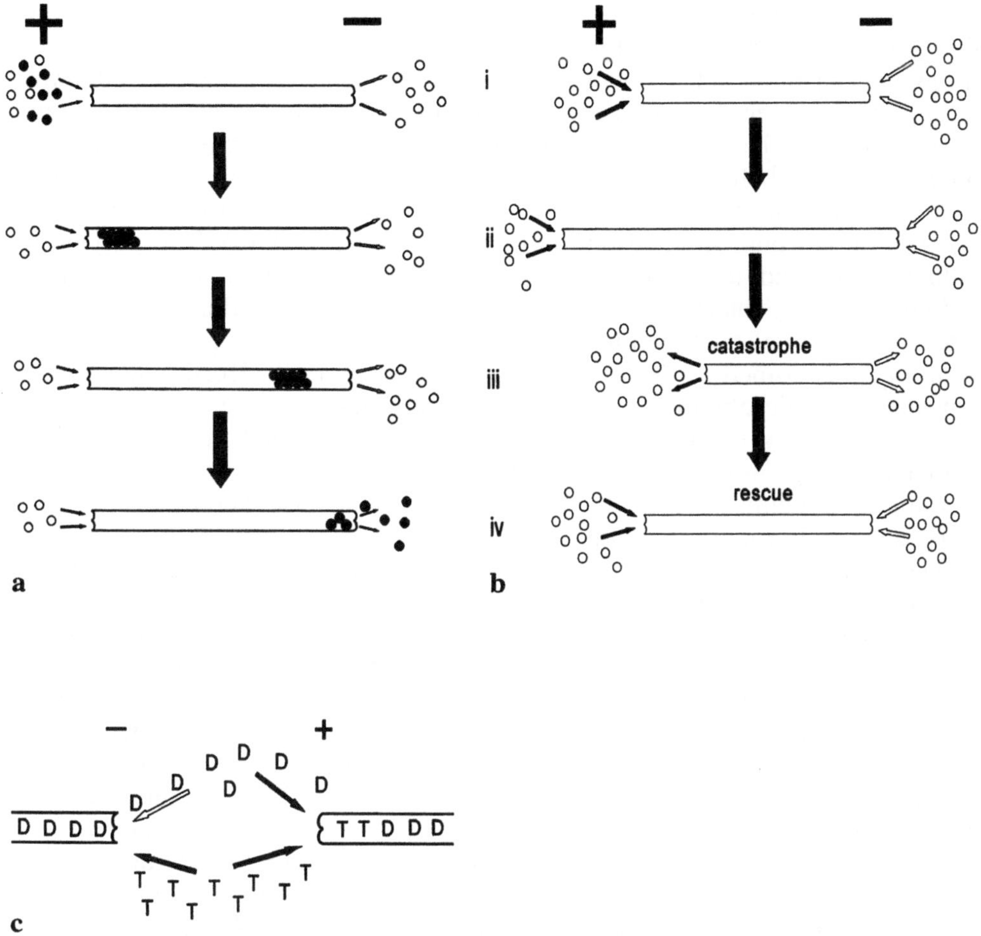

Fig. 1 a–c. Treadmilling, dynamic instability and the nucleotide cap theory. All arrows represent net movement of subunits. a, Schematic illustration of treadmilling, as applied to both microtubules and microfilaments. Net incorporation of subunits at the plus end and release at the minus end results in the movement of subunits (solid circles) through the polymer from the plus to minus end. b, Schematic diagram of the phenomenon of dynamic instability as applied to microtubules. i–ii, The increase in polymer is faster at the plus end. iii, A sudden net loss in subunits from the polymer results in "catastrophe"; the shrinking is much faster at the plus than the minus end. Also it may only occur at one end. iv, Catastrophe may result in total breakdown of the polymer, or "rescue" may occur in which the polymer reverts back to net growth. Fluctuation between growth and shrinking is random. c, Schematic illustration of the nucleotide cap theory as applied to microtubules and microfilaments. D, nucleoside diphosphates or nucleoside diphosphate + P_i (ADP for microfilaments, GDP for microtubules) at the exchangeable binding site on the free or incorporated subunits; T, equivalent nucleoside triphosphate. The range of arrow shading (light, medium, dark) represents the affinity between the free subunits and the polymer end. The highest affinity is between the free subunits with nucleoside triphosphate bound and the polymer end with the nucleotide cap not only the growth rates, but also the equilibrium constants for each end differed; hence the growth of one end was compensated by breakdown at the other and an overall equilibrium was achieved. This model was called treadmilling, because the effect would be a movement of individual subunits through the polymer from the net incorporating (+) end to the net dissociating (–) end (Fig. 1 a). However, although treadmilling does occur, it soon became apparent that the behaviour of microtubules was not as simple as this model implied. Mitchison and Kirschner (1984 a, b) were the first to measure real time length and number distribution of microtubules in vitro. The polymers could be observed to switch from growing to shrinking in an apparently random manner ("catastrophe" and "rescue"; see Fig. 1 b); the frequency of switching was quicker at the plus end than the minus end. The phrase *dynamic instability* was coined to describe this behaviour. In 1986, Horio and Hotani confirmed that microtubules never reach a steady state because of the randomness of the sudden changes from growing to shrinking and vice versa. This characteristic dynamic instability is also apparent in vivo. In animal cells microtubule dynamics are modulated by microtubule associated proteins, but the fundamental property remains.

The "GTP-cap" theory was proposed to explain dynamic instability (Carlier et al., 1987). After a tubulin unit incorporates into a polymer the GTP bound to the exchangeable site is hydrolysed into GDP and P_i. This hydrolysis lags behind the incorporation of the next subunit into the polymer, and the release of the P_i from the polymer is slower still. Hence, at a given moment during growth of a microtubule end there will be a number of tubulin units in the polymer which still have GTP or GDP and P_i bound (Fig. 1c). Bound GTP at the end of a polymer increases its stability (Davis et al., 1994). It is more kinetically favourable for a new tubulin unit to add on to the end of a polymer which has GTP bound at the exchangeable site, rather than GDP, thus when one end of a polymer is growing it is kinetically favourable for it to keep on growing, assuming no lack of available free tubulin. Because the polymers are inherently polar the addition of tubulin at the plus end of the polymer is more favoured than that at the minus end. This is also true for dissociation of the subunits from the polymer during shrinking, where tubulin comes off the microtubule much more readily at the plus end than at the minus end.

Therefore the plus end of a polymer will grow rapidly, but if free tubulin is unavailable, the GTP cap of this end will quickly disappear and the end will rapidly shrink (catastrophe). The more slowly growing minus end in the same circumstances, will shrink more slowly and it has a greater probability of starting to grow again (rescue). The rate of shrinking is faster than that of growing so in a solution of fixed tubulin concentration the net trend is towards longer and fewer microtubules. The difference in the way the two ends of the microtubule behave is explained by the fact that the tubulin subunits at the two ends of the polymer have different parts of the molecule exposed for new addition, and the binding of the GTP/GDP will affect the conformation of the parts of the molecule in a correspondingly different way. A great deal of effort has gone into proving the existence of the GTP cap, but the theory is still controversial.

The events which take place at the ends of polymerising and depolymerising microtubules still need to be elucidated in more molecular detail.

There are only a few isolated examples of direct phosphorylation of tubulin, and its regulatory significance is unclear. Koontz and Choi (1993) found that carrot suspension cells contained tubulin phosphorylating activity and that this was developmentally regulated, but the effect of phosphorylation was not investigated. Phosphorylation of specific tubulin isotypes on serine and tyrosine residues has been reported in animal cells (Ludueña et al., 1992) but its functional significance is unknown. Zhou et al. (1991) found that a protein kinase, $pp39^{mos}$, colocalised with microtubules at the spindle in transfected Swiss 3T3 cells, and in studies using immunoprecipitation β-tubulin was found to be the main substrate for the kinase. Crute et al. (1992) proposed that a casein kinase II-like activity which copurifies with tubulin is a MAP which phosphorylates β-tubulin.

B. Microfilaments

Polymerisation of actin into microfilaments is in many ways similar to polymerisation of tubulin into microtubules. The essential characteristics of subunit incorporation, and the involvement of nucleotide hydrolysis in the stabilisation of the polymer are the same (Korn et al., 1987; Carlier, 1989). As for microtubules, the overall polarity of the polymer results in different rates of incorporation of actin at different ends of the microfilament. However, slower rates of dissociation at the polymer end mean that the overall process of polymerisation is more stable, and the dramatic switching between growing and shrinking characteristic of microtubules is not observed. The theory of treadmilling is still considered a valid model for microfilaments (Fig. 1a) (Cleveland, 1982; Sheterline and Sparrow, 1994). The nucleotide cap theory has also been applied to microfilaments, with ATP rather than GTP bound to the subunits (Fig. 1c) (Korn et al., 1987).

Phosphorylation of actin does not seem to be of general importance in regulating microfilament dynamics. However, actin has been found to be phosphorylated in *Physarum polycephalum* when complexed with the associated protein fragmin (Furuhashi and Hatano, 1992). The effect of phosphorylation was to abolish the nucleating/capping activity of the fragmin/actin complex.

C. Intermediate Filaments

Intermediate filament protein polymerisation is very different from polymerisation of actin and tubulin. There is a less significant pool of unpolymerised monomer than for microtubules and microfilaments (Soellner et al., 1985). Early work on intermediate filament networks concluded that the polymers were not dynamic in vivo; that is, once polymerised the intermediate filaments stayed that way. This idea developed because the intermediate filament network was found to be highly insoluble when extracted from cells, and extreme conditions, such as 8 M urea, were required to solubilise the protein

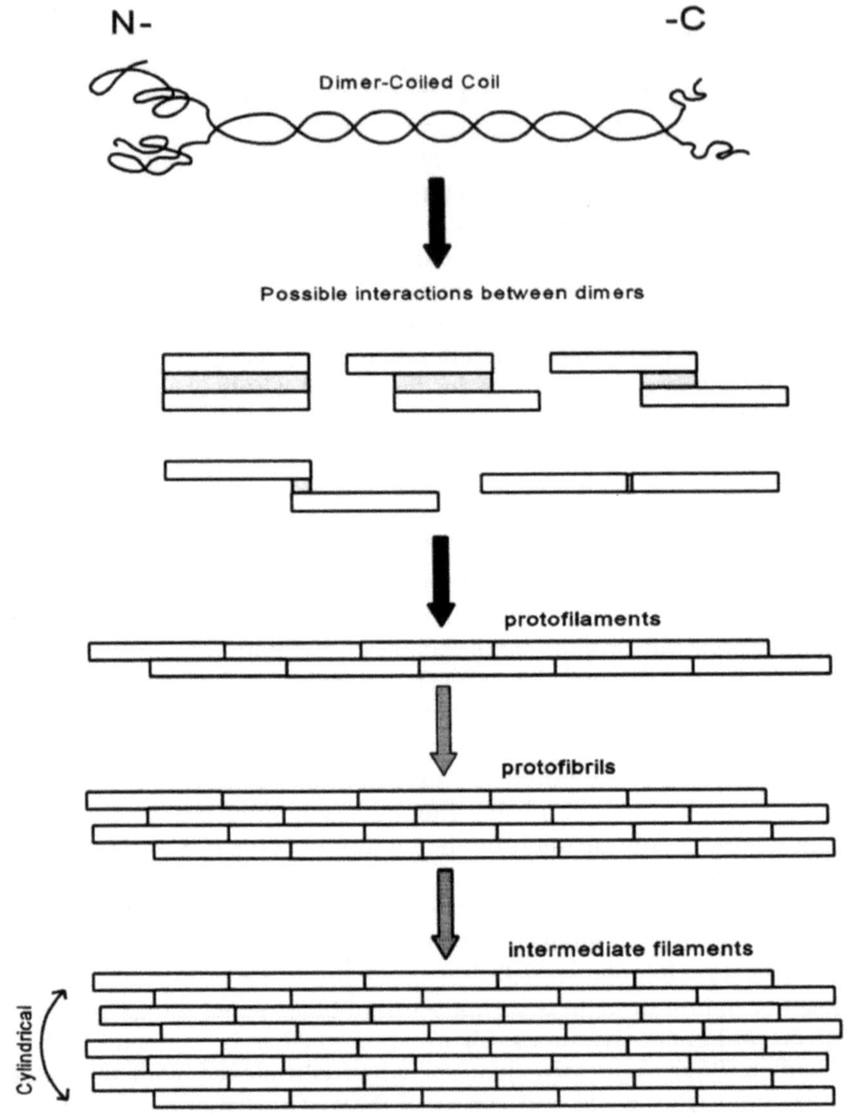
N-
-C
Dimer-Coiled Coil
Possible interactions between dimers
protofilaments
protofibrils
intermediate filaments
Cylindrical
a

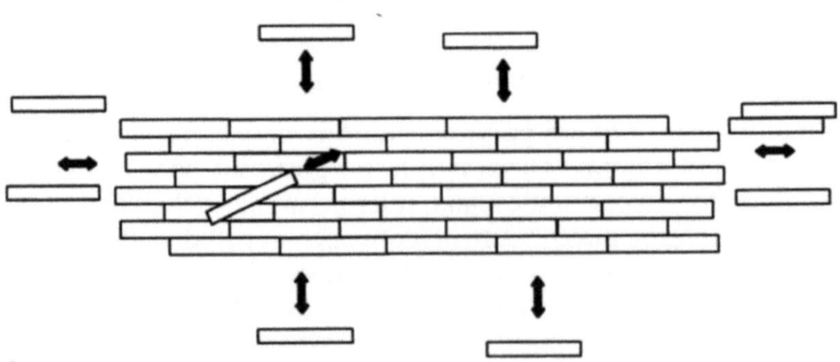
b

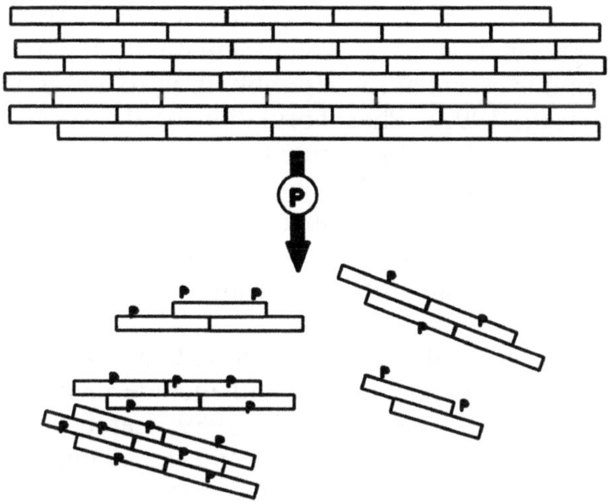

c

Fig. 2a–c. Intermediate filament dynamics. a, Schematic diagram of proposed stages of intermediate filament assembly (adapted from Heins and Aebi, 1994). b, Unlike in microtubules and microfilaments, subunits of intermediate filaments can exchange along the length of the filament. c, Schematic illustration of the effect of phosphorylation at the N-terminal domain on intermediate filament polymerisation

(Franke et al., 1979). The only known mechanism of breakdown of the networks was proteolysis. However when Vikstrom et al. (1989) microinjected labelled intermediate filament protein into animal cells it became apparent that the subunits of the intermediate filament networks were exchangeable, not static (Vikstrom et al., 1992). Intermediate filament disassembly is now generally believed to be caused by hyperphosphorylation of the component proteins. Vimentin and nuclear lamins are phosphorylated by $p34^{cdc2}$ kinase (Chou et al., 1990, 1991; see Sect. IV.A) Other protein kinases which are not specifically active during cell division have also been proposed to regulate the intermediate filament networks (Albers and Fuchs, 1992). When quiescent hepatocytes are treated with epidermal growth factor a rapid increase in phosphorylation and reorganisation of the keratin network is stimulated (Baribault et al., 1989). Microinjection of cAMP-dependent protein kinase into fibroblasts causes phosphorylation of vimentin and concomitant cytoskeletal rearrangements (Lamb et al., 1989). Figure 2 illustrates intermediate filament formation, and the way phosphorylation is thought to cause disassembly.

Eriksson et al. (1992) point out that there are examples of the involvement of phosphorylation in regulation of intermediate filament networks in which the rather simplistic model of phosphorylation leading to disassembly does not apply. The apparent contradictions can be explained, however, if the actual residues in the molecules which are phosphorylated are compared. Phosphorylation of a specific site in the N-terminal domain, which is known by deletion and point mutation studies to be vital for assembly, does cause disassembly of the intermediate filaments (Quinlan and Stewart, 1991). On the other hand, phosphorylation of a protein kinase C site in this domain of vimentin and desmin is essential for assembly in vitro and in vivo (Raats et al., 1990; Herrmann et al., 1992). Phosphorylation of the C-terminal domain of neurofilaments occurs when the protein is in a filamentous form; this influences the interaction of the filaments with microtubules (Hisanaga et al., 1991). Therefore although it is clear that phosphorylation is important for the dynamics of the networks, generalisations are dangerous unless the details of kinases, phosphatases and sites of modification are known (Heins and Aebi, 1994).

Protein kinases in animals have been found to phosphorylate the whole range of intermediate filament proteins and include cAMP-dependent protein kinase (Inagaki et al., 1987; Geisler et al., 1989; Lamb et al., 1989), protein kinase C (Geisler et al., 1989; Omary et al., 1992), calcium/calmodulin-dependent protein kinase II (Tokui et al., 1990; Ando et al., 1991; Yano et al., 1994), and a cytoskeleton associated kinase with properties distinct from other well known kinases (Harrison and Mobley, 1992). Foisner et al. (1991) present evidence that the protein plectin, which associates with intermediate filaments, actin associated proteins, and microtubule associated proteins, is phosphorylated by protein kinase C and this leads to its dissociation from intermediate filaments. In one plant study proteins from pea nuclei with lamin-like properties were found to be phosphorylated by a kinase with casein kinase II properties (Li and Roux, 1992).

Another significant difference between intermediate filaments and microtubules and microfilaments is that polymerisation is not accompanied by nucleotide hydrolysis. In microtubules and microfilaments this hydrolysis is thought to cause changes in the conformation of the incorporated molecules that increase the stability of the polymer (see Sect. II. A, B). It is not known what conformational changes, if any, occur when intermediate filament proteins are incorporated into filaments, but nucleotide hydrolysis is not involved in the process. Intermediate filament subunit proteins are varied in sequence and size of terminal domains, and there are therefore several possible bonds that can be formed between the dimers in the filaments (Fig. 2 a). Because of this the intermediate filaments are structurally much more varied than microtubules and microfilaments (Heins and Aebi, 1994).

The existence of intermediate filaments homologous to those found in animal cells is still a controversial matter in plant cell biology (see Shaw et al., 1991; Menzel, 1993). Proteins of a fibrillar nature can be extracted from plant cell tissue, and using the same conditions as those for animal intermediate filaments made to disassemble, and reassemble into filaments that are intermediate filament-like as judged by TEM (Hargreaves et al., 1989). These proteins also cross-react with animal intermediate filament antibodies, and the antibodies colocalise to the fibrillar structures in situ (Beven et al., 1991). But the antibodies may also label other elements in the cell; this leads some to question the conclusion that plants contain intermediate filaments homologous to those in animals (Menzel, 1993). The evidence for nuclear lamins, an intermediate filament protein that is closely related to the common ancestor shared by all the animal IF types (Albers and Fuchs, 1992) is nevertheless compelling, because the proteins have been identified in the higher plant nucleus (Beven et al., 1991; McNulty and Saunders, 1992; Minguez and Moreno Diaz de la Espina, 1993).

III. Phosphorylation of Cytoskeletal Associated Proteins During Interphase

A. Microtubule Associated Protein Phosphorylation

Microtubule associated proteins (MAPS) have long been known to enhance the stability of microtubules in vitro, and more recently evidence has emerged that this behaviour is relevant in vivo. Microinjection and transfection of MAPs has clearly demonstrated that these proteins alter the behaviour of microtubules. For example, Lee and Rook (1992) expressed the neuronal MAP, tau, in non-neuronal cells. The result was increased bundling of the microtubules. Whether this bundling is an indirect result of a stabilising effect of tau on microtubule dynamics, or directly carried out by the tau protein, is uncertain (for discussion, see Hirokawa, 1993). Weisshaar et al. (1992) expressed another neuronal MAP, MAP2c (the lower molecular mass form of MAP2), in non-neuronal cells. The result was an alteration of the microtubules to a form characteristic of neuronal cells, i.e., long and stiff bundles at the cortex. This study emphasises that a cell's complement of MAPs is likely to be highly specific, reflecting the functional and structural specialisation of the cytoskeleton in that cell type. An in vitro study illustrating a rather different effect of MAPs on microtubule dynamics was conducted by Pryer et al. (1992). Tau, MAP2, and a population of heat-stable MAPs were applied to tubulin and the effect on the dynamics of polymerisation assessed using video microscopy. Tau and MAP2 lowered the rate of catastrophe and increased the rate of rescue, but these events were still frequent, whereas the heat-stable MAPs suppressed catastrophe almost completely and promoted assembly, resulting in much more stable microtubules. Thus while different MAPs have the common property of altering the dynamics of microtubules, they can have a wide variety of effects; some may also be involved in bundling of microtubules by interaction between the side arm projections of MAPs associated with neighbouring microtubules (see Fig. 3).

Several proteins have been identified from plants which have the characteristics of MAPs (for review, see Chasan, 1993). A microtubule-binding protein has been isolated from carrot suspension cells; this binds to microtubules from both animal and plant sources (Cyr and Palevitz, 1989). A MAP has been isolated from the green alga *Dichotomosiphon tuberosus*, and has been shown to have microtubule bundling properties (Maekawa et al., 1990). Tobacco BY-2 cells also contain a MAP with microtubule bundling properties; this protein colocalised to microtubule arrays throughout the cell cycle (i.e., interphase

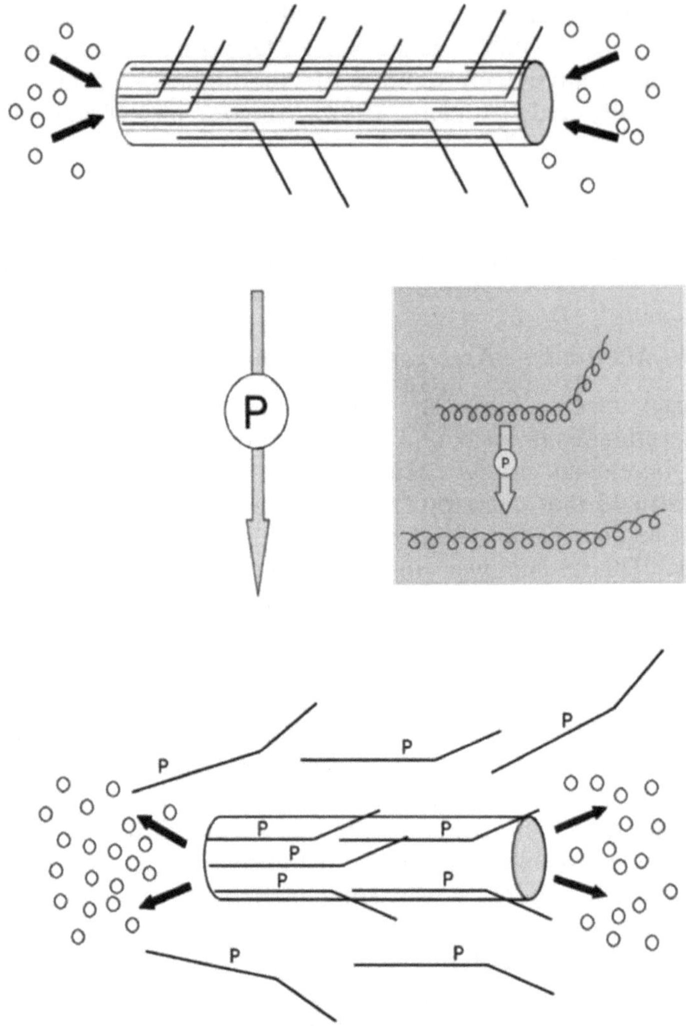

Fig. 3. MAP phosphorylation reduces binding and microtubule stability. Schematic illustration of MAPs on a microtubule showing that only part of each MAP molecule binds along the microtubule. (Based on studies carried out on neuronal MAPs: MAP2, Jensen and Smail, 1986; and tau, Hirokawa et al., 1988.) On phosphorylation the MAP molecule undergoes a conformational change (see inset) which reduces its affinity for the microtubule. The release of the MAP results in reduced microtubule stability, illustrated by a reduction in microtubule length

cortex, preprophase band, spindle, phragmoplast) (Chang-Jie and Sonobe, 1993). Vantard et al. (1991) have identified a 100 kDa putative plant MAP from maize suspension cells which crossreacts with an antibody to the neuronal MAP, tau. The properties of this and other MAPs isolated from maize suspension cells have been characterised in some detail (Schellenbaum et al., 1993). Most recently the properties of the 100 kDa MAP have been investigated further. The protein has been found to be heat stable, a characteristic of many animal MAPs. Antibodies have been raised to the 100kDa protein, and Western blotting against heat-stable fractions from carrot and tobacco cells shows cross reactivity with proteins of the same molecular weight in these plants. Significantly the 100kDa protein is phosphorylated in vitro by addition of maize cytosolic extract, indicating a further similarity to animal MAPs (Vantard et al., 1994).

A great deal of work has been done on phosphorylation of animal, in particular neuronal, MAPs. The results of particular relevance to plants are those which indicate the effects of phosphorylation on the interaction of the MAP with microtubules, and which involve protein kinases fundamental to eukaryotes or likely to have a functional equivalent in plants.

This latter category includes $p34^{cdc2}$, which is discussed in Sect. IV, and the mitogen-activated protein kinase. Mitogen-activated protein kinase homologues have been identified in alfalfa (Duerr et al., 1993; Jonak et al., 1993), and *Arabidopsis* (Mizoguchi et al., 1993). This kinase type was first identified because of its activation upon stimulation of animal cells by growth factors. It is now known that the kinase is involved, along with $p34^{cdc2}$ kinase, in the transition between G_2 and M phase (for reviews, see Cobb et al., 1991; Pelech and Sanghera, 1992; Ruderman, 1993). When isolated from M-phase cells and added to *Xenopus* oocytes at interphase, mitogen-activated protein kinase caused rearrangements of microtubules typical of the interphase–M-phase transition (Gotoh et al., 1991). This implies that the mitogen-activated protein kinase has its effect by phosphorylating a MAP. Shiina et al. (1992) worked with p220, a major MAP from *Xenopus* eggs, which they propose as an in vivo target for mitogen-activated protein kinase and maturation promotion factor (a complex of $p34^{cdc2}$ and cyclin). When p220 was extracted from interphase cells it bound potently to microtubules and stimulated tubulin polymerisation, whereas when it was taken from cells in M phase it showed virtually none of this activity. Staining patterns of p220 revealed its close association with microtubules during interphase, but a diffuse staining at M phase.

The following examples also suggest that MAP phosphorylation favours microtubule disassembly. Hagestedt et al. (1989), while attempting to purify the brain MAP, tau, found that phosphorylation by Ca^{2+}/calmodulin-dependent protein kinase changed the tau from a short elastic form to a long stiff form (Fig. 3 inset). This indicates that phosphorylation of the MAP, in this case, causes a conformational change in the molecule which prevents it from interacting with, and promoting assembly of, microtubules. Gliksman et al. (1992) used high resolution video microscopy to study microtubule dynamics in extracts of interphase sea urchin eggs. After the addition of the phosphatase inhibitor okadaic acid the microtubules became shorter, and their transition between growing and shrinking was more rapid. This overall effect reflected a modulation of rescue frequency (rather than alteration of actual rates of association/dissociation). The authors emphasise that this change in microtubule dynamics is characteristic of the interphase–M-phase transition, but the implication is clear for a more general model about the effect of phosphorylation state on MAP function and microtubule dynamics.

B. Actin Associated Protein Phosphorylation

The dynamics and organisation of microfilaments in vivo are tightly controlled by a large number of actin associated proteins (AAPs), so far numbering over 60 (see Kreis and Vale, 1993). Many of these proteins were originally thought to be species/cell type specific, but it is now evident that this is not the case.

The actin-associated proteins have been classified according to their effects on actin and microfilaments in vitro (Table 2). However, although the proteins in each group have common activities and regions of amino acid sequence, they also have features and domains in common with proteins from other groups (Vandekerckhove and Vancompernolle, 1992). The problem of how to classify these proteins is compounded by the fact that they can have different effects on actin/microfilaments in different circumstances. For example, profilin was originally proposed to act as an actin sequestering protein; that is, in binding to actin monomers it prevents their addition to microfilaments, hence altering microfilament dynamics by lowering the effective concentration of free actin. However, it was subsequently found that profilin probably does not have this role in vivo. Overexpression of the protein in CHO (Chinese hamster ovary) cells causes stabilization of the microfilament networks by reducing the dynamics of the actin polymerisation (Finkel et al., 1994). Its true function is probably to promote the exchange of ADP bound to actin monomers

Table 2. Classification of actin associated proteins according to binding domains (adapted from Vanderkerckhove and Vancompernolle, 1992)

AAP in which binding domain was originally identified	Group allocation according to in vitro function	Other proteins containing the domain	Identity with other AAP domains
Actobindin	monomer binding	fimbrin, plastin, villin, thymosin, myosin, actinin, tropomyosin	
Profilin	monomer binding, crosslinking		gelsolin family
Cofilin	severing	destrin, depactin, actophorin	gelsolin, profilin
Actinin	crosslinking	spectrin, fodrin, fimbrin, dystrophin, filamin, plastin, ABP-120, adducin, fascin	
Caldesmon	lateral binding		
Gelsolin	barbed end capping, severing	severin, fragmin	
Severin	barbed end capping, severing		
Villin	bundling		

(thus lowering the critical concentration for polymerisation); it also interacts with several signalling pathways in vivo (Theriot and Mitchison, 1993). Profilin has been identified in higher plants; initial studies on the protein show it binds to animal and plant actin, and the carboxy-terminal domain is important in binding, although the sequence identity with profilins from single-celled organisms and mammals is not high (Valenta et al., 1993). At least three genes for profilin have been identified in maize and the different forms show differential expression between anther and pollen tissue (Staiger et al., 1993). The recent reappraisal of the function of profilin illustrates the limited rele-

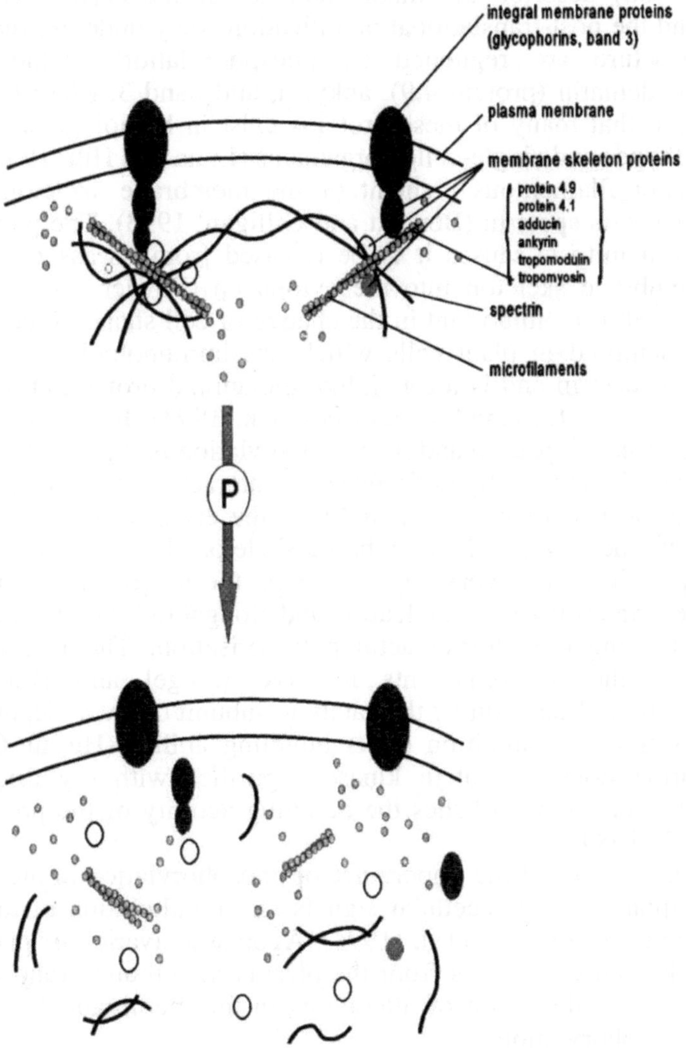

Fig. 4. Phosphorylation and the membrane skeleton. Schematic simplified model of the membrane skeleton. The proteins in parentheses are those identified in the erythrocyte. Phosphorylation causes a relaxation of tension created by the skeleton

vance that observations made in vitro can have for understanding of AAP function in vivo. Another complication with actin-binding proteins is that their association with actin may be incidental to other more important in vivo functions. An example of this is the ABP (actin-binding protein)-50 which was found to be an elongation factor (Yang et al., 1990).

The term membrane skeleton describes the proteins that underlie the plasma membrane and closely associate with cortical microfilaments (Fig. 4). The proteins involved are both globular and filamentous and they interact directly with integral and peripheral membrane proteins as well as microfilaments (Luna and Hitt, 1992; Hitt and Luna, 1994). Mische and Morrow (1988) describe the main elements of the erythrocyte membrane skeleton and the post-translational modifications they undergo; many elements of the structure are regulated by phosphorylation, including spectrin, protein 4.1, dematin (protein 4.9), ankyrin, and band 3. More recently it has been realised that many of these proteins exist in homologous form in other animal cell types and single-celled organisms (Luna and Hitt, 1992).

The major filamentous element of the membrane skeleton, apart from microfilaments, is spectrin (Bennett and Gilligan, 1993). The phosphorylation of spectrin at mitosis causes it to be released from its association with the plasma membrane skeleton into the cytoplasm (Fowler and Adams, 1992). This is thought to be important in the change of cell shape. Recently a protein has been identified in plant cells which has immunological cross-reactivity with human spectrin and is located, like the animal protein, at the cell cortex (Michaud et al., 1991; Faraday and Spaswick, 1993). It will be of interest to discover the role of spectrin and its phosphorylation in plant cells, in which cell shape is controlled by a very different set of interactions than in animal cells.

Dematin, also known as protein 4.9 in the erythrocyte, is another central protein in the network at the membrane skeleton. It is classified as an actin-bundling protein, but it works by increasing the length of the lag phase (the time between microfilament nucleation and elongation) and thereby decreasing the rate of elongation during actin polymerisation. The resulting effect is bundling of the microfilaments in vivo (Siegel and Branton, 1985). Phosphorylation of dematin by the catalytic subunit of cAMP-dependent protein kinase results in the abolition of its bundling ability (Husain-Chishti et al., 1988). Furthermore, a protein kinase copurifies with erythrocyte dematin which when activated abolishes the bundling activity of the protein (Husain-Chishti et al., 1989).

Another example of the importance of phosphorylation in the regulation of cellular responses to extracellular signals via the alteration of the membrane skeleton comes from Yano et al. (1994). Agonist activation of platelets results in the production of vesicles from the plasma membrane. Yano et al. showed that this effect is the result of alterations in the membrane skeleton brought about by phosphorylation.

The heat shock protein HSP27 is a well-known target for phosphorylation upon stimulation of animal cells receptive to growth factors, the result being alterations in cellular differentiation. Overexpression of this protein in fibroblasts causes an increased stability of the microfilaments and improved resistance to hyperthermia and the microfilament depolymerising agent cytochalasin D (Lavoie et al., 1993a). Moreover expression of this protein in CHO cells causes an increase in microfilament concentration at the cortex. When the cells were mitogen activated this caused a rapid increase in submembranous filament number. When cells were transfected with HSP27 which had point mutations at the phosphorylation sites, there was no effect on microfilament dynamics upon stimulation, and the basal level of microfilaments was lower (Lavoie et al., 1993). The avian homologue of HSP27, IAP (inhibitor of actin polymerisation), acts in vitro by binding to the cap region of the microfilament (Miron et al., 1988, 1991). Therefore there is good evidence that this protein is involved in the signal transduction pathway between mitogen binding and alterations in the microfilaments via this phosphorylation. The wide variety of functions of the HSPs so far characterised in animals (including roles not necessarily induced by heat shock, e.g., chaperone activity, protein folding, signal transduction; Gething and Sambrook, 1992) suggests that they may also have an important role in plant cells.

IV. Phosphorylation of Cytoskeletal Proteins During Cell Division

The most obvious reorganisation of the plant cytoskeleton occurs when cells divide. The cell contents rearrange in preparation for division and the arrays of cytoskeletal proteins go through well documented changes. The cortical, lateral microtubules condense before prophase to form a tight band across the middle of the cell, the preprophase band, which predetermines the site of future division (Goddard et al., 1994); cortical microfilaments coalign with the microtubules of the preprophase band, but they also remain throughout the cytoplasm during mitosis (McCurdy et al., 1991). The timing of the rearrangements must be precise and as mitosis progresses the specific arrays which form and breakdown must do so in harmony. Hence the regulation of the cytoskeleton during the process of cell division provides an excellent example of organisation in which to examine the role of phosphorylation.

It is now certain that phosphorylation and dephosphorylation play a central role in the regulation of the cell cycle. The $p34^{cdc2}$ complex is of particular importance. This kinase was first identified in yeast but it has been subsequently described in a large range of eukaryotes as a highly conserved protein (Norbury and Nurse, 1992). Plants in which $p34^{cdc2}$ has so far been identified include alfalfa (Hirt et al., 1991), maize (Colasanti et al., 1991), *Arabidopsis* (Martinez et al., 1992), and soybean (Miao et al., 1993). Staiger and Doonan (1993) have reviewed the developments of this specific topic in plants.

A. Intermediate Filaments

The kinase $p34^{cdc2}$ phosphorylates nuclear lamins (Dessev et al., 1991; Peter et al., 1990, 1991; see also Nigg, 1992), and cytoplasmic intermediate filament proteins, most notably vimentin and desmin (Chou et al., 1989, 1990, 1991; Ando et al., 1993; Kusubata et al., 1993) and also neurofilaments (Hisanaga et al., 1993) and glial filaments (Matsuoka et al., 1992). This phosphorylation causes the intermediate filament network to breakdown into protofilaments (Fig. 2 c) which do not reform full filaments again until interphase. Phosphorylation of an intermediate filament cross-linking protein IFAP-300 by $p34^{cdc2}$ occurs at a mitosis specific site (Skalli et al., 1992). The suggestion is therefore that $p34^{cdc2}$ coordinates the reorganisation of intermediate filament networks both by direct effects on the filaments and by phosphorylation-dependent breaking of the IFAP-300 crosslinks.

B. Microtubule Associated Proteins

The $p34^{cdc2}$ kinase also phosphorylates MAPs. It is thought that this decreases the affinity of the MAPs for microtubules and hence decreases microtubule stability (Fig. 3 a). Several approaches have been used to study this mode of regulation. Verde et al. (1990) added purified kinase to cell-free *Xenopus* extracts; this resulted in destabilization of the microtubules. Vandre et al. (1991) have shown that MAP4 is phosphorylated in a mitosis specific manner. Scott et al. (1993) found that phosphorylation of tau by $p34^{cdc2}$ decreases the microtubule assembly induced by tau. A mitosis specific kinase which may be related to $p34^{cdc2}$ has been isolated from mitotic spindles on the basis that it phosphorylates MAPs 4 and 1B (Tombes et al., 1991). Similarly a tau protein kinase has been purified and found to be a homologue of $p34^{cdc2}$ (Mawal-Dewan et al., 1992; Kobayashi et al., 1993; Hisanaga et al., 1993).

C. Actin Associated Proteins

There is evidence that some AAPs are phosphorylated by the $p34^{cdc2}$ family of kinases. For example, caldesmon, an AAP whose proposed functions include microfilament bundling, involvement in the regulation of actin/myosin contraction and regulation of actin polymerisation, is phosphorylated by the kinase during mitosis. This releases it from microfilaments (Hosoya et al., 1993; Ishikawa et al., 1992), and therefore enhances the potential for microfilament rearrangements. Myosin II is also phosphorylated by $p34^{cdc2}$ kinase at cell division (Satterwhite et al., 1992). This phosphorylation inhibits the ATPase activity of myosin II and may be important in delaying cytokinesis until chromosome separation has occurred. Spectrin is an important element of the membrane skeleton which interacts with actin in a complex with protein 4.1 and adducin (Sect. III.B; Table 2 and Fig. 4). It is phosphorylated by $p34^{cdc2}$ during mitosis and this causes it to redistribute to the cytosol (Fowler and Adams, 1992). Presumably the reason for this is that spectrin phosphorylation alters the interactions between microfilaments and the plasma membrane; this would permit the rearrangements that occur during mitosis. Although there is no current evidence that $p34^{cdc2}$ regulates any of the many microfilament severing, capping and crosslinking proteins this is probably because this topic has not been addressed. There is much more evidence available for the involvement of other kinases in the regulation of these proteins (see below for examples), but is is likely that among the many associated-proteins there are further substrates for $p34^{cdc2}$ or related kinases which bring about changes in microfilament dynamics necessary for successful cell division.

D. Indirect Evidence for Phosphorylation of the Plant Cytoskeleton During Cell Division

A $p34^{cdc2}$ homologue has been localised to the preprophase band (PPB) in dividing onion root tip cells (Mineyuki et al., 1991), dividing maize root tip and stomatal guard complex cells (Colasanti et al, 1993) (Figs. 5 and 6). In the latter study disruption of the microtubules of the PPB by oryzalin resulted in no PPB-like distribution of $p34^{cdc2}$. Colasanti et al. propose that because the PPB is known to define the future site of cell division (Wick, 1991), $p34^{cdc2}$ is involved in imprinting the signal defining the site while it is at the PPB.

In dividing animal cells the centrioles and the surrounding pericentriolar material together make up the microtubule organising centre (MTOC) (Brinkley, 1985). This rather amorphous structure is thought to be the central nucleating point for the microtubules of the spindle, as well as for cytoplasmic microtubules during interphase. In plant cells centrioles are not present, rather, regions on the surface of the nucleus during interphase and the presence of a microtubule spindle during mitosis suggest that a form of MTOC must exist during cell division (Lambert, 1993). Vandre et al. (1986) raised antibodies to phosphorylated proteins of the MTOC of animal PTK1 cells. These antibodies revealed that phosphorylated proteins exist in the MTOCs not only in the polar bodies of animal cells but also in onion root cells in a rather diffuse area around the dividing chromosomes, and associated with the microtubules of the plant cell phragmoplast. The antibody only gave a signal when introduced into dividing cells. This suggests that phosphorylation is involved in the regulation of microtubule dynamics during cell division in all eukaryotes. One of the antigens in animal cells has subsequently been identified as MAP4 (Vandre et al., 1991).

Traas et al. (1992) and Young et al. (1994) used the same antibody in immunofluorescence studies to investigate the spatial and temporal distribution of mitosis-specific phosphoproteins in more detail in plants. At interphase a low level of staining was visible in the nucleus. Once G_2 commenced (as indicated by the gradual condensation of microtubules into the preprophase band), immunofluorescent staining was evident throughout the nucleus and cytoplasm. The signal increased further in intensity during prophase and metaphase, and then rapidly returned to a low interphase level at telophase, as shown in Fig. 7. When Vandre et al. (1986) first used the antibody on onion root tip cells, the phragmoplast showed strong labelling. However, Traas et al. (1992) did not find staining of the phragmoplast in tobacco BY-2 cells and maize root meristem cells; Young et al. (1994) did not see staining of the phragmoplast in onion meristematic cells, but did find it in tobacco BY-2 sus-

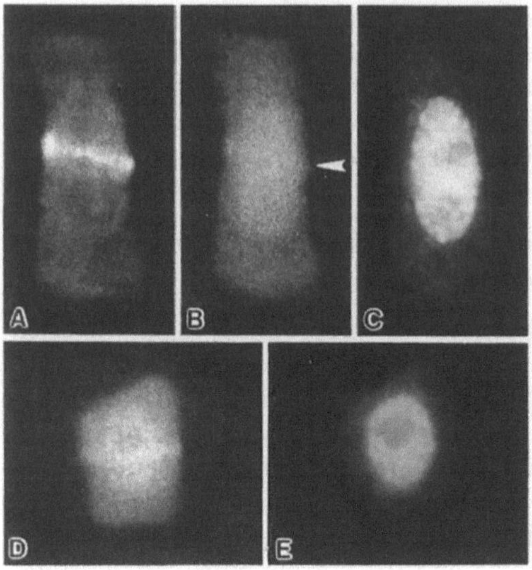

Fig. 5

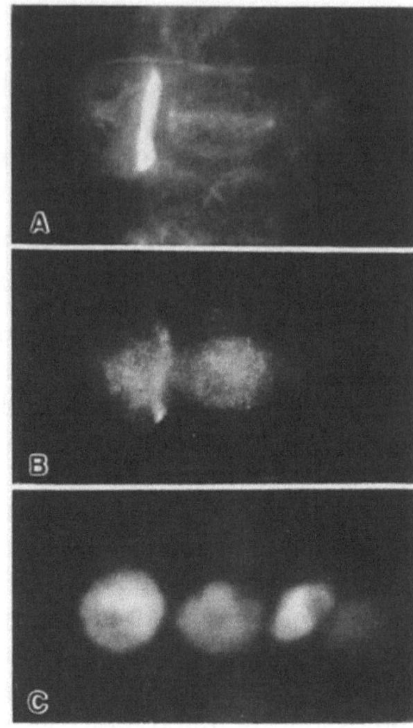

Fig. 6

Fig. 5. Immunofluorescence staining of maize root tip cells in early prophase. A, Cell stained with tubulin antibody: the preprophase band (PPB) is easily visible. B, The same cell stained with antibodies raised to a peptide based on the sequence of the maize cdc2 homologue ($p34^{cdc2}Zm$); arrowhead indicates position of the $p34^{cdc2}$ PPB. C, The same cell stained with DAPI. D, A different cell immunostained with the $p34^{cdc2}Zm$ antibody only; E, and with DAPI. × 850. Reproduced from Colasanti et al. (1993) by permission of the authors and the American Society of Plant Physiologists

Fig. 6. Immunofluorescence staining of maize leaf epidermis showing a developing stomatal complex. A, Stained with tubulin antibody, showing the subsidiary cell mother cell (SMC) on the left with a heavily stained PPB characteristic of this cell type in monocotyledons, and the guard mother cell on the right in which the PPB is also visible. B, The same two cells stained with anti-$p34^{cdc2}Zm$; the localisation of the antigen to the PPB is particularly evident in the SMC. C, The same two cells stained with DAPI. ×1300. Reproduced from Colasanti et al. (1993) by permission of the authors and the American Society of Plant Physiologists

pension cells. Although these results are contradictory there does seem to be a correlation between mitosis specific protein phosphorylation and cytoskeletal rearrangements during mitosis. A further interesting aspect of the study by Traas et al. (1992) was the isolation of one particular antigen in a cytoskeleton-enriched fraction. This protein has an apparent M_r of approximately 65 kDa and cross-reacted with a broad specificity antibody which binds to all known animal intermediate filaments.

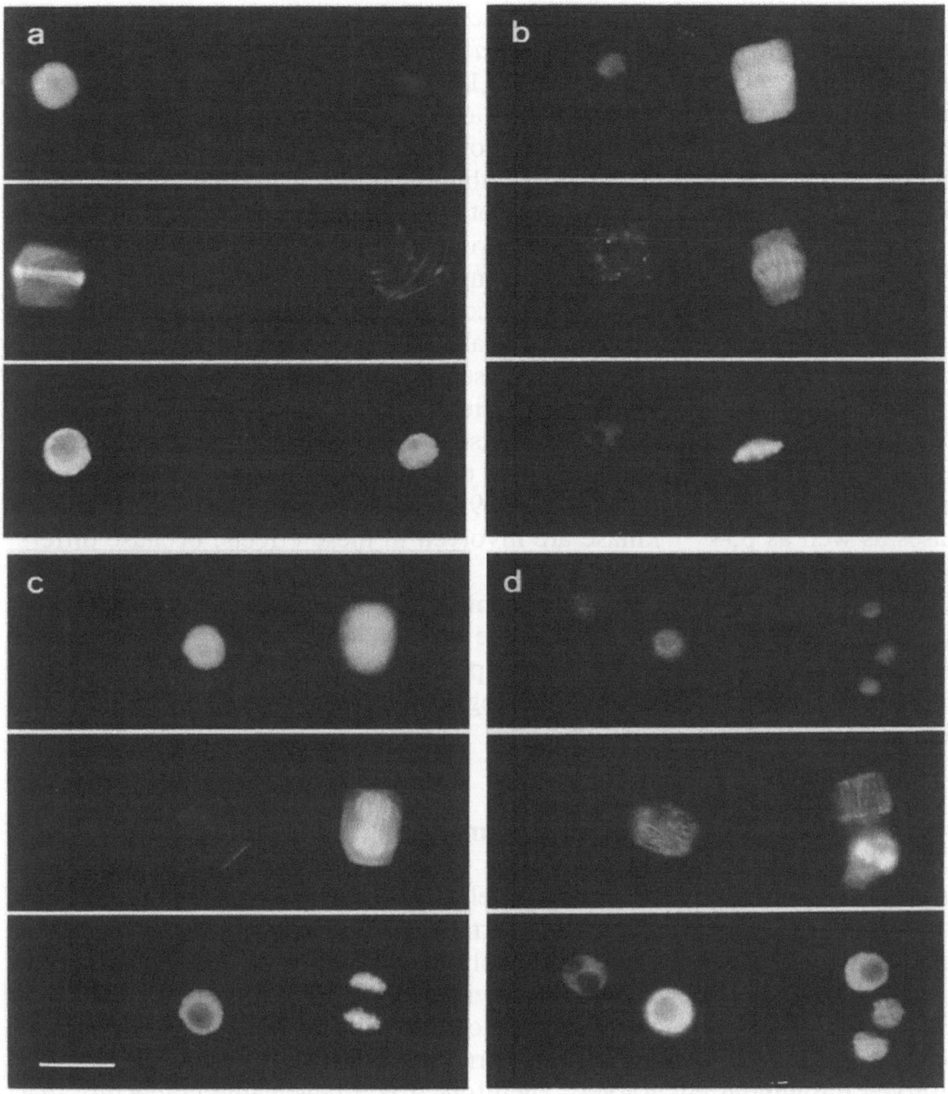

Fig. 7. Immunofluorescence staining of maize root meristem cells. Each panel in a–d shows groups of cells stained: top, with MPM-2 (antibody raised against animal mitosis specific phosphoproteins, see Sect. IV.D); center, antitubulin; bottom, DAPI. a, The left cell is at preprophase and the right one at interphase; the difference in the MPM-2 staining between interphase and early prophase is clear. b, The left cell is at interphase and the right cell at metaphase; the MPM-2 labelling is at its highest and is distributed throughout the cell, with stronger labelling around the spindle. c, The left cell is at early prophase, the right cell is at anaphase; the MPM-2 labelling is still high at this stage. d, Cells at telophase; the MPM-2 labelling is back down to interphase levels. ×940. Reproduced from Traas et al. (1992) by permission of the authors and Blackwell Scientific Publications Ltd.

The effect of the protein phosphatase inhibitors okadaic acid and microcystin during the division of *Tradescantia* hair cells has been studied (Wolniak and Larsen, 1992; Larsen and Wolniak, 1993). The timing of metaphase was altered, and when treatment was applied at metaphase, anaphase entry became asynchronous. However, chromosome movement was not affected once it had begun, and the structure of the microtubules appeared normal. Phosphatase activity therefore seems to be required at specific stages in mitosis to ensure the correct timing of cytoskeletal rearrangements. Hasezawa and Nagata (1992) have obtained similar results using this approach in the study of cell cycle progression in synchronized tobacco BY-2 cells. Okadaic acid and calyculin A blocked progression specifically at G_2, anaphase, and $M–G_1$. Furthermore, the introduction of the kinase inhibitors staurosporine and K252a into synchronised tobacco BY-2 cells during the onset of cell division caused a very large reduction in the number of visible PPBS, and the application of the inhibitors after the formation of the PPB caused a decrease in the rate of PPB disappearance (Katsuta and Shibaoka, 1992). These workers propose that other kinases as well as $p34^{cdc2}$ kinase are involved in these effects, because different inhibitors affect each step differently.

This evidence shows that in both animals and plants phosphorylation coordinates cytoskeletal dynamics during the cell cycle. It seems probable that a range of kinases (including $p34^{cdc2}$) and phosphatases act in concert to control cytoskeletal changes; we are only just beginning to understand the complexity of this control by phosphorylation.

V. Summary

It is clear that phosphorylation of cytoskeletal proteins, either the polymer subunits themselves (as is the case for intermediate filaments) or associated proteins that regulate the dynamics of the polymers, is an important mechanism in the regulation of the cytoskeleton. The evidence for this conclusion comes primarily from studies on animal cells. However, the basic proteins involved, tubulin, actin, and to a lesser extent intermediate filament proteins, are highly conserved throughout eukaryotes. Therefore it is a reasonable conjecture that the mechanisms by which the dynamics of the proteins are regulated have common themes in plants and animals. In the future, further cytoskeletal associated proteins are likely to be identified from animal and plant cells, and work on the plant cytoskeleton will reveal the extent to which its regulation by phosphorylation and other mechanisms differs from animals. Hugdahl et al. (1993) have carried out a preliminary study of the MAP binding domains of plant tubulin using neuronal MAP2. They found that this MAP enhances the polymerisation of taxol-stabilised plant microtubules by lowering the critical concentration approximately 8-fold. This is a significant effect, although this MAP lowers the critical concentration for animal microtubule polymerisation by 80-fold. Also, the stoichiometry of MAP binding to the plant microtubules is the same as that for animal microtubules, and the plant tubulin molecules possess, like the animal tubulin, one high and one low affinity binding site for the MAP. These studies indicate that although there are significant differences between the animal and plant proteins the basic characteristics and mechanisms are conserved. Intensive study of phosphorylation in the plant cytoskeleton is needed to provide a clear understanding of the role it plays in controlling cytoskeletal dynamics during plant cell growth and differentiation.

Acknowledgements

We are grateful to the BBSRC for a studentship to SMR, and to Paul Le Mière for drawing the diagrams.

VI. References

Albers K, Fuchs E (1992) The molecular biology of intermediate filament proteins. Int Rev Cytol 134: 243–279

Allen C, Borisy GG (1974) Structural polarity and directional growth of microtubules of *Chlamydomonas* flagella. J Mol Biol 90: 381–402

Ando S, Tokui T, Yamauchi T, Sugiura H, Tanabe K, Inagaki M (1991) Evidence that ser-82 is a unique phosphorylation site on vimentin for Ca^{2+}-calmodulin-dependent protein kinase II. Biochem Biophys Res Commun 175: 955–962

Ando S, Tsujimura K, Matsuoka Y, Tokui T, Hisanaga S, Okumura E, Uchiyama M, Kishimoto T, Yasuda H, Inagaki M (1993) Phosphorylation of synthetic vimentin peptides by cdc2 kinase. Biochem Biophys Res Commun 195: 837–843

Baribault H, Blouin R, Bourgon L, Marceau N (1989) Epidermal growth factor-induced selective phosphorylation of cultured rat hepatocyte 55-kD cytokeratin before filament organization and DNA synthesis. J Cell Biol 109: 1665–1676

Bennett V, Gilligan DM (1993) The spectrin-based membrane skeleton and micron-scale organization of the plasma membrane. Annu Rev Cell Biol 9: 27–66

Bergen LG, Borisy GG (1980) Head-to-tail polymerization of microtubules in vitro. J Cell Biol 84: 141–150

Beven A, Guan Y, Peart J, Cooper C, Shaw P (1991) Monoclonal antibodies to plant nuclear matrix reveal intermediate filament-related components within the nucleus. J Cell Sci 98: 293–302

Brinkley BR (1985) Microtubule organizing centres. Annu Rev Cell Biol 1: 145–172

Carlier MF (1989) Role of nucleotide hydrolysis in the dynamics of actin filaments and microtubules. Int Rev Cyt 115: 139–170

Carlier MF, Melki R, Pantaloni D, Hill TL, Chen Y (1987) Synchronous oscillations in microtubule polymerisation. Proc Natl Acad Sci USA 84: 5257–5261

Chang-Jie J, Sonobe S (1993) Identification and preliminary characterization of a 65 kDa higher-plant microtubule-associated protein. J Cell Sci 105: 891–901

Chasan R (1993) Of microtubules and MAPs. Plant Cell 5: 995–997

Chou Y-H, Rosevear E, Goldman RD (1989) Phosphorylation and disassembly of intermediate filaments in mitotic cells. Proc Natl Acad Sci USA 86: 1885–1889

Chou Y-H, Bischoff JR, Beach D, Goldman RD (1990) Intermediate filament reorganization during mitosis is mediated by p34cdc2 phosphorylation of vimentin. Cell 62: 1063–1071

Chou Y-H, Ngai K-L, Goldman R (1991) The regulation of intermediate filament reorganization in mitosis. J Biol Chem 266: 7325–7328

Cleveland DW (1982) Treadmilling of tubulin and actin. Cell 28: 689–691

Cobb M, Boulton TG, Robbins DJ (1991) Extracellular signal-regulated kinases: ERKs in progress. Mol Biol Cell 2: 965–978

Colasanti J, Tyers M, Sundaresan V (1991) Isolation and characterization of cDNA clones encoding a functional p34^{cdc2} homologue from *Zea mays*. Proc Natl Acad Sci USA 88: 3377–3381

Colasanti J, Cho S-O, Wick S, Sundaresan V (1993) Localization of the functional p34^{cdc2} homolog of maize in root tip and stomatal complex cells: association with predicted division sites. Plant Cell 5: 1101–1111

Crute BE, Van Buskirk RG (1992) A casein kinase-like kinase phosphorylates β-tubulin and may be a microtubule-associated protein. J Neurochem 59: 2017–2023

Cyr RJ, Palevitz BA (1989) Microtubule-binding proteins from carrot. Planta 177: 245–260

Davis A, Sage CR, Dougherty CA, Farrell KW (1994) Microtubule dynamics modulated by guanosine triphosphate hydrolysis activity of β-tubulin. Science 264: 839–842

Dentler WL, Granett S, Witman GB, Rosenbaum JL (1974) Directionality of brain microtubule assembly in vitro. Proc Natl Acad Sci USA 71: 1710–1714

Dessev G, Iovcheva-Dessev C, Beach D, Goldman RD (1991) A complex containing p34^{cdc2} and cyclin B phosphorylates the nuclear lamin and disassembles nuclei of clam oocytes in vitro. J Cell Biol 112: 523–533

Duerr B, Gawienowski M, Ropp T, Jacobs T (1993) MsERK1: a mitogen-activated protein kinase from a flowering plant. Plant Cell 5: 87–96

Eriksson JE, Brautigan DL, Vallee R, Olmstead J, Fujiki H, Goldman RD (1992) Cytoskeletal integrity in interphase cells requires protein phosphatase activity. Proc Natl Acad Sci USA 89: 11093–11097

Faraday CD, Spaswick RM (1993) Evidence for a membrane skeleton in higher plants. A spectrin-like polypeptide co-isolates with rice root plasma membranes. FEBS Lett 318: 313–316

Finkel T, Theriot JA, Dise KR, Tomaselli GF, Goldschmidt-Clermont PJ (1994) Dynamic actin structures stabilized by profilin. Proc Natl Acad Sci USA 91: 1510–1514

Foisner R, Traub P, Wiche G (1991) Protein kinase A- and protein kinase C-regulated interaction of plectin with lamin B and vimentin. Proc Natl Acad Sci USA 88: 3812–3816

Fowler VM, Adams EJH (1992) Spectrin redistributes to the cytosol and is phosphorylated during mitosis in cultured cells. J Cell Biol 119: 1559–1572

Franke WW, Schmid E, Winter S, Osborn M, Weber K (1979) Widespread occurrence of intermediate-sized filaments of the vimentin-type in cultured cells from diverse vertebrates. Exp Cell Res 123: 25–46

Furuhashi K, Hatano S (1992) Actin kinase: a protein kinase that phosphorylates actin of fragmin-actin complex. J Biochem 111: 366–370

Gething MJ, Sambrook J (1992) Protein folding in the cell. Nature 355: 33–45

Geisler N, Hatzfield M, Weber K (1989) Phosphorylation in vitro of vimentin by protein kinases A and C is restricted to the head domain. FEBS Lett 183: 441–447

Gliksman NR, Parsons SF, Salmon ED (1992) Okadaic acid induces interphase to mitotic-like microtubule dynamic instability by inactivating rescue. J Cell Biol 119: 1271–1276

Goddard RH, Wick SM, Silflow CD, Snustad DP (1994) Microtubule components of the plant cell cytoskeleton. Plant Physiol 104: 1–6

Gordon D, Eisenberg E, Korn E (1976) Characterization of cytoplasmic actin from *Acanthamoeba castellani* by a new method. J Biol Chem 251: 4778–4786

Gotoh Y, Nishida E, Matsuda S, Shiina N, Kosako H, Shiokawa K, Akiyama T, Ohta K, Sakai H (1991) In vitro effects on microtubule dynamics of purified *Xenopus* M phase-activated MAP kinase. Nature 349: 251–254

Hagrestedt T, Lichtenberg B, Wille H, Mandelkow E-M, Mandelkow E (1989) Tau protein becomes long and stiff upon phosphorylation: correlation between paracrystalline structure and degrees of phosphorylation. J Cell Biol 109: 1643–1651

Hargreaves AJ, Goodbody KC, Lloyd CW (1989) Reconstitution of intermediate filaments from a higher plant. Biochem J 261: 679–682

Harrison BC, Mobley PL (1992) Phosphorylation of glial fibrillary acidic protein and vimentin by cytoskeletal-associated intermediate filament protein kinase activity in astrocytes. J Neurochem 58: 320–327

Hasezawa S, Nagata T (1992) Okadaic acid as a probe to analyse the cell cycle progression in plant cells. Bot Acta 105: 63–69

Heins S, Aebi U (1994) Making heads and tails of intermediate filament assembly, dynamics and networks. Curr Opin Cell Biol 6: 25–33

Herrmann H, Hoffmann I, Franke WW (1992) Identification of a nonapeptide motif in the vimentin head domain involved in intermediate filament assembly. J Mol Biol 223: 637–650

Hirokawa N (1994) Microtubule organization and dynamics dependent on microtubule-associated proteins. Curr Opin Cell Biol 6: 74–81

Hirokawa N, Shiomura Y, Okabe S (1988) Tau proteins: the molecular structure and mode of binding on microtubules. J Cell Biol 107: 1449–1459

Hirt H, Pay A, Gyorgyey J, Bako L, Nemeth K, Bogre L, Schweyen RJ, Heberle-Bors E, Dudits D (1991) Complementation of a yeast cell cycle mutant by an alfalfa cDNA encoding a protein kinase homologue. Proc Natl Acad Sci USA 88: 1636–1640

Hisanaga S, Kusubata M, Okumura E, Kishimoto T (1991) Phosphorylation of neurofilament H subunit at the tail domain by CDC2 kinase dissociates the association to microtubules. J Biol Chem 266: 21798–21803

Hisanaga S, Ishiguro K, Uchida T, Okumura E, Okano T, Kishimoto T (1993) Tau protein kinase II has a similar characteristic to cdc2 kinase for phosphorylating neurofilament proteins. J Biol Chem 268: 15056–15060

Hitt AL, Luna EJ (1994) Membrane interactions with the actin cytoskeleton. Curr Opin Cell Biol 6: 120–130

Horio T, Hotani H (1986) Visualization of the dynamic instability of individual microtubules by dark-field microscopy. Nature 321: 605–607

Hosoya N, Hosoya H, Yamashiro S, Mohri H, Matsumura F (1993) Localization of caldesmon and its dephosphorylation during cell division. J Cell Biol 121: 1075–1082

Hugdahl JD, Bokros CL, Hanesworth WR, Aalund GR, Morejohn LC (1993) Unique functional characteristics of the polymerization and MAP binding domains of plant tubulin. Plant Cell 5: 1063–1080

Husain-Chishti A, Levin A, Branton D (1988) Abolition of actin-bundling by phosphorylation of human erythrocyte protein 4.9. Nature 334: 718–721

Husain-Chishti A, Faquin W, Wu CC, Branton D (1989) Purification of erythrocyte dematin (protein-4.9) reveals an endogenous protein-kinase that modulates actin-bundling activity. J Biol Chem 264: 8985–8991

Inagaki M, Nishi Y, Nishizawa K, Matsuyama M, Sato C (1987) Site-specific phosphorylation induces disassembly of vimentin filaments in vitro. Nature 328: 649–652

Ishikawa R, Kagami O, Hayashi C, Kohama K (1992) The binding of nonmuscle caldesmon from brain to microtubules. Regulation by Ca^{2+}-calmodulin and cdc2 kinase. FEBS Lett 299: 54–56

Jensen CG, Smaill BH (1986) Analysis of the spatial organization of microtubule-associated proteins. J Cell Biol 103: 559–569

Jonak C, Pay A, Bogre L, Hirt H, Heberle-Bors E (1993) The plant homologue of MAP kinase is expressed in a cell cycle-dependent and organ-specific manner. Plant J 3: 611–617

Katsuta J, Shibaoka H (1992) Inhibition by kinase inhibitors of the development and the disappearance of the preprophase band of microtubules in tobacco BY-2 cells. J Cell Sci 103: 397–405

Kobayashi S, Ishiguro K, Omori A, Takamatsu M, Arioka M, Imahori K, Uchida T (1993) A cdc2-related kinase PSSALRE/cdk5 is homologous with the 30 kDa subunit of tau protein kinase II, a proline-directed protein kinase associated with microtubules. FEBS Lett 335: 171–175

Koontz DA, Choi JH (1993) Evidence for phosphorylation of tubulin in carrot suspension cells. Physiol Plant 87: 576–583

Korn ED, Carlier M-F, Pantaloni D (1987) Actin polymerization and ATP hydrolysis. Science 238: 638–644

Kreis T, Vale R (1993) Guidebook to the cytoskeletal and motor proteins. Oxford University Press, London

Kusubata M, Matsuoka Y, Tsujimura K, Ito H, Ando S, Kamijo M, Yasuda H, Ohba Y, Okumura E, Kishimoto T, Inagaki N (1993) CDC2 kinase phosphorylation of desmin at three serine/threonine residues in the amino-terminal head domain. Biochem Biophys Res Commun 190: 927–934

Lamb NJC, Fernandez A, Feramisco JR, Welch WJ (1989) Modulation of vimentin containing intermediate filament distribution and phosphorylation in living fibroblasts by the cAMP-dependent protein kinase. J Cell Biol 108: 2409–2422

Lambert A-M (1993) Microtubule-organizing centers in higher plants. Curr Opin Cell Biol 5: 116–122

Larsen PM, Wolniak SM (1993) Asynchronous entry into anaphase induced by okadaic acid: spindle microtubule organization and microtubule/kinetochore attachments. Protoplasma 177: 53–65

Lavoie JN, Gingras-Breton G, Tanguay RM, Landry J (1993a) Induction of Chinese hamster HSP 27 gene expression in mouse cells confers resistance to heat shock. J Biol Chem 268: 3420–3429

Lavoie JN, Hickey E, Weber LA, Landry J (1993b) Modulation of actin microfilament dynamics and fluid phase pinocytosis by phosphorylation of heat shock protein 27. J Biol Chem 268: 24210–24214

Lee G, Rook SL (1992) Expression of tau protein in non-neuronal cells: microtubule binding and stabilization. J Cell Sci 102: 227–237

Li H, Roux SJ (1992) Casein kinase II protein kinase is bound to lamina-matrix and phosphorylates lamin-like protein in isolated pea nuclei. Proc Natl Acad Sci USA 89: 8434–8438

Lloyd CW (ed) (1991) The cytoskeletal basis of plant growth and form. Academic Press, London

Luna EJ, Hitt AL (1992) Cytoskeleton-plasma membrane interactions. Science 258: 955–964

Ludueña RF, Banerjee A, Khan IA (1992) Tubulin structure and biochemistry. Curr Opin Cell Biol 4: 53–57

McCurdy DW, Williamson RE (1991) Actin and actin-associated proteins. In: Lloyd CW (ed) The cytoskeletal basis of plant growth and form. Academic Press, London, pp 3–14

McNulty AK, Saunders MJ (1992) Purification and immunological detection of pea nuclear intermediate filaments: evidence for plant nuclear lamins. J Cell Sci 103: 407–414

Maekawa T, Ogihara S, Murofushi H, Nagai R (1990) Green algal microtubule-associated protein with a molecular weight of 90 kDa which bundles microtubules. Protoplasma 158: 10–18

Margolis RL, Wilson L (1978) Opposite end assembly and disassembly of microtubules at steady state in vitro. Cell 13: 1–8

Martinez MC, Jørgensen J-E, Lawton MA, Lamb CJ, Doerner PW (1992) Spatial pattern of *cdc2* expression in relation to meristem activity and cell proliferation during plant development. Proc Natl Acad Sci USA 89: 7360–7364

Matsuoka Y, Nishizawa N, Yano T, Shibata M, Ando S, Takahashi T, Inagaki M (1992) Two different protein kinases act on a different time schedule as glial filament kinases during mitosis. EMBO J 11: 2895–2902

Mawal-Dewan M, Sen PC, Abdel-Ghany M, Shalloway D, Racker E (1992) Phosphorylation of tau protein by purified $p34^{cdc28}$ and a related protein kinase from neurofilaments. J Biol Chem 267: 19705–19709

Menzel D (1993) Chasing coiled coils: intermediate filaments in plants. Bot Acta 106: 294–300

Miao G-H, Hong Z, Verma DPS (1993) Two functional soybean genes encoding $p34^{cdc2}$ protein kinases are regulated by different plant developmental pathways. Proc Natl Acad Sci USA 90: 943–947

Michaud D, Guillet G, Rogers PA, Charest PM (1991) Identification of a 220 kDa membrane-associated plant cell protein immunologically related to human β-spectrin. FEBS Lett 294: 77–80

Mineyuki Y, Yamashita M, Nagahama Y (1991) $p34^{cdc2}$ kinase homologue in the preprophase band. Protoplasma 162: 182–186

Minguez A, Moreno Diaz de la Espina S (1993) Immunological characterization of lamins in the nuclear matrix of onion cells. J Cell Sci 106: 431–439

Miron T, Wilchek M, Geiger B (1988) Characterization of an inhibitor of actin polymerization in vinculin-rich fraction of turkey gizzard smooth muscle. Eur J Biochem 178: 543–553

Miron T, Vancompernolle K, Vandekerckhove J, Wilchek M, Geiger B (1991) A 25 kDa inhibitor of actin polymerization is a low molecular mass heat shock protein. J Cell Biol 114: 255–261

Mische SM, Morrow JS (1988) Post-translational regulation of the erythrocyte cortical cytoskeleton. Protoplasma 145: 167–175

Mitchison T, Kirschner M (1984a) Dynamic instability of microtubule growth. Nature 312: 237–242

Mitchison T, Kirschner M (1984b) Microtubule assembly nucleated by isolated centrosomes. Nature 312: 232–237

Mizoguchi T, Hayashi N, Yamaguchi-Shinozaki K, Kamada H, Shinozaki K (1993) ATMPKs: a gene family of plant protein MAP kinases in *Arabidopsis thaliana*. FEBS Lett 3: 440–444

Nigg EA (1992) Assembly-disassembly of the nuclear lamina. Curr Opin Cell Biol 4: 105–109

Norbury C, Nurse P (1992) Animal cell cycles and their control. Annu Rev Biochem 61: 441–470

Omary MB, Baxter GT, Chou C-F, Riopel CL, Lin WY (1992) PKC_{ε}-related kinase associates with and phosphorylates cytokeratin 8 and 18. J Cell Biol 117: 583–593

Oosawa F, Kasai M (1962) A theory of linear and helical aggregations of macromolecules. J Mol Biol 4: 10–21

Pardee JD, Spudich JA (1982) Purification of muscle actin. Methods Enzymol 85: 164–181

Pelech SL, Sanghera JS (1992) MAP kinases: charting the regulatory pathways. Science 257: 1355–1356

Peter M, Nakagawa J, Dorée M, Lebbé J-C, Nigg EA (1990) In vitro disassembly of the nuclear lamina and M-phase specific phosphorylation of lamins by cdc2 kinase. Cell 61: 591–602

Peter M, Heitlinger E, Häner M, Aebi U, Nigg EA (1991) Disassembly of in vitro formed lamin head-to-tail polymers by cdc2 kinase. EMBO J 10: 1535–1544

Pryer NK, Walker RA, Skeen VP, Bourns BD, Soboeiro MF, Salmon ED (1992) Brain microtubule-associated proteins modulate microtubule dynamic instability in vitro. J Cell Sci 103: 965–976

Quinlan RA, Stewart M (1991) Molecular interactions in intermediate filaments. BioEssays 13: 597–600

Raats JMH, Pieper FR, Vree Egberts TMW, Verrijp KN, Ramaekers FCS, Bloemendal H (1990) Assembly of amino-terminally deleted desmin in vimentin-free cells. J Cell Biol 111: 1971–1985

Renner W, Franke WW, Schmid E (1981) Reconstitution of intermediate-filament sized filaments from denatured monomers of vimentin. J Mol Biol 149: 285–306

Ruderman JV (1993) MAP kinase and the activation of quiescent cells. Curr Opin Cell Biol 5: 207–213

Satterwhite LL, Lokha MJ, Wilson KL, Scherson TY, Cisek LJ, Corden JL, Pollard TD (1992) Phosphorylation of myosin-II regulatory light chain by cyclin-p34cdc2: a mechanism for the timing of cytokinesis. J Cell Biol 118: 595–605

Schellenbaum P, Vantard M, Peter C, Fellows A, Lambert A-M (1993) Coassembly properties of higher-plant microtubule-associated proteins with purified brain and plant tubulins. Plant J 3: 253–260

Scott CW, Vuillet PR, Caputo LB (1993) Phosphorylation of tau by proline-directed-kinase ($p34^{cdc2}/p58^{cyclin\text{-}a}$) decreased tau induced microtubule assembly and antibody SM133 reactivity. Brain Res 611: 237–242

Shaw PJ, Fairbairn DJ, Lloyd CW (1991) Cytoplasmic and nuclear intermediate filament antigens in higher plants. In: Lloyd CW (ed) The cytoskeletal basis of plant growth and form. Academic Press, London, pp. 69–81

Shelanski ML, Gaskin F, Cantor CR (1973) Microtubule assembly in absence of added nucleotide. Proc Natl Acad Sci USA 70: 765–768

Siegel DL, Branton D (1985) Partial purification and characterization of an actin-bundling protein, band 4.9, from human erythrocytes. J Cell Biol 100: 775–785

Sheterline P, Sparrow JC (1994) Actin. Protein Profile 1: 1–121

Shiina N, Moriguchi T, Ohta K, Gotoh Y, Nishida E (1992) Regulation of major microtubule-associated proteins by MPF and MAP kinase. EMBO J 11: 3977–3984

Skalli O, Chou Y-H, Goldman RD (1992) Cell cycle-dependent changes in the organization of an intermediate filament-associated protein: correlation with phosphorylation by $p34^{cdc2}$. Proc Natl Acad Sci USA 89: 11959–11963

Soellner P, Quinlan RA, Franke WW (1985) Identification of a distinct soluble subunit of an intermediate filament protein: tetrameric vimentin from living cells. Proc Natl Acad Sci USA 82: 7929–7933

Staiger C, Doonan J (1993) Cell division in higher plants. Curr Opin Cell Biol 5: 226–231

Staiger CJ, Goodbody KC, Hussey PJ, Valenta R, Drobak BK, Lloyd CW (1993) The profilin multigene family of maize: differential expression of three isoforms. Plant J 4: 631–641

Theriot JA, Mitchison TJ (1993) The three faces of profilin. Cell 75: 835–838

Tokui T, Yamauchi T, Yano T, Nishi Y, Kusegawa M, Yatani R, Inagaki M (1990) Ca^{2+}, calmodulin-dependent protein kinase II phosphorylates various types of non-epithelial intermediate filament proteins. Biochem Biophys Res Commun 169: 896–904

Tombes RM, Peloquin JG, Borisy GG (1991) Specific association of an M-phase kinase with isolated mitotic spindles and identification of two substrates as MAP4 and MAP1B. Cell Regul 2: 861–874

Traas JA, Bevan AF, Doonan JH, Shaw PJ (1992) Cell cycle dependent changes in labelling of specific phosphoproteins by the monoclonal antibody MPM-2 in plant cells. Plant J 2: 723–732

Valenta R, Ferreira F, Grote M, Swoboda I, Vrtala S, Duchêne M, Deviller P, Meagher RB, McKinney E, Heberle-Bors E, Kraft D, Scheiner O (1993) Identification of profilin as an actin-binding protein in higher plants. J Biol Chem 268: 22777–22781

Vandekerckhove J, Vancompernolle K (1992) Structural relationships of actin-binding proteins. Curr Opin Cell Biol 4: 36–42

Vandré DD, Davis FM, Rao PN (1986) Distribution of cytoskeletal proteins sharing a conserved phosphorylation epitope. Eur J Cell Biol 41: 72–81

Vandré DD, Centonze VE, Peloquin J, Tombes RM, Borisy GG (1991) Proteins of the mammalian mitotic spindle: phosphorylation/dephosphorylation of MAP-4 during mitosis. J Cell Sci 98: 577–588

Vantard M, Schellenbaum P, Fellous A, Lambert A-M (1991) Characterization of maize microtubule-associated proteins, one of which is immunologically related to tau. Biochemistry 30: 9334–9340

Vantard M, Peter C, Fellous A, Schellenbaum P, Lambert A-M (1994) Characterization of a 100 kDa heat-stable microtubule-associated protein from higher plants. FEBS Lett 220: 847–853

Verde F, Labbe J-C, Doree M, Karsenti E (1990) Regulation of microtubule dynamics by cdc2 protein kinase in cell-free extracts of *Xenopus* eggs. Nature 343: 233–238

Vikstrom KL, Borisy GG, Goldman RD (1989) Dynamic aspects of intermediate filament networks in BHK-21 cells. Proc Natl Acad Sci USA 86: 549–553

Vikstrom KL, Lim S-S, Goldman RD, Borisy GG (1992) Steady state dynamics of intermediate filament networks. J Cell Biol 118: 121–129

Weisshaar B, Doll T, Matus KA (1992) Reorganization of the microtubular cytoskeleton by embryonic microtubule-associated protein 2 (MAP2c). Development 116: 1151–1161

Wick SM (1991) The preprophase band. In: Lloyd CW (ed) The cytoskeletal basis of plant growth and form. Academic Press, London, pp 231–244

Wolniak SM, Larsen PM (1992) Changes in the metaphase transit times and the pattern of sister chromatid separation in stamen hair cells of *Tradescantia* after treatment with protein phosphatase inhibitors. J Cell Sci 102: 691–715

Yang F, Demma M, Watten V, Dharmawardhane S, Condeelis J (1990) Identification of an actin-binding protein from *Dictyostelium* as elongation factor 1a. Nature 347: 494–496

Yano S, Fukunaga K, Ushio Y, Miyamoto E (1994) Activation of Ca^{2+}/calmodulin-dependent protein kinase II and phosphorylation of intermediate filament proteins by stimulation of glutamate receptors in cultured rat cortical astrocytes. J Biol Chem 269: 5428–5439

Yano Y, Kambayashi J-I, Shiba E, Sakon M, Oiki E, Fukuda K, Kawasaki T, Mori T (1994) The role of protein phosphorylation and cytoskeletal reorganization in microparticle formation from the platelet plasma membrane. Biochem J 299: 303–308

Young T, Hyams JS, Lloyd CW (1994) Increased cell cycle-dependent staining of plant-cells by the antibody MPM-2 correlates with PPB formation. Plant J 5: 279–284

Zhou R, Oskarsson M, Paules RS, Schulz N, Cleveland D, Vande Woude GF (1991) Ability of the c-mos product to associate with and phosphorylate tubulin. Science 251: 671–675

The Regulatory Phosphorylation of C_4 Phospho*enol*pyruvate Carboxylase: a Cardinal Event in C_4 Photosynthesis

Jean Vidal, Jean-Noël Pierre, and Cristina Echevarria

Contents

I. Introduction

Phospho*enol*pyruvate (PEP) carboxylase (EC 4.1.1.31, PEPC) is a widely distributed enzyme in plant tissues (embryophytes), green algae (chlorobionts) and microorganisms, but it is not present in animals (O'Leary, 1982; Andreo et al., 1987). It catalyses the exergonic β-carboxylation of PEP ($\Delta G = -7$ kcal/mol) in the presence of a bivalent cation, generally Mg^{2+}:

$$PEP + HCO_3^- \rightarrow OAA + P_i$$

The reaction proceeds via a stepwise mechanism that involves the reversible, rate-limiting formation of carboxyphosphate and enolate of PEP (O'Leary, 1982). CO_2 may be formed within the active site from carboxyphosphate to react with the enolate species (Janc et al., 1992). In vivo, the reaction product oxaloacetate (OAA) is rapidly transformed to stable aspartate and/or L-malate by coupled transaminase or NAD/NADP-dependent L-malate dehydrogenases, respectively. In plants, no less than eleven different physiological roles for PEPC have been reported and the list is presumably not exhaustive (Latzko and Kelly, 1983). The ubiquitous nature of this cytosolic enzyme is presumably due to the metabolic fate of the end-product, L-malate, which can be stored, transported and used as a source of CO_2 and reducing power in a number of metabolic pathways. Due to its very low K_m for the substrate bicarbonate (micromolar range), the enzyme serves an obvious general role in the carbon economy of cells by recapturing respiratory CO_2 and more specific functions, namely: a "housekeeping" anaplerotic function (consisting of the replenishment of Krebs cycle intermediates for use as biosynthetic precursors), pH stat (for cellular homeostasis of pH), nitrogen assimilation in legume root-nodules, cation absorption and transport, stomatal movements, grain formation and germination, fruit maturation, style/pollen tube interactions and CAM C_4 photosynthesis. Consistently, several isoenzymes have been charaterized that presumably fulfill specific tasks in a variety of tissues (for a review, see Lepiniec et al., 1994).

The discovery of C_4 photosynthesis in the mid-1960s and the involvement of a specific form of PEPC in this pathway has considerably boosted interest in the enzyme (Hatch, 1992). During the last 10 years, different aspects of this topic (O'Leary, 1982; Andreo et al., 1987), including PEPC phosphorylation (Ranjeva and Boudet, 1987; Budde and Chollet, 1988; Jiao and Chollet, 1991; Nimmo, 1993; Huber et al., 1994) and molecular biology (Lepiniec et al., 1994; Rajagopalan et al., 1994) have been comprehensively reviewed, and the reader should consult this literature for more detailed information. Here, focusing on the photosynthetic PEPC isoenzyme, we analyze the progress made in studying the mechanism of its reversible phosphorylation and identifying the components of the signal transduction cascade that controls this process. An assessment of the role of this post-translational modulation of PEPC with respect to the functioning and regulation of C_4 photosynthesis is proposed. For convenience and brevity, in the text reference is made often to reviews rather than the plethora of related articles even though this precludes citing many individual authors who have contributed to studies on plant PEPC.

II. C_4 PEPC, a Highly Regulated Plant Enzyme

A. *PEPC in C_4 Photosynthesis*

C_4 plants exhibit specific anatomical and biochemical features that comprise the C_4 syndrome. Their leaf architecture more or less conforms to the classical "Kranz" anatomy characterized by concentrically organized photosynthetic tissues, i.e., outer mesophyll (M) surrounding inner bundle sheath (BS) cells. In terms of metabolic adaptation, there exist 3 basic types of C_4 plants, but the general metabolic scheme of division of labor is conserved; i.e., two cycles, C_4

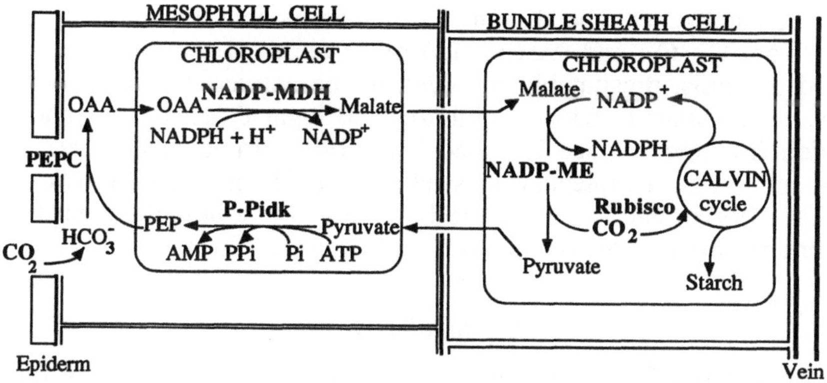

Fig. 1. Spatial organization of carbon flow during C_4 photosynthesis in *Sorghum*, maize, and *Digitaria*. PEPC, Phospho*enol*pyruvate carboxylase; NADP-MDH, NADP-dependent L-malate dehydrogenase; NADP-ME, NADP-dependent malic enzyme; P-Pidk, pyruvate P_i dikinase; RuBisCo, ribulose bisphosphate carboxylase/oxygenase

and Calvin cycles, working in concert to achieve CO_2 assimilation by the C_4 leaf. In "L-malate formers", the primary fixation of carbon dioxide (in its hydrated form) is carried out by PEPC in the cytoplasm of M-cells to form OAA that will be reduced to L-malate in M-chloroplasts; diffusive transport of L-malate to BS-cells and its decarboxylation by NADP-malic enzyme in the chloroplast stroma, making CO_2 available for incorporation by RuBisCo in the Calvin cycle (Fig. 1).

In C_3 plants, O_2 competes with CO_2 at the active site of RuBisCo leading to the production of phosphoglycollate, the substrate of the photo respiratory pathway. This causes more than 30% of the carbon initially fixed to be lost by the C_3 leaf. In C_4 plants, the C_4 cycle acts as a pump that concentrates CO_2 at the site of RuBisCo, thereby building up a large increase in the CO_2/O_2 ratio that favours the enzyme's carboxylation over oxygenation activity. This mechanism largely prevents the wasteful production of CO_2 by photorespiration and loss of carbon from C_4 leaves that is observed with C_3 plants. This explains why under arid environments C_4 plants have a better water use efficiency and a higher productivity than C_3 plants (Hatch, 1977, 1987, 1992). Such a complex and highly integrated pathway, involving different cell types and compartments, needs a high degree of coordination and enzymatic control. We describe hereafter what we have learned about the different regulatory mechanisms which act on PEPC in the M cytosol to adjust its activity to the carbon demand of the Calvin cycle in the BS.

B. Structure

During the last decade, thanks to the development of molecular approaches and DNA sequencing, a growing number of complete aminoacid sequences of PEPCs from various sources has been deduced from corresponding full length cDNAs and cloned genes. The alignment of the aminoacid sequences of 14 PEPC proteins (from C_3, C_4, and CAM plants) made apparent the presence of well conserved domains, presumably involved in the active site, interaction with effectors, i.e, L-malate (feedback inhibitor) and sugar-P (activators), and phosphorylation (for reviews, see Lepiniec et al., 1994; Rajagopalan et al., 1994). The PEPC subunit has a molecular mass close to 110 kDa (108 kDa in *Sorghum*; Crétin et al., 1991b) and is believed to be arranged in a homotetramer to form the quaternary structure of the enzyme (Andreo et al., 1987).

C. Regulation

Light is obviously of cardinal importance in that it provides energy for anabolic processes and as a regulatory signal. In the context of C_4 photosynthesis, PEPC is subject to an elaborate regulation at both the transcriptional and post-transcriptional levels.

1. Transcriptional Level

De-etiolation of a C_4 leaf is characterised by the appearance of an immunologically distinct and predominant PEPC isozyme that accumulates very specifically in the cytosol of M-cells (Thomas et al., 1987a). This tissue-specific expression is preceded by a very large increase in quantity and translational capacity of a specific mRNA (Thomas et al., 1987b, 1990). Consistently, "run on" experiments performed with nuclei purified from light- or dark-adapted C_4 leaves demonstrated a much higher capacity of the former to transcribe PEPC mRNA (Thomas et al., 1990). Photoreversibility-based (red/red-far red) experiments revealed that the chromoprotein phytochrome is the light sensor that stimulates PEPC gene activity (Thomas et al., 1990). In green *Sorghum* leaves, but not in maize leaves (Taylor, 1989), the intriguing observation was made that the production of mRNA, but not of the enzyme subunit, was rhythmic, being the highest during the light phase (Thomas et al., 1990). This peculiarity also pertains to some other light-regulated genes like the wheat *cab-1* gene, for which it was suggested that a circadian clock modulates phytochrome-mediated transcription (Nagy et al., 1988; Taylor, 1989). On the other hand, nothing is known about how the photoreceptor is functionally linked to the transcriptional efficiency of the PEPC gene. Along these lines, recent work (Neuhaus et al., 1993; Bowler et al., 1994) has demonstrated the role of calcium, as well as G proteins, calmodulin and cGMP, in the transduction of light signals to photoregulated gene promoters by at least two different pathways in a phytochrome-deficient mutant (*aureus*) of tomato. These results strongly argue that the DNA-binding activity of certain nuclear regulatory proteins is modulated by phosphorylation. For example, this is the case for the protein factor AT-1 from pea whose DNA-binding ability is reversibly modulated via phosphorylation by a protein kinase which is functionally related to a NII-type casein kinase (Datta and Cashmore, 1989). It will be particularly interesting to see whether the transcriptional activity of other photodependent genes, including the PEPC gene, is controlled in a similar way in C_4 plants. Structural analyses of the promoter of the maize PEPC gene revealed the presence of CpG islands, consistent with possible regulation via methylation/ demethylation of specific sites (Langdale et al., 1991; Ngernprasirtsiri et al., 1989), and putative cis-elements that could be implicated in the light-dependent transcriptional response (Schäffner and Sheen, 1992; Becker et al., 1992). Consistently, gel-shift experiments using cloned fragments of the PEPC promoter detected three types of leaf-specific nuclear factors, MNF1, MNF2a and MNF2b, two of which have been indentified by screening an expression library with synthetic oligonucleotides but their involvement in the transduction of light signals is not as yet clear (Yanagisawa and Izui, 1993).

Finally, both nitrogen (Sugiharto et al., 1992b) and the plant hormones cytokinins (Sugiharto et al., 1992a) and ABA (Amzallag et al., 1990) also stimulate the transcriptional capacity of the maize PEPC gene. The expression of this PEPC gene in C_4 plants is therefore complex and results from the integration of different external and internal stimuli (reviewed in Lepiniec et al., 1994; Rajagopalan et al., 1994).

2. Post-transcriptional Level

Metabolite control. PEPC accumulates to an extremely high level in the M-cell cytoplasm, reaching up to 50% of the soluble M proteins (Pierre et al., 1992). In vitro, this allosteric enzyme is subject to diverse regulatory influences by C_4 photosynthesis metabolites, i.e., negative feedback by the end-products L-malate, OAA and/or aspartate, and allosteric activation by glucose-6P (G-6P) as well as sugar-P like dihydroxyacetone-P (Doncaster and Leegood, 1987; Huber et al., 1986). L-malate seems to interact with the enzyme at different sites, producing competitive, non-competitive or mixed inhibition depending on pH, phosphorylation state, and concentration (Andreo et al., 1987; Duff et al., 1995). G-6P increases V_m of PEPC and, more importantly, its affinity for PEP, such that, by lowering the apparent K_m value for this substrate, it improves its ability to compete with L-malate (Andreo et al., 1987). This is particularly marked and varies greatly with pH around pH 7, a value which is likely to be close to physiological and which presumably shows diurnal fluctuation (see Sect. III.C). Hence, the in vivo activity of PEPC should depend on the actual balance of negative/positive C_4 photosynthesis-related effector metabolites, and the magnitude of their effect is modulated further by cytosolic pH (pH_c). When L-malate accumulates, due to reduced consumption by the Calvin cycle, negative feedback will tend to diminish its synthesis through PEPC; conversely, any increase in G-6P, reflecting accumulation of C_3 metabolites, will in turn stimulate the activity of the enzyme.

Covalent control. C_4 PEPC is reversibly modulated by a light-dependent phosphorylation process. This post-translational modification was discovered in 1984 on a CAM plant PEPC (Nimmo et al., 1984) and soon observed with C_4 plant PEPC (Budde and Chollet, 1986). It has become evident that it modulates the above-mentioned metabolite regulation of C_4 PEPC, in a light/dark manner, so that the resulting catalytic activity of the enzyme will meet the carbon demand of the Calvin cycle. We also note the possibility of a metabolite control of the susceptibility of PEPC to phosphorylation via a target, rather than a protein kinase effect. The phosphorylation process is slow and presumably confers to PEPC hysteretic properties. In this respect, it is interesting to note that other C_4 cycle enzymes, chloroplastic NADP-dependent L-malate dehydrogenase and pyruvate-P_i dikinase, are rapidly light-activated by redox and dephosphorylation cascades, respectively, that adjust their catalytic activity to the incident light energy (Chollet et al., 1990). The transduction pathway that leads to PEPC phosphorylation is one of the paradigms in plant signalling.

Other control mechanisms. Changes in the aggregation or redox state of the maize enzyme could provide a basis for its reversible light-regulation (Walker et al., 1986; Andreo et al., 1987; McNaughton et al., 1989; Willeford et al., 1990; Chardot and Wedding, 1992). However, to date there is no in vivo experimental support in favour of this hypothesis (Jiao and Chollet, 1991).

III. Regulatory Phosphorylation of PEPC and Its Role in C_4 Photosynthesis

A. The Target Residue

PEPC phosphorylation is complete after a C_4 leaf has been illuminated for approximately 1–2 h, and the final ratio of phosphorylated/dephosphorylated enzyme depends on the light energy (Jiao and Chollet, 1988; Echevarria et al., 1990; Bakrim et al., 1992). Upon return to darkness, dephosphorylation, likely by a type 2A protein phosphatase (Carter et al., 1990; McNaughton et al., 1991), follows with a similar time course (Echevarria et al., 1990; Bakrim et al., 1992). When measured at suboptimal, but near-physiological pH (7.3) and PEP concentration (2.5 mM), this covalent modulation mainly increases the enzyme's IC_{50} for L-malate (3–6-fold) and catalytic activity (2–3-fold) (Jiao and Chollet, 1988, 1991; Arrio-Dupont et al., 1992; Duff et al., 1995); the latter increase was found to be even higher at pH 7–7.1 (6–7-fold) (unpubl.). The actual L-malate concentration in the M-cell cytosol is not known with certainty, but in order to account for the observed rate of C_4 photosynthesis in full sunlight, and based on the hypothesis of a gradient-driven diffusive transport of this metabolite to the adjacent BS-cells, it was calculated that it should reach concentrations as high as 15–20 mM (Stitt and Heldt, 1985; Leegood and Osmond, 1990). Given an IC_{50} value of 0.2–0.4 mM of the dephosphorylated PEPC for L-malate, a tremendous inhibition is expected. Therefore, it is not surprising that most of the regulatory mechanisms acting on PEPC are aimed to protect the enzyme against this central C_4 photosynthesis metabolite. In this respect, based on concentrations of the effector metabolites believed to be physiological (Stitt and Heldt, 1985; Leegood and Osmond, 1990), a recent reinvestigation of the kinetic and regulatory properties of purified, recombinant *Sorghum* PEPC indicated that (1) phosphorylation decreases K_a for G-6P (approximately 4-fold at pH 7.3, with the maize enzyme) (Duff et al., 1995), (2) G-6P causes a large decrease in the apparent K_m for the substrate PEP, (3) PEP in turn reduces the affinity of the enzyme for L-malate. Consistent with this scheme is the observation that the effect (protection against L-malate) of phosphorylation and G-6P on PEPC was not simply additive; rather there is some substantial amplification, e.g., IC_{50} for L-malate is 0.2–0.4 mM for the dephosphorylated PEPC; it increases 3–5-fold upon phosphorylation or in the presence of 4 mM G-6P, but 50–75-fold when both factors are combined (the final value is 15 mM at pH 7.3, 2.5 mM PEP) (unpubl.). This would clarify how the enzyme is apt to cope with such high concentrations of metabolically active L-malate in the illuminated C_4 leaf. These observations highlight the deep interaction between the different regulatory mechanisms that affect PEPC activity and point to the critical role of PEPC phosphorylation for the control of carbon flux in C_4 photosynthesis (see Sect. III.D, for a discussion).

The sequence and location of the phosphorylation site have been determined following in vivo labeling of maize and *Sorghum* leaf PEPC, after feeding excised leaves [^{32}P]phosphate and rapid isolation of the protein from illuminated leaves by FPLC or immunoaffinity chromatography (Jiao et al., 1991b). In vitro labeling of the purified dephosphorylated enzyme in reconstituted systems by an endogenous PEPC kinase (Jiao and Chollet, 1990) or by the catalytic subunit of mammalian cAMP-dependent protein kinase (cAMP-PK) has also been performed (Jiao and Chollet, 1990; Terada et al., 1990). From trypsin-generated peptides, a single ^{32}P-labelled nonapeptide was isolated and sequenced by automated Edman degradation analysis. Comparing this aminoacid sequence,

His-His-*Ser(P)*-Ile-Asp-Ala-Gln-Leu-Arg,

to the nucleotide sequence of the corresponding DNA made it apparent that the phosphorylation domain is very close to the protein's N-terminus and contains a single seryl residue (serine 15 and 8 for maize and *Sorghum* PEPC, respectively) whose phosphorylation state undergoes changes in response to light/dark transitions (Jiao et al., 1991b) (Fig. 2). A survey of the deduced aminoacid sequences of PEPC subunits from diverse sources (Lepiniec et al., 1994; Rajagopalan et al., 1994) has led to the conclusion that this phosphorylation site is conserved in all plant enzymes, but is found in neither cyanobacterial nor bacterial PEPCs.

Conserved motifs flanking the phosphorylatable serine have been suggested to represent structural determinants for substrate recognition by PEPC kinase. Further investigation of structure/function relationships of the target residue with respect to phosphorylation has been carried out by recombinant protein technology based on the complementation of PEPC-deficient mutants of *Escherichia coli* with a plasmid bearing a full-length cDNA encoding *Sorghum* C_4 PEPC. The protein was purified close to electrophoretic homogeneity from bacterial extracts by using a fast immunochromatography procedure and

Maize	MASTKAPGPGEKHHSIDAQLRQLVP
Sorghum	MAS-------------ERHHSIDAQLRALAP
Flaveria trinerva	MANRNV------EKLASIDAQLRLLVP

Fig. 2. Amino-acid sequence of the phosphorylation site of C_4-type PEPC. Maize: Jiao and Chollet, 1990; Terada et al., 1990; Jiao et al., 1991b; *Sorghum*: Jiao et al., 1991b; Lepiniec et al., 1992; Wang et al. 1992; *Flaveria trinervia*: Hermans and Westhoff, 1992

identified as a genuine, phosphate-free, in vitro phosphorylatable C_4 plant PEPC (Crétin et al., 1991a). The experimental achievement opened the way to site-directed mutagenesis of PEPC. Changing serine 8 to aspartate (S8D) or cysteine (S8C) showed that phosphorylation can be functionally mimicked by the introduction of a negative charge in the N-terminal domain of the protein (Wang et al., 1992). Convincingly, engineering S-carboxymethylation of the cysteinyl residue in the S8C mutant by iodoacetic acid, which adds a negative charge, or iodoacetamide, which does not, further confirmed the previous observations, as only the former-SH modification was found to display the expected change in properties (Duff et al., 1993). These results also provided the unequivocal proof that there is cause-and-effect relationship between phosphorylation of a single serine and modification of enzymatic properties. Additional supportive evidence came from the use of affinity-purified antibodies directed against a synthetic peptide encompassing the phosphorylation site of *Sorghum* PEPC; upon binding epitopes of the enzyme's N-terminus, these antibodies impaired the access of the protein kinase and caused a concomitant change in PEPC properties mimicking phosphorylation (Pacquit et al., 1995). Thus, the addition of a negative charge (phosphate group, mutagenesis or chemical modification) or binding specific antibodies at the PEPC phosphorylation site result in increasing its IC_{50} for L-malate and velocity. This effect is presumably due to a modification of enzyme conformation; however, to date, there is no clue as to the precise molecular mechanism underlying this process. X-ray diffraction studies are now awaited to elucidate how PEPC phosphorylation modifies the enzyme's architecture.

Along these lines, using the hanging-drop vapor diffusion technique, protein crystals of *E. coli* PEPC and preliminary crystallographic data have been reported (Inoue et al., 1989). A similar set of data has been produced with the purified dephosphorylated recombinant *Sorghum* PEPC, but this research aera is still in its infancy.

B. PEPC Kinases: Identification and Properties

Protein kinases play a key role in the regulation of many cellular functions and for this reason are intensely studied. They are classified according to the nature of the phosphorylated amino acid residue (serine/threonine or tyrosine kinases) and further distinguished by their mode of regulation (Hunter, 1987, 1991; Hanks et al., 1988; Kemp and Pearson, 1990). A central problem is the question of how these enzymes recognize their diverse substrate proteins in vivo, since in reconstituted in vitro assays it is difficult to distinguish meaningful from gratuitous phosphorylation. Therefore, since there is no strict specificity of the converter enzyme for its target, functional/regulatory criteria should be met before it can be accepted that a given pair of protein substrate/protein kinase is physiologically relevant (Nimmo and Cohen, 1977). Indeed, plant protein kinases, like their animal counterparts, are implicated in the regulation of a variety of physiological events. During the last decade a growing number of plant protein kinases and targets has been reported (for reviews, see Ranjeva and Boudet, 1987; Budde and Chollet, 1988; Budde and Randall, 1989; Roberts and Harmon, 1992; Huber et al., 1994), among which PEPC is one of the best characterized plant proteins that is regulated by phosphorylation. However, PEPC does not escape the rule and different protein kinases are capable of phosphorylating this protein in vitro (see below). In view of the well established molecular characteristics of the target, it can be anticipated that the authentic PEPC kinase must be a serine/threonine protein kinase that phosphorylates the target serine of the N-terminal domain in the conserved motif acidic-basic-X-X-Ser-Ile, and that incorporation of phosphate at this serine residue elicits changes in the catalytic and regulatory properties of PEPC. Estimation of PEPC kinase activity in vitro relied on the detection of radiochemical labeling and/or, more accurately, the determination of the changes in functional (V; catalytic activity test, pH 7.1) and regulatory (IC_{50} for L-malate; L-malate test, pH 7.3) properties of highly purified PEPC. Controls to estimate site-specificity of in vitro phosphorylation (Ser 8 or 15 for *Sorghum* or maize PEPC, respectively) were based on the use of the recombinant mutant form of PEPC (S8D), which should not be radio labeled, and/or of specific phosphorylation site antibodies, which should impair recognition by PEPC kinase and block phosphorylation.

1. Calcium-independent PEPC Kinases

The first demonstration of the in vitro phosphorylation of purified, dephosphorylated maize PEPC by a protein fraction purified from illuminated C_4 leaves (by blue dextran agarose chromatography) was reported by Jiao and Chollet (1989a). Extensive purification of this very low abundance enzyme has been achieved with a purification yield of 4000-fold, leading to the identification of a 30 kDa polypeptide which catalyses target phosphorylation (serine 15 and 8 for maize and *Sorghum* PEPC, respectively) and concomitantly elicits the expected changes in PEPC properties (Wang and Chollet, 1993b). In vitro, at optimal pH, this protein kinase had an apparent K_m for PEPC (subunit) of 2.5 μM and needed Mg^{2+} and ATP (K_m total ATP, 40 μM) to be active; previous work had shown that neither EGTA, Ca^{2+}, calmodulin antagonists, P_iP, fructose-2,6-P nor cytosolic thioredoxin h had any effect on its activity. In contrast,

L-malate was found to be an inhibitor of the phosphorylation reaction (Jiao and Chollet, 1989b; Chollet et al., 1990). The optimum pH of the protein kinase was close to 8, and its catalytic activity varied sharply with pH between 7 and 8 (Wang and Chollet, 1993b). Similar findings were reported for a calcium-independent PEPC kinase present in *Sorghum* leaves (Vidal et al., 1990; Bakrim et al., 1992). Terada et al. (1990) and Jiao and Chollet (1990) showed that the catalytic subunit of the mammalian cAMP-PK also achieves a bona fide phosphorylation of maize PEPC. In this respect, it should be noted that no homolog of cAMP-PK has been found in plants. Similarly, a polycation-stimulated protein kinase which copurifies with and phosphorylates a highly purified preparation of phytochrome from etiolated *Avena* leaves was found to use maize PEPC as a substrate, but the phosphorylation site was not determined (Wong et al., 1989). Advantage was taken of these observations to demonstrate that, in vitro, L-malate diminishes the rate of PEPC phosphorylation in a pH and concentration-dependent manner and that this effect reflected an interaction of the effector with the target enzyme rather than with the protein kinase per se (Wang and Chollet, 1993b). Furthermore it was seen, that the photosynthesis-related metabolites G-6P and PEP were able to relieve L-malate inhibition (using physiological concentrations of these metabolites and cAMP-PK in a reconstituted assay at pH 7.3) of the phosphorylation rate of *Sorghum* PEPC (Echevarria et al., 1994). It is important to note that not only do these metabolites affect the activity of the target enzyme but they also modulate its phosphorylatability in vitro and possibly in vivo. This indirect way of regulating protein phosphorylation has been characterized in animal cells (Kenelly and Krebs, 1991) and may allow an individual target protein to be specifically controlled by a multisubstrate protein kinase.

2. Calcium-dependent PEPC Kinases

Crude, desalted protein extracts from *Sorghum* leaves catalyse the in vitro phosphorylation of PEPC, but this reaction was dependent on calcium and inhibited by anti-calmodulin antibodies (Echevarria et al., 1988). However, while a PEPC kinase activity was consistently recovered after affinity chromatography on calmodulin-Sepharose, subsequent in vitro phosphorylation of PEPC was insensitive to both calmodulin (CaM) and calmodulin antibodies, yet still required calcium. Although the phosphorylation site and the molar stoichiometry of phosphate incorporated/PEPC subunit were not known with certainty, this leaf protein kinase did not seem to induce any change in the sensitivity of PEPC to L-malate (Vidal et al., 1990; Bakrim et al., 1992). On the other hand, applying the purification procedure described above (Jiao and Chollet, 1989a) but with several alterations, Ogawa et al. (1992) purified a Ca^{2+}-dependent protein kinase able to phosphorylate maize PEPC on serine 15, and decrease its sensitivity to L-malate. This enzyme has a monomeric molecular mass of 50–60 kDa and was found to be highly Ca^{2+}-activated and inhibited by the calmodulin antagonist W7 [*N*-(6-aminohexyl)-5-chloro-1-naphthalenesulfonamide] and a potent inhibitor of myosin light chain kinase

[KT5926 (IC_{50}, 2.5 μM); Yabuta et al., 1994], an enzyme belonging to the Ca^{2+}/CaM-dependent protein kinase family. The Ca^{2+}-dependent PEPC kinase is reminiscent of plant CDPK (calcium-dependent protein kinase), a protein kinase not found in animals and which possesses an intrinsic, calmodulin-like calcium-binding regulatory domain linked to the catalytic domain (Roberts and Harmon, 1992). These findings cast some doubt on the identity of PEPC kinase and prompted further investigation.

3. Applying Physiological Criteria to Identify PEPC Kinase

The demonstration that a given protein kinase does, indeed, modify the catalytic properties of PEPC in vitro is a necessary but not unequivocal criterion to conclude that it is the genuine converter enzyme of the light-transduction pathway. Different leaf protein kinases, and even mammalian cAMP-PK, achieve a correct phosphorylation of the regulatory serine in reconstituted assays, resulting in an increase in velocity and IC_{50} (L-malate) of PEPC. To be physiologically relevant, the candidate protein kinase should also be subject to in vivo photoregulation. Thus, a major goal was to first unravel what mechanism(s) underlies the light control of PEPC kinase activity and second, which of the PEPC kinases detected in the C_4 leaf by the various former experiments is stimulated by light. Many possibilities were investigated but it was only when cycloheximide (CHX), a cytosolic protein synthesis inhibitor, was fed to excised *Sorghum* or maize leaves, that PEPC phosphorylation was blocked during the subsequent illumination period (Jiao et al., 1991a; Bakrim et al., 1992). Confirmation of this observation was reported by Carter et al. (1991) following the use of CHX and puromycin, another cytosolic protein synthesis inhibitor, which blocked the nocturnal appearance of the PEPC kinase activity in detached leaves of the CAM plant *Bryophyllum fedtschenkoi*. Interestingly, a subsequent IRGA-based experiment showed that uptake of CHX by an excised *Sorghum* or maize leaf performing steady state photosynthesis caused a progressive and well-correlated decrease in its CO_2 assimilation rate and PEPC phosphorylation status (Bakrim et al., 1993). The demonstration that this inhibitor-based observation was not due to perturbation of the Calvin cycle, photosynthetic electron flow, photophosphorylation or stomatal functioning strengthened the view that the light-dependent enhancement of PEPC phosphorylation state reflected stimulation of the rapid synthesis of the PEPC kinase itself, or alternatively, the synthesis of an unknown protein factor required for its activation. This interpretation involved protein turnover as a primary component in the transduction cascade and predicted the occurrence of a light/dark modulation of the translational capacity of specific mRNAs encoding these proteins in C_4 M-cells.

The most convincing attempt to identify C_4 PEPC kinase to date was performed recently by Li and Chollet (1993); they have attempted to sort out the different PEPC kinases in a green maize leaf based on the criterion of a photoregulation event involving protein synthesis (see above). Concentrated soluble-protein extracts from maize leaves rapidly prepared in the presence of protease inhibitors were analysed by SDS-PAGE and a subsequent in situ renaturation method was applied. Protein kinases were detected in the gel by their property to autophosphorylate. With this strategy, at least seven bands were revealed, but the ~30 kDa protein kinase was not radio labeled. Comparable gels were renatured sliced, pulverized and protein kinases assayed in vitro in reconstituted assays in the presence of purified PEPC as a substrate and [γ-^{32}P]ATP. Three proteins of molecular mass of ~57, ~37 and ~30 kDa exhibited PEPC phosphorylation activity on the target serine 15 and, interestingly, only the ~57 kDa protein kinase required calcium to catalyse this reaction whereas the two others did not. To assess which of them, if any, was light-responsive, leaves were light/dark adapted or treated with CHX or with methyl viologen (MV) in the light (see Sect. III.C.1) and the extracted soluble proteins analysed similarly. The activity of the calcium-independent PEPC kinases (~37 and ~30 kDa proteins) was strongly inhibited in darkened leaves, or in those illuminated leaves that were previously treated with CHX or MV, two compounds previously shown to block the in vivo light-dependent stimulation of C_4 PEPC kinase, while the calcium-dependent ~57 kDa protein kinase remained unaffected, similar to that of control (light) leaves. These data provided critical in vivo evidence which argues that C_4 PEPC kinase per se is calcium-independent. The exact nature of the two calcium-independent proteins able to carry out in vitro PEPC phosphorylation and the question of how the photoregulation mechanism works remain to be solved; are they true isozymes or structurally related and generated by proteolysis from a parent protein? The answer will come from the future isolation and molecular characterization of these PEPC kinase polypeptides.

C. The Signal Transduction Cascade

1. PEPC Phosphorylation in the Intact C_4 Leaf

SDS-PAGE analysis of immunoprecipitated PEPC from extracts of illuminated or darkened *Sorghum* leaves fed [^{32}P]phosphate showed that the 110 kDa carboxylase subunit was substantially more labeled during the light period (Bakrim et al., 1992). Consistently, PEPC kinase activity in desalted crude extracts of light-adapted maize leaves was several-fold greater than that from the corresponding dark tissue (Echevarria et al., 1990; McNaughton et al., 1991). This light-induced phosphorylation of PEPC was correlated with an increase in both IC_{50} for L-malate and V of PEPC, when measured at suboptimal levels of pH (7.3) and PEP (2.5 mM). Kinetic data indicated that (1) both the phosphorylation and dephosphorylation reactions plateaued after the leaf had been placed for about 100 min in the light or dark, respectively, and (2) increasing the light intensity step-wise from darkness to 735 μE/m^2·s at leaf level was paralleled by a progressive decrease in L-malate sensitivity (Bakrim et al., 1992). These observations suggested that phosphorylation determines the target enzyme's sensitivity towards its allosteric inhibitor, L-malate, and that the light-dependency of the phosphorylation process is coupled to the functioning of C_4 photosynthesis. The latter assumption was verified following illumination of detached *Sorghum* or maize leaves in the presence of various inhibitors of photosynthetic electron flow, ATP synthesis and the Calvin cycle; diuron and methyl viologen, that inhibit photosystem II and I, respectively, the uncoupler gramicidin which blocks ATP synthesis and DL-glyceraldehyde, a Calvin cycle inhibitor, markedly decreased the in vivo phosphorylation of PEPC (Bakrim et al., 1992; Jiao and Chollet, 1992). Obviously, the sensing and transduction of the light signal that initiates the phosphorylation cascade was by the photosynthetic apparatus and probably involved the electron transport chain and the Calvin cycle as intermediate steps. Well-known regulatory redox systems, like NADPH or reduced-thioredoxin, appeared not to be involved (Jiao and Chollet, 1989b; Chollet et al., 1990). A consistent hypothesis was that light-transduction might be mediated through changes in the levels of photosynthetic metabolites and/or energy charge; as a consequence, a transcellular message formed in BS-cells during C_4 photosynthesis would be delivered to adjacent M-cells where the physiological response is observed.

2. Light-dependent Phosphorylation of PEPC in C_4 Mesophyll Protoplasts and Cells

With respect to the identification of the components of the light-dependent phosphorylation cascade, especially those steps located upstream of the protein kinase, the intact leaf is of limited use. Thus, a cellular approach based on the production of M-cell protoplasts has been developed. In general, isolated intact protoplasts suspended in liquid medium constitute a suitable and homogeneous experimental system for studies of biochemical, physiological and cytochemical processes in plants because (1) they can be uniformly exposed to equal concentrations of various compounds in the suspending medium, and (2) they retain enzymes and organelles such that they represent functional entities capable of active photosynthesis and other pathways of cell metabolism. Convenient methods for obtaining reproducibly viable M-cell protoplasts from *Sorghum*, maize or *Digitaria* leaves, devoid of contamination by BS and organelles have been devised (Pierre et al., 1992; Devi and Raghavendra, 1992).

The objective was to find favourable experimental conditions that allowed in situ expression of PEPC phosphorylation. In this respect, the response of the protoplast preparation was checked under light or dark conditions, using the L-malate-sensitivity test at pH 7.3 or the catalytic-activity test at pH 7.1. If protoplasts were kept in a minimal medium (400 mM sorbitol and 10 mM Hepes-KOH, pH 7), these tests indicated very little modification in PEPC properties in the light. In contrast, when M-cell protoplast were suspended in incubation medium supplemented with Murashige and Skoog (MS) basal medium (0.44%), light markedly decreased L-malate sensitivity (35%) and increased catalytic activity (200%) of the endogenous PEPC (Fig. 3). Consistently, these changes were due to a substantial, light-induced increase in the activity of PEPC kinase (Fig. 4); in rapidly prepared protein extracts from these protoplasts, the

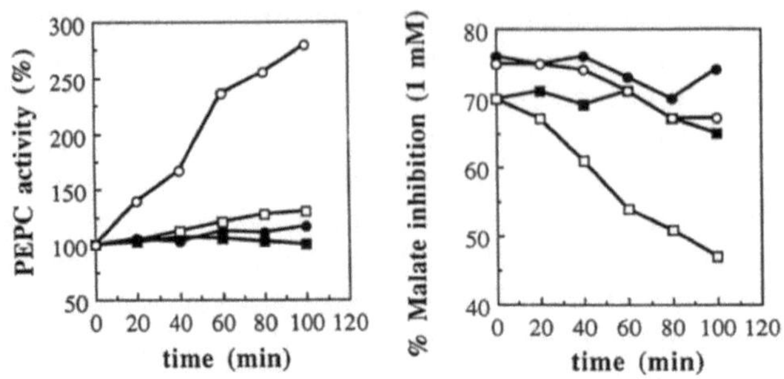

Fig. 3. Influence of light and NH_4Cl on the activity and malate sensitivity of PEPC in *Sorghum* M-cell protoplasts. Intact protoplasts were kept in darkness or illuminated at 300 µE/m^2·s at 25 °C in the presence or absence of NH_4Cl (20 mM). Aliquots of the protoplast suspension were removed at the indicated times and mixed directly with reaction medium. PEPC activity was recorded spectrophotometrically at 340 nm, in a 1 ml medium containing subsaturing PEP (1.5 mM) at pH 7.1 (activity test) or pH 7.3, and with or without 1 mM L-malate (malate test). pH 7.1: ● light, dark or dark and 20 µM NH_4Cl; ○ light and 20 µM NH_4Cl. pH 7.3: ■ light, dark or dark and 20 µM NH_4Cl; □ light and 20 µM NH_4Cl

protein kinase did not require calcium for activity. Treatment of intact protoplasts with CHX suppressed PEPC phosphorylation in a dose-dependent manner (Pierre et al., 1992). Once again, these data established at the cellular level that protein turnover is a component of the light-activated cascade leading to the stimulation of the activity of a Ca^{2+}-independent PEPC kinase and phosphorylation of its target enzyme. Similar data have been obtained with intact C_4 M-cells from *Digitaria sanguinalis* (S. Duff et al., unpubl.). Since, during the mechanical isolation procedure, these cells are released with their plasmodes-

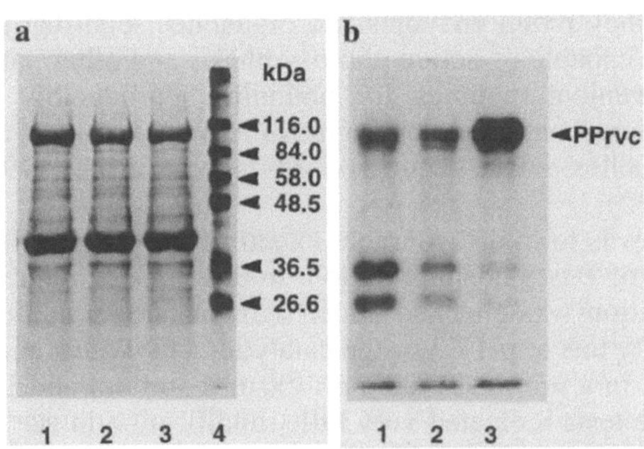

Fig. 4. PEPC kinase activity in crude extracts of illuminated or darkened *Sorghum* M-cell protoplasts. SDS-PAGE analysis of phosphorylation assays supplemented with [γ-^{32}P]ATP-Mg, 0.2 U of exogenous immunopurified dephosphorylation PEPC and dark (lanes 1 and 2, 0 and 60 min, respectively) or light (lane 3, 60 min) protoplast extracts. a, Coomassie Blue stained gel; b, corresponding autoradiograph. PPrvc, PEPC

mata open, this experimental system provides the means for studying the effect of those compounds that permeate poorly through the protoplast plasma membrane. Because isolated M-cells and protoplasts from the C_4 leaf possess the intrinsic ability (i.e, the whole signalling machinery) to perform PEPC phosphorylation in a light-dependent manner, they represent an ideal tool to unravel elements of the transduction chain.

3. Cytosolic pH and PEPC Phosphorylation

The above-mentioned inhibitor-based experiments performed in planta have suggested that a functional Calvin cycle (bundle sheath) is required for the effective phosphorylation of C_4 PEPC (mesophyll), thus implicating that the transduction chain involves the participation of these neighbouring cell types and intercellular communication. Since M-cell protoplasts have been physically separated from the Calvin cycle-containing BS-cells, they are expected not to receive the message. What, then, in the suspension medium is apt to mimick the physiological inducer? Among the different salts that comprise the MS medium, the active compound that triggers the in situ phosphorylation of PEPC was identified as NH_4NO_3 (Table 1). It was shown further that NH_4Cl and methylamine were good substitutes. These 3 compounds are weak bases which permeate into cells in their neutral form and therefore tend to increase cytosolic pH (pH_c) (Sanders et al., 1981; Bertl et al., 1984; Felle, 1987). Consistent with these findings are the data from Yin et al. (1990, 1993), Devi and Raghavendra (1992), Raghavendra et al. (1993), and Rajagopalan et al. (1993) that indicate that a cytosolic alkalization occurs in M-cells during photosynthesis in C_3 and C_4 plants. Weak base-induced alkalization of pH_c was verified

Table 1. Effect of Murashige and Skoog (MS) medium, NH_4Cl and methylamine on the L-malate sensitivity of PEPC in *Sorghum* M-cell protoplasts

Treatment	Malate sensitivity (% inhibition)
Dark	72
Light	
none	67
MS (0.44%)	42
NH_4NO_3	44
NH_4Cl	40
KNO_3	70
methylamine	40

Malate sensitivity of PEPC was measured in aliquots of protoplasts suspended in incubation medium and supplemented with the components (20 mM each) listed. Incubation was for 100 min at 27 °C after loading C_4 M-cell protoplasts with the fluorescent probe BCECF-AM and performing in situ fluorescence imaging by confocal microscopy. A rapid and marked increase in fluorescence, which reflects an increase in pH_c, was observed soon after either of the weak base inducers were added to the suspension medium (N. Giglioli et al., unpubl.). Despite the use of neutral red, a fluorescence quenching agent in the vacuoles, it was difficult to assess and further quantify the variation in fluorescence that occurs in the M-cell cytosol. On the other hand, applying the "null-point" method (Van der Veen et al., 1992) and monitoring the induced changes in protoplast fluorescence by flow cytometry led to quite satisfactory results. In this experiment the basal pH_c in C_4 M-cell protoplasts was found to be around 6.8–7.0 a value which is in good agreement with those already reported for plant cells (Felle, 1989; Kurdjian and Guern, 1989), and the addition of 10 mM weak base to the suspension medium increased pH_c by 0.3–0.4 pH-unit. In fact, the magnitude of the pH_c shift was expectedly dependent on the concentration of the exogenous inducer and well correlated with the corresponding changes in the activity of PEPC kinase, as measured in a reconstituted assay (N. Giglioli et al., unpubl.).

Therefore, intracellular alkalization in M-cell protoplasts is an early event in the transduction chain leading to phosphorylation of PEPC. This observation gave a clue for understanding how cells communicate in the illuminated C_4 leaf and for clarifying the nature of the putative intercellular message. It led to the proposal of a working hypothesis as follows: during C_4 photosynthesis, metabolite trafficking between the two photosynthetic cell-types takes place; a good candidate to represent the message is 3-phosphoglyceric acid (3-PGA), a Calvin cycle intermediate that diffuses to M-cells for its subsequent reduction in the M-chloroplast stroma (Hatch, 1987, 1992). As this process proceeds via the protonated form of 3-PGA, pumping of protons into the vacuole will ensue and a net alkalization of M-cell pH_c should result (Yin et al., 1990; Devi and Raghavendra, 1992; Raghavendra et al., 1993). This working model also predicts a quantitative coupling between light, the phosphorylation process and ultimately the rate of CO_2 assimilation, as mediated by a pH signal. Strongly supporting this view are the recent results by S. Duff et al. (unpubl.) and N. Giglioli et al. (unpubl.), showing that incubation of intact C_4 M-cells of *Digitaria sanguinalis* in a suspension medium containing 5 mM 3-PGA led to a significant decrease in L-malate sensitivity of PEPC and an increase in PEPC kinase activity, while 2-PGA or other C_4 photosynthesis-related metabolites had no effect.

4. Calcium and PEPC Phosphorylation

Various protein kinases have been found to phosphorylate C_4 PEPC in vitro. The Ca^{2+}-independent PEPC kinase initially reported by Chollet and coworkers (Jiao and Chollet, 1989a; Wang and Chollet 1993b) to occur in illuminated maize leaves is likely to represent the physiological protein kinase because it exhibits the requisite properties that conform to diagnostic physiological criteria (Li and Chollet, 1993), i.e., clearly its activity in C_4 leaf tissue is light/dark modulated and it confers on the target enzyme the changes in regulatory properties characteristic of a bona fide phosphorylation on the N-terminal serine of the phosphorylation domain. A homolog of this PEPC kinase has also been found in illuminated *Sorghum* leaves (Bakrim et al., 1992). In the C_4 leaf, this protein kinase should be found in the same cell as its target, namely M-cells. However, in the presence of the calcium ionophore A23187 (calcimycin) and EGTA, illuminated M-cell protoplasts did not show any tendency to phosphorylate PEPC in the presence of either of the weak-base inducers, but specific recovery was observed if Ca^{2+} was added back to the suspension medium. This finding supported the view that Ca^{2+} is involved upstream of the C_4 PEPC protein kinase in the light-transduction pathway.

Changes in cytosolic free Ca^{2+} play a central role in stimulus–response coupling in plant cells (Bush, 1993). The dominant source of Ca^{2+} in higher plants appears to be the central vacuole, rather than the external medium (Johannes et al., 1991). Diminution of the pH gradient across the tonoplast by pH changes on either side of the membrane results in a transient increase in cytosolic Ca^{2+} (Bush and Jones, 1987; Felle, 1988). However, this increase could also be due to the activation of a pH-dependent channel (Johannes et al., 1992), leading to a prolonged efflux of vacuolar Ca^{2+}. Evidence for such a mechanism has been obtained by using TMB-8 [8-(*N*,*N*-diethylamino)-octyl-3,4,5-trimethoxybenzoate hydrochloride], an internal Ca^{2+} channel blocker (Malagodi and Chiou, 1974), which severely inhibits PEPC phosphorylation

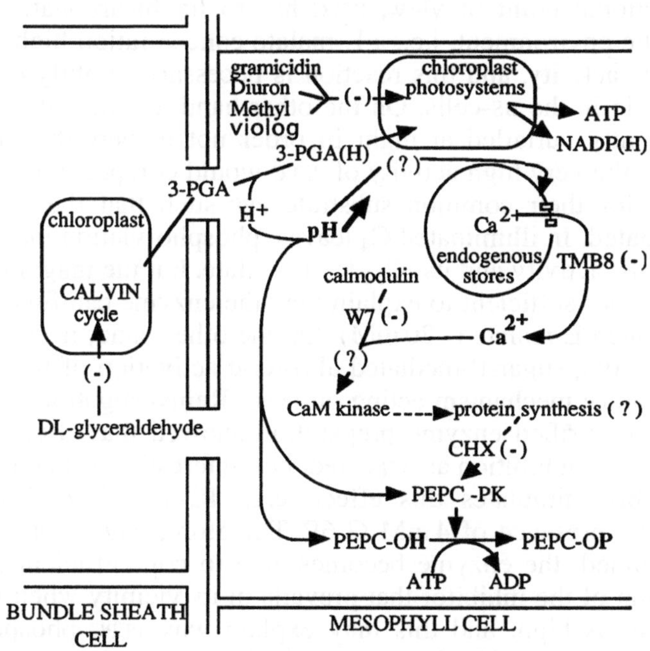

Fig. 5. Proposed scheme for the light-transduction pathway leading to the regulatory phosphorylation of C_4 PEPC

and the activity of the protein kinase when illuminated C_4 M-cell protoplasts are fed this compound in the presence of the inducer. Furthermore, W7 [*N*-(6-aminohexyl)-5-chloro-1-naphthalenesulfonamide], an inhibitor of Ca^{2+}/CaM-regulated protein kinases (Hidaka et al., 1981), had a similar, marked inhibitory effect. Collectively, these in situ data suggest that the transduction chain is presumably multicyclic, involving more than one protein kinase and requiring calcium to carry out an upstream step in the signalling pathway (Boon Chock et al., 1980).

An integrated working model of the transduction chain that controls PEPC phosphorylation and its relationship with the C_4-photosynthesis pathway is depicted in Fig. 5.

D. The Physiological Significance of the Regulatory Phosphorylation of C_4 PEPC

As described above, feeding an illuminated C_4 leaf the cytosolic protein synthesis inhibitor CHX markedly perturbs the activity of PEPC kinase, the phosphorylation status of PEPC, and ultimately C_4 photosynthesis in a highly correlated manner. The inhibitory effect of CHX is also observed in illuminated C_4 M-cell protoplasts treated with a suitable inducer of PEPC phosphorylation (Pierre et al., 1992). These results favour the view that PEPC phosphorylation has a critical regulatory impact on the overall functioning of C_4 photosynthesis.

From a functional point of view, PEPC has to fix bicarbonate in a very "hostile" cytosolic environment, i.e., a L-malate concentration high enough to severely block its activity, and this reaction is necessarily tightly coordinated to the Calvin cycle in the BS-cells. On the other hand, the activity of C_4 PEPC should be efficiently curtailed at night in order not to perturb normal dark metabolism, e.g., the very high activity of PEPC would compete with glycolytic pyruvate kinase for their common substrate PEP such that the Krebs cycle could not be initiated. In illuminated C_4 leaves, phosphorylation increases substantially both PEPC activity and its IC_{50} for L-malate, but the magnitude of this effect is probably not sufficient to explain how the enzyme can function in the presence of so much L-malate (~20 mM). On the other hand, it has long been proposed that in vivo, sugar-P-mediated allosteric activation of the enzyme is part of the regulatory mechanism acting on PEPC. Reinvestigation of this view by using a highly purified enzyme preparation showed that G-6P efficiently antagonizes L-malate inhibition as expected, but, interestingly, that phosphorylation considerably stimulates this effect, e.g., IC_{50} is 15 mM at pH 7.3, 2.5 mM PEP in the presence of 4 mM G-6P. Therefore, when both regulatory factors are combined, the enzyme becomes able to cope efficiently with the high concentration of the inhibitor that prevails in its vicinity when the rate of C_4 photosynthesis is high, and this may explain why PEPC phosphorylation appears to be so indispensable for the functioning of C_4 photosynthesis. The collective data also implicate a central role for pH_c and its light-induced variation in M-cells. If, as suggested by the aforementioned cell biology approach,

pH_c in M-cells is increased by 0.2–0.3 unit when leaves are exposed to high light, then a further modulation of the metabolic control of C_4 PEPC is expected; we speculate that this shift would be related to the rate of the Calvin cycle (see Fig. 5), and not only constitutes the message that initiates and transduces the signal through the chain leading to the synthesis of PEPC kinase and the phosphorylation of PEPC, but also modulates its sugar-P and phosphorylation-induced desensitization to L-malate and its increase in velocity. It is thus apparent that the metabolite and enzymatic protagonists in this regulatory network are interdependent and ultimately connected to light, and there lies the regulatory device that ensures coordination of the two metabolic cycles involved in C_4 photosynthesis. In the dark, dephosphorylation of PEPC and a decrease in pH to its original level would account for a more complete deactivation of the enzyme, as needed in a physiological context.

In closing, we would stress the following two points: (1) phosphorylation of C_4 PEPC is a slow process (approximately 100 min to be complete after switching on or off the light) and can by no means afford rapid regulation of the enzyme in response to abrupt fluctuations in its microenvironment. Thus, phosphorylation would constitute a coarse control of PEPC, whereas the fine-tuning of its activity would be the responsibility of metabolites produced during C_4 photosynthesis; and (2) the post-translational modification of PEPC is obviously slower than its catalytic rate and consequently, a hysteretic behavior of the enzyme is expected. Enzymes of this class are believed to possess a buffering effect on metabolites at the crossing of different pathways, i.e., they minimize sudden and large variations in the concentration of these metabolites (Neet and Ainslie, 1980). This would be modulated further by the photosynthetic metabolites (e.g., L-malate) that affect PEPC phosphorylatability via a target rather than converter-enzyme effect. In the present case, a hysteresis of PEPC would be of critical importance during the build-up and decline of photosynthetic PEP in the C_4 M-cell cytosol.

IV. Conclusions and Outlook

The intense research performed during the last decade has led to the view that PEPC phosphorylation is a cardinal regulatory event in C_4 photosynthesis. The increasingly complex pathway that links the light stimulus to PEPC kinase has the photosynthetic apparatus, cytosolic protein synthesis, pH, and Ca^{2+} as components and presumably more than one protein kinase is involved. Questions to be addressed in the future deal with (1) the unequivocal indentification and isolation of the genuine, light/dark-modulated PEPC kinase and the determination of its functional, regulatory and molecular properties; (2) the elucidation of the mechanism underlying the light-dependent increase in activity of the Ca^{2+}-independent PEPC kinase; is this protein, or protein factor, light-controlled at the level of translation, as suggested by the in vivo and in situ inhibitor-based experiments? If this proves to be the case, this protein kinase would appear to be a most unusual cascade enzyme with respect to its mode of regulation; (3) the identification of a putative Ca^{2+}-dependent protein kinase intervening in the upstream steps of the cascade; and (4) the role of Ca^{2+} and calcium channels in this process, with respect to pH_c. Resolution of these points will involve the improvement and development of cell pharmacological and molecular approaches. Although some insightful break throughs have been achieved, much work remains to be done before this very complex, multicyclic signal transduction cascade can be deciphered entirely. A good deal of structural data has become available on plant PEPC in relation to its regulatory phosphorylation. The study of its structure/function relationships involving site-directed mutagenesis of the phosphorylation domain will be continued and extended to domains forming the active and allosteric sites. Particularly awaited are X-ray crystallographic data and new insights on the three-dimensional structures of plant and microbial PEPCs.

Finally, (1) all plant PEPCs possess a highly conserved N-terminal phosphorylation domain, (2) the phosphorylation of C_3-type PEPCs from *Sorghum* roots (Pacquit et al., 1993), wheat (Van Quy et al., 1991), and tobacco (Wang and Chollet, 1993a) leaves, guard cells from *Vicia faba* leaves (Schnabl et al., 1992), and soybean root-nodules (Schuller and Werner, 1993) has been observed in reconstituted assays, and (3) the presence of endogenous protein kinases able to correctly phosphorylate these targets has been detected. It will be of particular interest to clarify in each of these specific cases the exact mechanism underlying PEPC phosphorylation and the regulatory impact on the corresponding physiology of the tissue. Research in this direction is becoming especially active.

Acknowledgements

The authors' research is supported in part by a grant from the Centre National de la Recherche Scientifique/U.S. National Science Foundation (in collaboration with Prof. Chollet's laboratory, University of Nebraska–Lincoln). We would like to thank Nathalie Giglioli and Valérie Pacquit for communicating unpublished results. Thanks are also due to Dr. Charles White for critical reading of the manuscript.

V. References

Amzallag GN, Lerner HR, Poljakoff-Mayber A (1990) Exogenous ABA as a modulator of the response of *Sorghum* to high salinity. J Exp Bot 41: 1529–1534

Andreo CS, Gonzalez DH, Iglesias AA (1987) Higher plant phospho*enol*pyruvate carboxylase, structure and regulation. FEBS Lett 213: 1–8

Arrio-Dupont M, Bakrim N, Echevarria C, Gadal P, LeMaréchal P, Vidal J (1992) Compared properties of PEPC from dark- and light-adapted *Sorghum* leaves: use of a rapid purification technique by immunochromatography. Plant Sci 81: 37–46

Bakrim N, Echevarria C, Crétin C, Arrio-Dupont M, Pierre J-N, Vidal J, Chollet R, Gadal P (1992) Regulatory phosphorylation of *Sorghum* leaf phospho*enol*pyruvate carboxylase, identification of the protein-serine kinase and some elements of the signal-transduction cascade. Eur J Biochem 204: 821–830

Bakrim N, Prioul J-L, Deleens E, Rocher J-P, Arrio-Dupont M, Vidal J, Gadal P, Chollet R (1993) Regulatory phosphorylation of phospho*enol*pyruvate carboxylase, a cardinal event influencing the photosynthesis rate in *Sorghum* and maize. Plant Physiol 101: 891–897

Becker TW, Templeman TS, Viret JF, Bogorad L (1992) The cab-m7 gene: a light-inducible, mesophyll-specific gene of maize. Plant Mol Biol 20: 49–60

Bertl A, Felle H, Bentrup FW (1984) Amine transport in *Riccia fluitans*. Cytoplasmic and vacuolar pH recorded by a pH-sensitive microelectrode. Plant Physiol 76: 75–78

Boon Chock P, Goo Rhee S, Stadtman ER (1980) Interconvertible enzyme cascades in cellular regulation. Annu Rev Biochem 49: 813–843

Bowler C, Neuhaus V, Yamagata H, Chua N-H (1994) Cyclic GMP and calcium mediate phytochrome phototransduction. Cell 77: 73–81

Budde RJA, Chollet R (1986) In vitro phosphorylation of maize leaf phospho*enol*pyruvate carboxylase. Plant Physiol 82: 1107–1114

Budde RJA, Chollet R (1988) Regulation of enzyme activity in plants by reversible phosphorylation. Physiol Plant 72: 435–439

Budde RJA, Randall DD (1989) Protein kinases in higher plants. In: Morré DJ, Boss WF, Loewus F (eds) Plant biology, vol 7, inositol metabolism in plants. AR Liss, New York, pp 344–351

Bush DS (1993) Regulation of cytosolic calcium in plants. Plant Physiol 103: 7–13

Bush DS, Jones RL (1987) Cytoplasmic calcium and α-amylase secretion from barley protoplasts. Eur J Cell Biol 46: 466–469

Carter PJ, Nimmo HG, Fewson CA, Wilkins MB (1990) *Bryophyllum fedtschenkoi* protein phosphatase type 2A can dephosphorylate phospho*enol*pyruvate carboxylase. FEBS Lett 263: 233–236

Carter PJ, Nimmo HG, Fewson CA, Wilkins MB (1991) Circadian rhythms in the activity of a plant protein kinase. EMBO J 10: 2063–2068

Chardot TP, Wedding R (1992) Role of cysteine in activation and allosteric regulation of maize phospho*enol*pyruvate carboxylase. Plant Physiol 98: 780–783

Chollet R, Budde RJA, Jiao J-A, Roeske CA (1990) Light/dark-regulation of C_4-photosynthesis enzymes by reversible phosphorylation. In: Baltscheffsky M (ed) Current research in photosynthesis, vol 4. Kluwer, Dordrecht, pp 135–142

Crétin C, Bakrim N, Keryer E, Santi S, Lepiniec L, Vidal J, Gadal P (1991a) Production in *Escherichia coli* of active *Sorghum* phospho*enol*pyruvate carboxylase which can be phosphorylated. Plant Mol Biol 17: 83–88

Crétin C, Santi S, Keryer E, Lepiniec L, Tagu D, Vidal J, Gadal P (1991b) The phospho*enol*pyruvate carboxylase gene family in *Sorghum*: promoter structure, amino acid sequences and expression of genes. Gene 99: 87–94

Datta N, Cashmore AR (1989) Binding of a pea nuclear protein to promoters of certain photoregulated genes is modulated by phosphorylation. Plant Cell 1: 1069–1077

Devi MT, Raghavendra AS (1992) Light activation of phospho*enol*pyruvate carboxylase in maize mesophyll protoplasts. J Plant Physiol 138: 435–439

Doncaster HD, Leegood RC (1987) Regulation of phospho*enol*pyruvate carboxylase activity in maize leaves. Plant Physiol 84: 82–87

Duff SMG, Lepiniec L, Crétin C, Andreo CS, Condon SA, Sarath G, Vidal J, Gadal P, Chollet R (1993) An engineered change in the L-malate sensitivity of a site-directed mutant of *Sorghum* phospho*enol*pyruvate carboxylase: the effect of sequential mutagenesis and S-carboxymethylation at position 8. Arch Biochem Biophys 306: 272–276

Duff SMG, Andreo CS, Pacquit V, Lepiniec L, Sarath G, Condon S, Vidal J, Gadal P, Chollet R (1995) Kinetic analysis of the non-phosphorylated, in vitro phosphorylated, and phosphorylation site mutant (Asp8) forms of intact recombinant C_4 phospho*enol*pyruvate carboxylase from *Sorghum*. Eur J Biochem 228: 92–95

Echevarria C, Vidal J, LeMaréchal P, Brulfert J, Ranjeva R, Gadal P (1988) The phosphorylation of *Sorghum* leaf phospo*enol*pyruvate carboxylase is a Ca^{2+}-calmodulin-dependent process. Biochem Biophys Res Commun 155: 835–840

Echevarria C, Vidal J, Jiao J-A, Chollet R (1990) Reversible light-activation of the phospho*enol*pyruvate carboxylase protein-serine kinase in maize leaves. FEBS Lett 275: 25–28

Echevarria C, Pacquit V, Bakrim N, Osuna L, Delgado B, Arrio-Dupont M, Vidal J (1994) The effect of pH on the covalent and metabolic control of C_4 phospho*enol*pyruvate carboxylase from *Sorghum* leaf. Arch Biochem Biophys 315: 425–430

Felle H (1987) Proton transport and pH control in *Sinapis alba* root hairs. A study carried out with double barrelled pH microelectrodes. J Exp Bot 38: 340–354

Felle H (1988) Cytoplasmic free calcium in *Riccia fluitans* L. and *Zea mays* L.: interaction of Ca^{2+} and pH? Planta 176: 248–255

Felle H (1989) pH as a second messenger in plants. In: Boss WF, Morré DJ (eds) Second messengers in plant growth and development. AR Liss, New York, pp 145–166

Hanks SK, Quinn A-M, Hunter T (1988) The protein kinase family: conserved features and deduced phylogeny of the catalytic domain. Science 241: 42–52

Hatch MD (1977) C_4 pathway of photosynthesis: mechanism and physiological function. Trends Biochem Sci 2: 199–202

Hatch MD (1987) C_4 photosynthesis: a unique blend of modified biochemistry, anatomy and ultrastructure. Biochim Biophys Acta 895: 81–106

Hatch MD (1992) C_4 photosynthesis: an unlikely process full of surprises. Plant Cell Physiol 33: 333–342

Hermans J, Westhoff P (1992) Homologous genes for the C_4 isoform of phospho*enol*pyruvate carboxylase in a C_3 and a C_4 *Flaveria* species. Mol Gen Genet 234: 275–284

Hidaka H, Sasaki Y, Tanaka T, Endo T, Ohno S, Fujii Y, Nagata T (1981) *N*-(6-Aminohexyl)-5-chloro-1-naphthalenesulfonamide, a calmodulin antagonist, inhibits cell proliferation. Proc Natl Acad Sci USA 78: 4354–4357

Huber SC, Sugiyama T, Akazawa T (1986) Light modulation of maize leaf phospho*enol*pyruvate carboxylase. Plant Physiol 82: 550–554

Huber SC, Huber JL, McMichael RW Jr (1994) Control of plant enzyme activity by reversible protein phosphorylation. Rev Cytol 149: 47–98

Hunter T (1987) A thousand and one protein kinases. Cell 50: 823–829

Hunter T (1991) Protein kinases classification. Methods Enzymol 200: 3–38

Inoue M, Hayashi M, Sugimoto M, Harada S, Kai Y, Kasai N, Terada K, Izui K (1989) First crystallization of a phospho*enol*pyruvate carboxylase from *Escherichia coli*. J Mol Biol 208: 509–510

Janc JW, Urbauer JL, O'Leary MH, Cleland WW (1992) Mechanistic studies of phospho*enol*pyruvate carboxylase from *Zea mays* with (Z)- and (E)-3-fluorophospho*enol*pyruvate as substrates. Biochemistry 31: 6432–6440

Jiao J-A, Chollet R (1988) Light/dark regulation of maize phospho*enol*pyruvate carboxylase by in vivo phosphorylation. Arch Biochem Biophys 261: 409–417

Jiao J-A, Chollet R (1989a) Regulatory seryl-phosphorylation of C_4 phospho*enol*pyruvate carboxylase by a soluble protein kinase from maize leaves. Arch Biochem Biophys 269: 526–535

Jiao J-A, Chollet R (1989b) Regulatory phosphorylation of C_4 phospho*enol*pyruvate carboxylase (PEPcase): effects of thioredoxin h, calcium, calmodulin (CaM) and metabolite effectors. Plant Physiol [Suppl] 89: 191

Jiao J-A, Chollet R (1990) Regulatory phosphorylation of serine-15 in maize phospho*enol*pyruvate carboxylase by a C_4-leaf protein-serine kinase. Arch Biochem Biophys 283: 300–305

Jiao J-A, Chollet R (1991) Posttranslational regulation of phospho*enol*pyruvate carboxylase in C_4 and crassulacean acid metabolism plants. Plant Physiol 95: 981–985

Jiao J-A, Chollet R (1992) Light activation of maize phospho*enol*pyruvate carboxylase protein-serine kinase activity is inhibited by mesophyll and bundle sheath-directed photosynthesis inhibitor. Plant Physiol 98: 152–156

Jiao J-A, Echevarria C, Vidal J, Chollet R (1991a) Protein turnover as a component in the light/dark regulation of phospho*enol*pyruvate carboxylase protein-serine kinase activity in C_4 plants. Proc Natl Acad Sci USA 88: 2712–2715

Jiao J-A, Vidal J, Echevarria C, Chollet R (1991b) In vivo regulatory phosphorylation site in C_4-leaf phospho*enol*pyruvate carboxylase from maize and *Sorghum*. Plant Physiol 96: 297–301

Johannes E, Brosnan JM, Sanders D (1991) Calcium channels and signal transduction in plant cells. Bioassays 13: 331–336

Johannes E, Brosnan JM, Sanders D (1992) Parallel pathways for intracellular Ca^{2+} release from the vacuole of higher plants. Plant J 2: 97–102

Kemp BC, Pearson RB (1990) Protein kinase recognition sequence motifs. Trends Biochem Sci 15: 342–346

Kenelly PJ, Krebs EG (1991) Consensus sequences as substrate specificity determinants for protein kinases and protein phosphatases. J Biol Chem 266: 15555–15558

Kurkdjian A, Guern J (1989) Intracellular pH_i measurement and importance in cell activity. Annu Rev Plant Physiol Plant Mol Biol 40: 271–303

Langdale JA, Taylor WC, Nelson T (1991) Cell-specific accumulation of maize phospho*enol*pyruvate carboxylase is correlated with demethylation at a specific site >3kb upstream of the gene. Mol Gen Genet 225: 49–55

Latzko E, Kelly GJ (1983) The many-faceted function of phospho*enol*pyruvate carboxylase in C_3 plants. Physiol Veg 21: 805–815

Leegood RC, Osmond CB (1990) The flux of meabolites in C_4 and CAM plants. In: Dennis DT, Turpin DH (eds) Plant physiology, biochemistry and molecular biology. Longman, Harlow, pp 274–298

Lepiniec L, Keryer E, Philippe H, Gadal P, Crétin C (1992) The phospho*enol*pyruvate carboxylase gene family of *Sorghum*: structure, function and molecular evolution. Plant Mol Biol 21: 487–502

Lepiniec L, Vidal J, Chollet R, Gadal P, Crétin C (1994) Phospho*enol*pyruvate carboxylase: structure, regulation and evolution. Plant Sci 99: 111–124

Li B, Chollet R (1993) Resolution and identification of phospho*enol*pyruvate carboxylase protein-kinase polypeptides and their reversible light activation on maize leaves. Arch Biochem Biophys 307: 416–419

McNaughton GAL, Fewson CA, Wilkins MB, Nimmo HG (1989) Purification, oligomerization state and L-malate sensitivity of maize leaf phospho*enol*pyruvate carboxylase. Biochem J 261: 349–355

McNaughton GAL, MacKintosh C, Fewson CA, Wilkins MB, Nimmo HG (1991) Illumination increases the phosphorylation state of maize leaf phospho*enol*pyruvate carboxylase by causing an increase in the activity of a protein kinase. Biochim Biophys Acta 1093: 189–195

Malagodi MH, Chiou CY (1974) Pharmacological evaluation of a new Ca^{2+} antagonist 8-(N, N-diethylamino)-octyl-3,4,5-trimethoxybenzoate hydrochloride (TMB-8): studies in smooth muscles. Eur J Parmacol 27: 25–33

Nagy F, Kay SA, Chua N-H (1988) A circadian clock regulates transcription of the wheat *Cab-1* gene. Genes Dev 2: 376–382

Neet KE, Ainslie GR Jr (1980) Hysteretic enzymes; historical background and development of the hysteretic concept of nonrapid enzyme transitions. Methods Enzymol 64: 192–226

Neuhaus G, Bowler C, Kern R, Chua N-H (1993) Calcium/calmodulin-dependent and -independent phytochrome signal transduction pathways. Cell 73: 937–952

Ngernprasirtsiri J, Chollet R, Kobayashi H, Sugiyama T, Akazawa T (1989) DNA methylation and the differential expression of C_4 photosynthesis genes in mesophyll and bundle sheath cells of greening maize leaves. J Biol Chem 264: 8241–8248

Nimmo HG, Cohen P (1977) Hormonal control of protein phosphorylation. Adv Cyclic Nucleotide Res 8: 145–266

Nimmo GA, Nimmo HG, Fewson CA, Wilkins MB (1984) Diurnal changes in the properties of phospho*enol*pyruvate carboxylase in *Bryophyllum* leaves: a possible covalent modification. FEBS Lett 178: 199–203

Nimmo HG (1993) The regulation of PEPC by reversible phosphorylation. In: Battey NH, Dickinson HG, Hetherington AM (eds) Post-translational modifications in plants. Cambridge University Press, Cambridge, pp 161–170

Ogawa N, Okumura S, Izui K (1992) A Ca^{2+}-dependent protein kinase phosphorylates phospho*enol*pyruvate carboxylase in maize. FEBS Lett 302: 86–88

O'Leary MH (1982) Phospho*enol*pyruvate carboxylase; an enzymologist's view. Annu Rev Plant Physiol 33: 297–315

Pacquit V, Santi S, Crétin C, Bui V-L, Vidal J, Gadal P (1993) Production and properties of recombinant C_3-type phospho*enol*pyruvate carboxylase from *Sorghum vulgare*: in vitro phosphorylation by leaf and root PyrPC protein serine kinases. Biochem Biophys Res Commun 197: 1415–1423

Pacquit V, Giglioli N, Crétin C, Pierre J-N, Vidal J, Echevarria C (1995) Regulatory phosphorylation of C_4 phospho*enol*pyruvate carboxylase from *Sorghum*: an immunological study using specific anti-phosphorylation site antibodies. Photosynth Res 43: 283–288

Pierre J-N, Pacquit V, Vidal J, Gadal P (1992) Regulatory phosphorylation of phospho*enol*pyruvate carboxylase in protoplasts from *Sorghum* mesophyll cells and the role of pH and Ca^{2+} as possible components of the light-transduction pathway. Eur J Biochem 210: 531–537

Raghavendra AS, Yin Z, Heber U (1993) Light-dependent pH changes in leaves of C_4 plants. Planta 189: 278–287

Rajagopalan AV, Tirumala Devi M, Raghvendra AS (1993) Patterns of phospho*enol*pyruvate carboxylase activity and cytosolic pH during light activation and dark deactivation in C_3 and C_4 plants. Photosynth Res 38: 51–60

Rajagopalan AV, Devi M, Raghavendra AS (1994) Molecular biology of C_4 phospho*enol*pyruvate carboxylase: structure, regulation and genetic engineering. Photosynth Res 39: 115–135

Ranjeva R, Boudet AM (1987) Phosphorylation of proteins in plants: regulatory effects and potential involvement in stimulus/response coupling. Annu Rev Plant Physiol 38: 73–93

Roberts DM, Harmon A (1992) Calcium-modulated proteins: targets of intracellular calcium in higher plants. Annu Rev Plant Physiol Plant Mol Biol 43: 375–414

Sanders D, Hansen U-P, Slayman CL (1981) Role of the plasma membrane proton pump in pH regulation in non-animal cells. Proc Natl Acad Sci USA 78: 5903–5907

Schäffner AR, Sheen J (1992) Maize C_4 photosynthesis involves differential regulation of phospho*enol*pyruvate carboxylase genes. Plant J 2: 221–232

Schnabl H, Denecke M, Schulz M (1992) In vitro and in vivo phosphorylation of stomatal phospho*enol*pyruvate carboxylase from *Vicia faba* L. Bot Acta 105: 367–367

Schuller K, Werner D (1993) Phosphorylation of soybean nodule phospho*enol*pyruvate carboxylase in vitro decreases sensitivity to inhibition by L-malate. Plant Physiol 101: 1267–1273

Stitt M, Heldt H (1985) Generation and maintenance of concentration gradients between the mesophyll and bundle sheath in maize leaves. Biochim Biophys Acta 808: 400–414

Sugiharto B, Burnell JN, Sugiyama T (1992a) Cytokinin is required to induce the nitrogen-dependent accumulation of mRNAs for phospho*enol*pyruvate carboxylase and carbonic anhydrase in detached maize leaves. Plant Physiol 100: 153–156

Sugiharto B, Susuki I, Burnell JN, Sugiyama T (1992b) Glutamine induces the N-dependent accumulation of mRNAs encoding phospho*enol*pyruvate carboxylase and carbonic anhydrase in detached maize leaf tissue. Plant Physiol 100: 2066–2070

Taylor WC (1989) Transcriptional regulation by a circadian rhythm. Plant Cell 1: 259–264

Terada K, Kai T, Okuno S, Fujisawa H, Izui K (1990) Maize leaf phospho*enol*pyruvate carboxylase: phosphorylation of Ser15 with a mammalian cyclic AMP-dependent protein kinase diminishes sensitivity to inhibition by L-malate. FEBS Lett 2: 241–244

Thomas M, Crétin C, Keryer E, Vidal J, Gadal P, Bidart JM, Bohuon C (1987a) Photoregulation process of *Sorghum* leaf phospho*enol*pyruvate carboxylase. Study with monoclonal antibodies. Biochem Biophys Res Commun 143: 170–177

Thomas M, Crétin C, Keryer E, Vidal J, Gadal P (1987b) Photocontrol of *Sorghum* leaf phospho*enol*pyruvate carboxylase: characterization of messenger RNA and of photoreceptor. Plant Physiol 85: 243–246

Thomas M, Crétin C, Vidal J, Keryer E, Gadal P, Monsinger E (1990) Light regulation of phospho*enol*pyruvate carboxylase mRNA in leaves of C_4 plants: evidence for phytochrome control on transcription during greening and for rhythmicity. Plant Sci 69: 65–78

Van der Veen R, Heimovaara-Dijkstra S, Wang M (1992) Cytosolic alkalinization mediated by abscisic acid is necessary, but not sufficient, for abscisic acid-induced gene expression in barley aleurone protoplasts. Plant Physiol 100: 699–705

Van Quy L, Foyer C, Champigny ML (1991) Effect of light and NO_3^- on wheat leaf PEPC activity: evidence for covalent modulation of the C_3 enzyme. Plant Physiol 97: 1476–1482

Vidal J, Echevarria C, Bakrim N, Arrio-Dupont M, Pierre J-N, Crétin C, Gadal P (1990) Regulation of phospho*enol*pyruvate carboxylase in *Sorghum* leaves: study of the phosphorylation process of the enzyme. In: Randall DD, Blevins DG (eds) Current topics in plant biochemistry and physiology, vol 9, The Interdisciplinary Plant Biochemistry and Physiology Program, University of Missouri, Columbia, MO, pp 365–370

Walker GH, Ku MSB, Edwards GE (1986) Catalytic activity of maize leaf phospho*enol*pyruvate carboxylase in relation to oligomerization. Plant Physiol 80: 848–855

Wang TH, Duff SMG, Lepiniec L, Crétin C, Sarath G, Condon SA, Vidal J, Gadal P, Chollet R (1992) Site-directed mutagenesis of the phosphorylatable serine (Ser 8) in C_4 phospho*enol*pyruvate carboxylase from *Sorghum*: the effect of negative charge at position 8. J Biol Chem 267: 16759–16762

Wang YH, Chollet R (1993a) In vitro phosphorylation of purified tobacco leaf phospho*enol*pyruvate carboxylase. FEBS Lett 328: 215–218

Wang YH, Chollet R (1993b) Partial purification and characterization of phospho*enol*pyruvate carboxylase protein-serine kinase from illuminated maize leaves. Arch Biochem Biophys 304: 496–502

Willeford KO, Wu MX, Meyer CR, Wedding RT (1990) The role of oligomerization in regulation of maize phospho*enol*pyruvate carboxylase activity. Influence of Mg-PEP and malate on the oligomeric equilibrium of PEP carboxylase. Biochem Biophys Res Commun 168: 778–785

Wong YS, McMicheal RW Jr, Lagarias JC (1989) Properties of a polycation-stimulated protein kinase associated with purified *Avena* phytochrome. Plant Physiol 91: 709–718

Yabuta N, Ogawa N, Izui K (1994) Partial purification and molecular properties of protein kinase phosphorylation C_4-type phospho*enol*pyruvate carboxylase in maize. Plant Cell Physiol [Suppl] 35: 95

Yanagisawa S, Izui K (1993) Molecular cloning of two DNA-binding proteins of maize that are structurally different but interact with the same sequence motif. J Biol Chem 268: 16028–16036

Yin Z, Neimanis S, Wagner U, Heber U (1990) Light-dependent pH changes in leaves of C_3

plants. I. Recording pH changes in various cellular compartments by fluorescent probes. Planta 182: 244–252
Yin Z, Heber U, Raghavendro AS (1993) Light-induced pH changes in leaves of C_4 plants. Planta 189: 267–277

Salicylic Acid—an Important Signal in Plants

Kristin K. Wobbe and Daniel F. Klessig

Contents

I. Introduction

Plants are a rich source of compounds used to treat the spectrum of human ailments—from headaches to heart disease to cancer. Aspirin, or acetylsalicylic acid, one of the most widely used drugs today, is plant based. Salicylic acid (SA) or closely related salicylates including methyl salicylate, saligenin (the alcohol of SA) and their glucosides, are found in many plants including the willow. As early as the 4th century B.C., Hippocrates purportedly prescribed willow bark to relieve the pain of childbirth (Rainsford, 1984; Weissman, 1991). In the 19th century acetylsalicylic acid was found to cause less gastrointestinal irritation than SA yet has similar therapeutic properties. Despite their long history, the mechanisms by which SA and other salicylates achieve their wide range of clinical effects (reduction of pain, fever, swelling, blood clotting and the risk of heart attacks and strokes) remain unresolved. In contrast, recent breakthroughs in plant pathology indicate that SA is a key signal molecule in disease resistance and suggest a new mode of action of SA.

SA is a member of a class of small molecules called phenolics, defined as an aromatic ring substituted with either a hydroxyl group or its derivative. It has been speculated that some phenolics function as plant growth regulators (Åberg, 1981) since when applied to plants, phenolics can produce a wide range of effects on many plant processes. For example, early studies demonstrated that exogenously applied SA affects seed germination, flowering, fruit yield, stomatal closure and glycolysis (for reviews, see Cutt and Klessig, 1992;

Raskin, 1992). However, it was difficult to determine if endogenous SA played a significant role in these processes since other phenolics could produce the same results. In addition, some of the effects could be due to the general chemical properties of SA: its acidity and its ability to chelate iron. Only recently has SA been conclusively demonstrated to play a role as a regulatory molecule.

II. SA in Thermogenesis

The first conclusive evidence that SA serves as an endogenous regulatory molecule came from studies by Raskin and coworkers on the thermogenic voodoo lily (*Sauromatium guttatum*) (Raskin et al., 1987, 1989). During flowering, the spadix of the voodoo lily exhibits two temperature increases of up to 12 °C. Raskin and coworkers showed that levels of endogenous SA increased dramatically just prior to both temperature spikes. In addition, exogenously applied SA induced the temperature increases and the concomitant production of aromatic compounds. Of 33 derivatives tested, only acetylsalicylic acid and 2,6-dihydroxybenzoic acid were capable of inducing the temperature rise, indicating the high level of specificity of the process.

The elevation in temperature is caused by increased utilization of the alternative respiratory pathway. Electrons are diverted from the cytochrome respiratory pathway, where the energy flow is conserved chemically, to the alternative respiratory pathway, where energy is released as heat (Meeuse, 1975). Rhoades and MacIntosh (1992) have demonstrated that expression of the gene encoding alternative oxidase, the terminal electron acceptor of the alternative respiratory pathway, is induced by SA in voodoo lilies. SA has also been shown to increase both the use of the alternative respiratory pathway (Kapulnik et al., 1992) and the amount of alternative oxidase in tobacco, a non-thermogenic plant (Rhoades and MacIntosh, 1993).

III. Disease Resistance

The second plant process shown to utilize SA as a signal molecule is disease resistance. This exciting area of research has developed rapidly in recent years and has suggested fascinating parallels between human and plant defense systems.

When a plant is infected by a pathogen to which it is resistant, a wide variety of biochemical and physiological responses are induced. The end result of these responses is the restriction of the pathogen and therefore, the damage which it may cause. A susceptible plant, in contrast, cannot restrict pathogen replication, in many cases allowing the pathogen to spread throughout the plant, with the potential to cause severe damage or even death of the plant. Susceptibility may be due to an inability of the plant either to recognize the pathogen or to successfully mount a defense response.

The defense response is frequently manifested as restriction of the pathogen to a small region surrounding the initial entry site, followed by necrosis of the tissue at the infection site. These two processes together characterize the hypersensitive response (HR) (Matthews, 1991). Frequently accompanying the HR is the development of acquired or induced resistance. Following the initial infection, the plant becomes, over a several day period, more resistant to a secondary infection by the same or even unrelated pathogens, both near the initial site of infection (local acquired resistance) and in distal, uninoculated portions of the plant (systemic acquired resistance, SAR).

At the biochemical level, a wide variety of processes are activated during a resistance response. Several of these are involved in strengthening the plant's passive defense system, mainly by fortifying the cell wall. These include synthesis and incorporation of hydroxyproline-rich glycoproteins, cellulose, and lignin into the cell wall and deposition of callose and suberin onto the cell wall. In addition, there is enhanced peroxidase activity which is necessary for lignification and may be involved in crosslinking cell wall proteins. A number of active defense mechanisms are also induced including synthesis of proteinase inhibitors which block the activity of both microbial and insect proteases, production of phytoalexins which have anti-microbial activity, induction of hydrolytic activities such as chitinases and β-1,3-glucanases that degrade the cell walls of microbes, and production of anti-viral activities.

Frequently used markers to study gene activation during plant defense are the phenylalanine ammonia lyase (PAL) gene and the pathogenesis-related (PR) protein genes. The PAL gene is induced relatively early after infection (Fritig et al., 1973; Legrande et al., 1976; Duchesne et al., 1977; Dong et al., 1991) and encodes a key enzyme in the phenylpropanoid pathway. This pathway is responsible for production of a range of defense molecules including phytoalexins, SA and other phenolic compounds. Expression of the PR protein genes correlates well, though not completely (Ye et al., 1989), with resistance responses to a wide variety of viral, fungal and bacterial pathogens. In particular, their expression in uninoculated tissue correlates with the onset of SAR (Ward et al., 1991). PR proteins are divided into at least five families (PR-1 through PR-5), each of which has been demonstrated to have anti-fungal activity, either in vitro (Mauch et al., 1988; Vigers et al., 1991; Woloshuk et al., 1991; D. Liu et al., 1994; Ponstein et al., 1994) or in vivo (Broglie et al., 1991; Alexander et al., 1993; Yoshikawa et al., 1993; D. Liu et al., 1994). The number of PR protein families is expanding and it is likely that some of the new members will be found to have anti-viral and anti-bacterial activity. A number of reviews are available which provide a more detailed discussion of plant defense responses (Bowles, 1990; Dixon and Harrison, 1990; Linthorst, 1991; Madamanchi and Kuć, 1991; White and Antoniw, 1991; Cutt and Klessig, 1992; Ryals et al., 1992).

The complexity of the plant response to pathogen attack suggests that there may be multiple signals operating through several pathways to produce the various defense reactions. In addition to the induction of local defenses, there is also the induction of SAR which requires a signal(s) capable of long distance movement from the initial infection site to the distal portions of the plant. Recent work has identified several potential candidates for long distance signal molecules. These include systemin (McGurl et al., 1992), jasmonates (Farmer et al., 1992), salicylic acid, ethylene (Pegg, 1976; Yang and Pratt, 1978), and electrical potentials (Wilson et al., 1992). For reviews, see Enyedi et al. (1992a), Malamy and Klessig (1992), and Dempsey and Klessig (1995).

IV. SA as a Signal in Plant Defense

The first suggestion that SA might be involved in inducing the plant's defense was provided by White and coworkers. They found that injecting aspirin or SA into tobacco leaves led to enhanced resistance to TMV (White, 1979) and induction of PR protein synthesis (Antoniw and White, 1980). This initial observation has been expanded to include protection against many other pathogens: viral, bacterial and fungal (Weete, 1992; Palva et al., 1994; reviewed by Malamy and Klessig, 1992). SA also induces PR genes in many other plants, both monocots and dicots (for review, see Klessig and Malamy, 1994). However, not all plant-pathogen systems respond to SA (Roggero and Pennazio, 1988; Ye et al., 1989; Roggero and Pennazio, 1991).

More direct proof of SA's involvement was provided a decade later by studies in tobacco (Malamy et al., 1990) and cucumber (Métraux et al., 1990; Rasmussen et al., 1991). After infection of resistant tobacco with TMV, SA levels increased in both the inoculated and uninoculated leaves. The increase in SA levels preceded or paralleled both the HR and induction of PR genes in the inoculated and uninoculated leaves. In contrast, SA levels did not increase in a susceptible cultivar of tobacco. In cucumber infected with tobacco necrosis virus (TNV), *Colletotrichum lagenarium* (Métraux et al., 1990) or *Pseudomonas syringae* pv. *syringae* (Rasmussen et al., 1991; Smith et al., 1991), SA levels were seen to increase in phloem exudates. The increase in SA levels preceded induction of SAR and peroxidase activity. While SA itself has been found to be toxic to some pathogens, i.e., *Colletotrichum falcatum, Fusarium oxysporum* (Singh, 1978) and *Agrobacterium tumefaciens* (Saint-Pierre et al., 1984), it is not toxic to TMV (Malamy and Klessig, unpubl.) or *C. lagenarium* (Mills and Wood, 1984). Therefore, the increased resistance associated with elevated SA levels cannot be due simply to SA's toxicity to the incoming pathogen. Later studies have demonstrated increases in endogenous SA in a number of other plant-pathogen systems including *Arabidopsis thaliana* inoculated with turnip crinkle virus (Uknes et al., 1993b; Dempsey and Klessig, unpubl.) and *Pseudomonas syringae* (Summermatter et al., 1994), and tobacco inoculated with tobacco necrosis virus, *P. syringae, Peronospora tabacina* (Silverman et al., 1993) or *Erwinia carotovora* ssp. *carotovora* (Palva et al., 1994).

Temperature shift experiments also support the hypothesis that SA plays a role in defense signaling. Resistance to TMV in tobacco is temperature sensitive. At temperatures greater than 28 °C, PR protein synthesis and the HR are blocked and the infection becomes systemic (Kassanis, 1952; Gianinazzi, 1970). At these high temperatures, there was also no increase in endogenous SA levels (Yalpani et al., 1991; Malamy et al., 1992). When resistance was restored by moving the plants to lower temperature, SA levels rose dramatically. The elevation of SA levels preceded PR-1 gene activation and formation of necrotic lesions (Malamy et al., 1992).

Additional support for SA as a signal molecule was provided by gene expression studies (Ward et al., 1991). In tobacco, nine gene families coordinately induced during establishment of SAR were also induced by SA treatment. Four other defense-related genes, induced only in pathogen-inoculated tissues, were not activated by SA. Further evidence that SA plays a role in establishment of SAR comes from transgenic tobacco experiments by Gaffney et al. (1993). They transformed *xanthi*-nc (resistant) tobacco with the *nah G* gene of *Pseudomonas putida*. This gene encodes salicylate hydroxylase, an enzyme that converts SA to catechol. Catechol does not induce SAR or PR gene expression. In plants expressing high levels of salicylate hydroxylase, TMV infection caused only a 2–3-fold induction of SA in inoculated leaves, compared to a 180-fold induction of SA in non-transformed control plants. Despite this large difference in SA levels, PR-1 gene expression was activated in the TMV-inoculated leaves of the *nah G* transgenic plants as well as in the non-transformed control plants. This result suggests that either the small increase in SA was sufficient or that another signal was involved. In contrast, PR-1 gene activation (Vernooij et al., 1994) and SAR development were inhibited in the upper uninoculated leaves of the infected transgenic plants (Gaffney et al., 1993). If the only effect of salicylate hydroxylase is destruction of SA, this result strongly implicates SA as a signal for induction of SAR.

SA may also be involved in the initial restriction of pathogen spread. While SA levels increase throughout the plant, the highest levels are seen in the infected leaf (Malamy et al., 1990), particularly immediately around the lesion site (Enyedi et al., 1992b). In addition, the primary lesions on *nah G* transformed tobacco are larger than on the non-transformed tobacco (Gaffney et al., 1993), consistent with a reduced ability to restrict pathogen replication and/or spread. In addition, *nah G* transformed tobacco and *Arabidopsis* showed increased susceptibility to primary infection by viral, fungal, and bacterial pathogens (Delaney et al., 1994).

In sum, the above studies argue strongly that SA is necessary for establishment of at least part of the plant's defenses. Application of exogenous SA induces PR gene expression and acquired resistance. Elevated levels of endogenous SA are associated with resistance and precede activation of defense-related genes; these increases in SA levels appear to be required as reduction or elimination of endogenous SA results in a decrease in the resistance response. However, it is still unknown which defense processes are directly affected by SA. Genetic experiments involving SA synthesis and response mutants should prove useful in delineating the SA-dependent processes.

V. SA as the Translocated Signal

Initial studies indicated that SA might be the translocated signal which initiates SAR. SA was found in the phloem in both pathogen-inoculated cucumber (Métraux et al., 1990; Rasmussen et al., 1991) and tobacco (Yalpani et al., 1991). In addition, increases in SA levels either in phloem or in uninoculated leaves of infected plants preceded establishment of SAR (Métraux et al., 1990) and induction of defense-related gene expression (Malamy et al., 1990; Meuwly et al., 1994).

Subsequent work, however, suggests that the translocated signal is not SA and production of the translocated signal precedes increases in SA. Hammerschmidt and coworkers found that the translocated signal(s) required for induction of SAR and associated peroxidase activity, as well as for elevated SA levels in upper, uninoculated leaves of *P. syringae*-infected cucumber appeared prior to detectable increases in SA levels in the phloem of the inoculated leaf. Despite the detachment of the inoculated leaf two to four hours before SA levels were detectable in the phloem, the upper uninoculated leaves still exhibited increases in SA levels (Rasmussen et al., 1991), peroxidase gene expression and induced resistance (Smith et al., 1991).

The knowledge that the systemic mobile signal can pass through a graft junction (Gianinazzi and Ahl, 1983) and the availability of *nah G* transgenic tobacco formed the basis for a second study that argues against SA being the mobile signal. Vernooij et al. (1994) found that when the upper grafted portion of the plant (scion) was derived from the *nah G* plants, neither SAR nor PR-1 gene induction occurred following infection of the rootstock, regardless of whether the rootstock was derived from the *nah G* transgenic or from the nontransformed *xanthi*-nc parental plants. However, a nontransformed *xanthi*-nc scion grafted onto *nah G* rootstock expressed PR-1 genes and SAR in response to TMV infection of the rootstock. These results suggest that SA is neither the translocated signal nor is it necessary for generation of the translocated signal. In contrast, SA appears to be required for mediation of the translocated signal in distal portions of the plant.

A third study using chemicals to help dissect the SA signal transduction pathway leading to acquired resistance also suggests that SA is not the mobile signal. It has long been known that chemicals other than SA can induce resistance in plants. A short list of these includes L-α-aminobutyric acid, thiamine-HCl (Asselin et al., 1985), polyacrylic acid (Gianinazzi and Kassanis, 1974), and $BaCl_2$ (White et al., 1986). Although treatment of tobacco with the above chemicals led to increases in total SA levels, enhanced resistance to TMV and PR-1 gene expression was induced only in the treated leaves (Malamy et al., unpubl.). This is consistent with experiments which showed that injection of SA induced PR protein accumulation and enhanced resistance only in the treated leaf (Van Loon and Antoniw, 1982). Even though SA can be detected in the phloem (Métraux et al., 1990; Rasmussen et al., 1991; Yalpani et al., 1991), the above experiments argue that SA is not the translocated signal for induction of SAR, and further support the theory that the production of the translocated signal is independent of SA levels, perhaps because it is produced earlier in the signaling pathway than is SA. However, the recent demonstration that in tobacco approximately 70% of the SA in an upper, uninoculated leaf has been translocated from the TMV-infected lower leaf suggests that SA could serve as the translocated signal for SAR (Shulaev et al., 1995) Perhaps there are redundant translocated signals with either SA or a yet unidentified second chemical serving to induce SAR.

VI. SA Metabolism

In plants, SA is probably derived from phenylalanine (El-Basyouni et al., 1964; Chadha and Brown, 1974; Yalpani et al., 1993a). This amino acid is converted to trans-cinnamic acid by PAL, which, as mentioned previously, is rapidly induced after infection of resistant plants and is a key enzyme in the phenylpropanoid biosynthetic pathway. SA can be produced from cinnamic acid via one of two intermediates: benzoic acid (BA) or *o*-coumaric acid.

Yalpani et al. (1993a) provided evidence that in tobacco the bulk of SA is produced from BA. Using ^{14}C-labeled precursors, they demonstrated that injection of labeled trans-cinnamic acid led to production of labeled BA and trace amounts of labeled SA; no label was found in *o*-coumaric acid. Large amounts of radiolabeled SA were detected following injection of ^{14}C-BA. In contrast, after injection of labeled *o*-coumaric acid, no labeled SA was detected. Concurrent work (León et al., 1993) demonstrated that the enzymatic activity responsible for the conversion of BA to SA, BA-2-hydroxylase (BA2H), was induced by increases in BA concentration and by TMV infection. This induction of BA2H activity was cycloheximide sensitive. Since the increase in BA levels and BA2H activity paralleled the increase in SA content, the rate-limiting step in SA synthesis may be the formation of BA.

Not all plants, however, utilize the same pathway. *Primula acaulis* and *Gaultheria procumbens* synthesize SA via *o*-coumaric acid (El-Basyouni et al., 1964). Chadha and Brown (1974) showed that in healthy tomato seedlings, SA is synthesized via BA. In contrast, after infection by *Agrobacterium tumefaciens* ^{14}C-labeled trans-cinnamic acid was converted to *o*-coumaric acid. In addition, incorporation of ^{14}C into SA from BA was also reduced, suggesting that the biosynthetic pathway switched after infection. Alternatively, reduced levels of ^{14}C-labeled SA produced from BA may have been due to dilution of the ^{14}C-BA by the release of endogenous BA from a pool of conjugated BA (see below). It is possible that some of the newly synthesized *o*-coumaric acid was destined for production of coumarins, rather than SA. Coumarins, like SA, are induced during disease resistance.

High concentrations of phenolic acids such as SA are phytotoxic. Therefore many are found to exist as sugar conjugates (Harborne, 1964; Towers, 1964; Cooper-Driver et al., 1972; Ben-Tal and Cleland, 1982; Umetani et al., 1990). Both Malamy et al. (1992) and Enyedi et al. (1992b) discovered that after TMV infection of tobacco, a large pool of SA existed in a conjugated form. After removal of free SA from infected leaf extracts, it was found that acid hydrolysis of the remaining material released a large amount of additional SA. Treatment of this material with β-glucosidase, instead of acid, released a similar amount of SA, demonstrating that the majority of conjugated SA exists as SA-β-glucoside (SAG), though approximately 10% of the total SA was not released by enzymatic treatment. Time course experiments showed that SA produced after infection is rapidly conjugated to form SAG in both inoculated (Enyedi et al., 1992b, 1993; Malamy et al., 1992) and uninoculated tissue (Malamy et al., 1992), with SAG becoming the predominant form. In *Arabidopsis*, SA is also converted to an acid hydrolyzable conjugate; again over 90% can be released by treatment with β-glucosidase (Bowling et al., 1994).

The enzymatic activity responsible for conversion of SA to SAG, UDP-glucose: SA glucosyltransferase, has been characterized from several plant species including oats (Yalpani et al., 1992), *Mallotus japonicus* (Tanaka et al., 1992), and tobacco (Enyedi and Raskin, 1993). In all these species, the glucosyltransferase activity is induced by SA. In tobacco, TMV infection increases the activity seven-fold above basal levels by two to three days post inoculation, paralleling the increase in SA levels.

To determine if SAG is bioactive in the absence of SA, SAG was chemically synthesized and injected into tobacco leaves (Hennig et al., 1993b). While PR-1 gene expression was rapidly induced, isolation of SA and SAG from injected tissue demonstrated that the SAG was rapidly degraded to free SA in the extracellular spaces. Further experiments with cultured tobacco cells demonstrated that only free SA was transported into the cells; SAG was not. In addition, a cell wall-associated, non-specific β-glucosidase active in releasing SA from SAG has been detected (Chen et al., 1995). These complications make determination of SAG's activity difficult. However, comparison with other sugar conjugates suggests that SAG is inactive. Many bioactive small molecules such as phenolics and phytohormones have conjugated forms that are inactive (Cohen and Bandurski, 1982; Letham and Palni, 1988; Conn, 1984; Reinecke and Bandurski, 1988).

It has been suggested that SAG serves as a storage form of SA, allowing for rapid release of SA in the event of a secondary infection (Hennig et al., 1993b; Klessig and Malamy, 1994). During this second infection, damage to the cell may allow the normally intracellular SAG to come in contact with the extracellular β-glucosidase. Release of SA would provide a signal for reactivation of defense responses more rapidly than in the case of a primary infection. Presumably, this would allow the plant to restrict the pathogen at an earlier stage in the infection, resulting in, for example, decreased lesion size.

These studies have provided a much clearer picture of the complex regulation of SA levels in tobacco (see Fig. 1). Initially, in healthy leaves there is very little SA, either free or in conjugated form. Within 24 h of TMV infection, free SA levels begin to rise and by 2 days p.i. a large percentage of total SA is in the form of SAG. Total SA levels continue to increase, with SAG becoming the predominant form. We now know that in addition to induction of PAL, an early

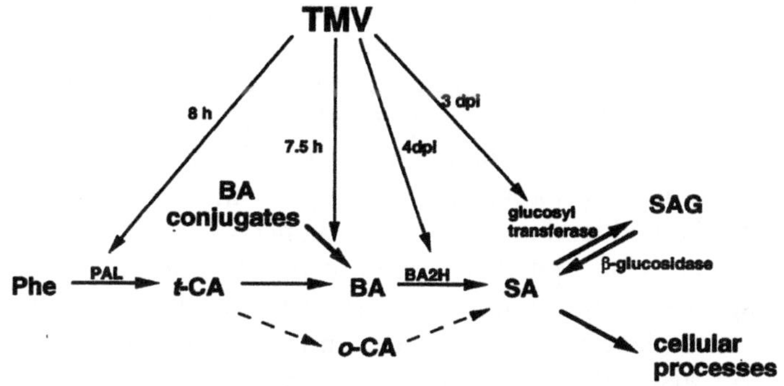

Fig. 1. Effect of TMV infection on SA metabolism in tobacco. TMV infection induces many of the enzymatic processess involved in SA production and conjugation. PAL activity is induced by 8 h pi. The pool of BA conjugate is substantially decreased by 7.5 h after moving infected tobacco from 32 °C to 24 °C (it is not known whether the decrease in the pool of BA conjugate is as rapid during nomal TMV infection at 22–24 °C). BA2H and the UDP-glucose:SA glucosyl transferase activities are both induced above basal levels by 2 d pi, with BA2H activity peaking at 4 d pi, and the glucosyl transferase by 3 d pi. SA levels increase detectably by 24 h pi. SA then affects cellular processes such as inducing PR gene expression, which becomes detectable by 36 h pi. The dotted line represents the possible alternative pathway for SA synthesis. This pathway does not appear to be used in TMV-infected tobacco. Phe, phenylalanine; *t*-CA, trans-cinnamic acid; *o*-CA, ortho-coumaric acid

enzyme in the biosynthetic pathway, TMV infection regulates SA metabolism at a number of other steps. There is a large pool of BA existing in an acid-hydrolyzable, conjugated form in healthy tissue. Yalpani et al. (1993a) observed that a transient decrease in the size of the pool of BA conjugates parallels the initial rise in BA and SA, suggesting that release of BA may provide an initial source for immediate conversion to SA. This would allow for more rapid production of SA, by bypassing several preceding steps in the biosynthetic pathway. In addition, increases in BA concentration induce BA2H activity (León et al., 1993), further stimulating production of SA. Increases in SA levels, in turn, induce the activity of the conjugating enzyme, UDP-glucose:SA glucosyltransferase (Enyedi and Raskin, 1993), thereby stimulating SAG production. Thus, the entire system appears to be tightly regulated to allow efficient production, and perhaps maintenance, of bioactive concentrations of SA while reducing the possibility that SA would reach phytotoxic levels. Recently, Raskin and coworkers (pers. comm.) discovered that a large amount of SA synthesized after TMV infection is converted to methyl salicylate in addition to SAG.

VII. Mechanisms of Action of SA

The discovery of a salicylic acid binding protein (SABP) in tobacco has shed considerable light on the role of SA in plant defense (Chen and Klessig, 1991;

Chen et al., 1993a, b). SABP is a soluble protein of 240–280 kDa, apparently composed of four identical or similar 57 kDa subunits. The specificity (it binds only analogs of SA that induce PR gene expression or resistance) and binding affinity for SA (K_d of 14 μM) suggest a physiological role for SABP in transduction of the SA signal. A cDNA clone encoding the 57 kDa subunit of SABP was obtained by screening an expression library with monoclonal antibodies against the SABP. The deduced amino acid sequence of the cDNA and the partial amino acid sequence of the purified SABP are highly homologous to catalase, the enzyme responsible for conversion of H_2O_2 to H_2O and O_2. In fact, purified SABP has catalase activity in vitro and this activity is inhibited upon binding of SA. SA also inhibits the activity of catalase in crude tobacco leaf extracts and in tobacco suspension cell (Conrath et al., 1995).

Moreover, there is a strong correlation between the effectiveness of the different SA analogs to inhibit catalase and their abilities both to bind SABP and to induce PR genes and resistance (Chen et al., 1993a, b). This correlation has been further strengthened by the recent demonstration that three additional analogs of SA (4-chlorosalicylic acid, 5-chlorosalicylic acid, and 3,5-dichlorosalicylic acid) are highly active in (*i*) inducing PR-1 gene expression and enhanced resistance to TMV in tobacco, (*ii*) competing with SA for SABP binding, and (*iii*) inhibiting catalase activity in vitro and in vivo (Conrath et al., 1995).

H_2O_2 production is an ongoing process in plants. Thus, as anticipated, inhibition of catalase by SA was shown to elevate levels of H_2O_2 in vivo as did treatment with a known inhibitor of plant and animal catalases, 3-amino-1,2,4-triazole (3AT). Furthermore, when H_2O_2 levels were raised artificially by injecting leaves with H_2O_2 or 3AT, PR-1 gene expression was activated. Expression of PR-1 was also induced by injecting compounds that promote H_2O_2 generation in vivo such as paraquat. In contrast, treatment with 3-hydroxybenzoic acid did not induce elevated H_2O_2 levels or PR-1 gene expression, consistent with its inability to bind SABP or inhibit catalase activity (Chen et al., 1993b). The above results argue that SA acts by inhibiting catalase activity which leads to elevated H_2O_2 levels. H_2O_2 or another reactive oxygen species (ROS) derived from it, then activates defense-related genes such as PR-1, perhaps by acting as a second messenger.

To determine if this mechanism is specific to tobacco or is a more general mechanism for activation of defense responses, several other plant species were tested for SA-binding activity (Sánchez-Casas and Klessig, 1994). High levels of SABP were found in cucumber, tomato and *Arabidopsis*, while little, if any, were detected in soybean, maize and rice. The amount of SA-binding activity correlated with the extent to which catalase activity was inhibited by SA. In addition, treatment of cultured rose cells with 1 mM SA increased internal H_2O_2 levels by 50% (Murphy et al., 1993). These results suggest that SA-inhibitable catalases are not unique to tobacco.

2,6-Dichloroisonicotinic acid (INA) has been shown in both tobacco and *Arabidopsis* to induce resistance to a variety of pathogens and to activate a set of genes triggered during SAR (Ward et al., 1991; Uknes et al., 1992). However,

INA treatment does not lead to increases in SA or SAG (Vernooij et al., 1995; Malamy et al., unpubl.) suggesting either that it enters the signal transduction pathway downstream of SA or that there is a separate pathway. NAD(P)H (nicotinamide adenine [phosphate] dinucleotide) has been shown to bind to mammalian catalase, though it is neither required for nor inhibitory to the enzyme's activity (Kirkman and Gaetani, 1984). INA contains a nicotinic acid moiety and may therefore also bind to plant catalase. In fact, INA was found both to compete with SA for binding SABP and to potently inhibit catalase activity of cultured tobacco suspension cells in vivo (Conrath et al., 1995). These results, in addition to the previous findings, argue that SA and a wide variety of SAR-including compounds activate the defense response by modulating catalase activity.

The immediate consequence of inhibiting catalase is an increase in intracellular H_2O_2 levels. H_2O_2 is one of several reactive oxygen species, others being $O_2^{\cdot -}$ (superoxide anion), HO_2 (perhydroxyl radical) and $OH^{\cdot}$ (hydroxyl radical), all of which are interconvertible. Large increases in the level of ROS, termed an oxidative or respiratory burst, have previously been associated with early stages of the plant defense response (for review, see Sutherland, 1991). Treatment of several species of cultured cells with elicitors (Doke, 1983a; Apostol et al., 1989; Legendre et al., 1993; Degousée et al., 1994) or of plants with pathogens (Doke, 1983b; Chai and Doke, 1987; Doke and Ohashi, 1988) results in an immediate oxidative burst. In incompatible interactions (plant is resistant to the pathogen) there is frequently a second rise in ROS levels that occurs within several minutes to hours after infection (Keppler and Baker, 1989; Keppler et al., 1989; Baker et al., 1991; Orlandi et al., 1992). The rapid appearance of the ROS suggests that SA is not involved in these early oxidative bursts as SA levels do not increase until considerably later.

Production of H_2O_2 and other ROS appears to serve multiple purposes in the plant. The ROS can act directly to neutralize the pathogen. Since ROS can be toxic, the initial increase may effectively diminish the pathogen's ability to establish an infection either by damaging the pathogen or reducing its numbers (Keppler and Baker, 1989). In addition, H_2O_2 is directly involved in crosslinking cell wall proteins and increasing lignification (Bradley et al., 1992), known components of the defense arsenal. There is also evidence that ROS cause host plasma membrane damage (Keppler and Novacky, 1986; Degousée et al., 1994) and cell death during the HR (Doke, 1983a, b; Doke and Ohashi, 1988).

ROS also appear to play a signalling role in the defense reaction and other plant responses. ROS are well suited for such a role; they are small, readily diffusable and can be rapidly synthesized and destroyed. There are several systems for both their synthesis and destruction, allowing precise control of their levels. They have been implicated in a number of processes. Induction of phytoalexin biosynthesis by fungal elicitor treatment of soybean suspension cells is preceded by production of ROS (Apostol et al., 1989; Legendre et al., 1993; Degousée et al., 1994). Recently, Levine and coworkers (1994) reported that in elicitor or pathogen challenged soybean cells, H_2O_2 functions as a local trigger of programmed cell death and as a diffusible signal for the activation of genes encoding cellular protectants such as glutathione S-transferase, glutathione reductase and polyubiquitin. ROS have also been associated with chilling tolerance in maize (Prasad et al., 1994), utilization of the alternative respiratory pathway in yeast (Minagawa et al., 1992), and induction of the alternative oxidase gene in petunia (Wagner, 1995).

Use of ROS as signal molecules is well documented in animal systems (Schreck and Baeuerle, 1991). ROS are involved in vertebrate immune responses mediated by the transcription factor NF-κB (Schreck et al., 1991; Xanthoudakis, 1992; Schenk et al., 1994). The transcription factor AP-1 regulates genes involved in growth, differentiation and UV damage repair. AP-1 synthesis and activation is also induced by ROS (Stein, 1989; Abate et al., 1990; Devary, 1991). There is evidence that ROS trigger apoptosis (programmed cell death; Hockenbery et al., 1993). It has been suggested that H_2O_2 also mediates many of the metabolic changes associated with binding of insulin to its receptor (May and deHaven, 1979).

Very recently, the involvement of catalase inhibition by SA and elevated H_2O_2 levels in plant defense responses has been called into question. While H_2O_2 and H_2O_2-inducing chemicals activated PR-1 genes in wild type tobacco, PR-1 induction was strongly suppressed in *nah G* transgenic plants in which SA is converted to catechol by the *nah G*-encoded salicylate hydroxylase (Neuenschwander et al., 1995; Bi et al., 1995). In addition, no detectable increase of H_2O_2 levels was found during the onset of SAR (Neuenschwander et al., 1995) and Bi and coworkers (1995) were unable to detect reductions in catalase activity after *P. syringae* infection of tobacco or pretreatment of leaf disks with SA. Moreover, H_2O_2 was found to induce accumulation of SA and activate BA2H, the enzyme which converts BA to SA (León et al., 1995; Neuenschwander et al., 1995). Taken together, these results suggest that H_2O_2 may act upstream of SA in the signal transduction pathway rather than, or in addition to, acting downstream of SA. It should be noted that extremely high levels of H_2O_2 (150–1000 mM) were required to induce SA accumulation.

We (Chen et al., unpubl.) have also found that PR-1 gene activation by prooxidants such as H_2O_2 is suppressed in *nah G* transformed tobacco; but induction of PR-1 genes by INA was also depressed in *nah G* versus wild type tobacco. Since INA does act through SA (Vernooij et al., 1995; Malamy et al., unpubl.), these results suggest that *nah G* and wild type plants may differ in more than just the loss of the SA signal. Since catechol, which is produced from SA by salicylate hydroxylase, is a potent antioxidant and antioxidants can suppress SA or INA activation of PR-1 genes (Conrath et al., 1995; Chen et al., unpubl.), these results argue that PR-1 gene activation may be modulated through the plant's redox state and an altered redox state may account for some of the difference in responsiveness to prooxidants and INA in *nah G* versus wild type plants.

It has recently been discovered that SA has several additional effects in plants. In addition to blocking catalase activity, SA (as well as INA) can also inhibit ascorbate peroxidase (APX; Durner and Klessig, 1995). APX is a critical enzyme in the other major pathway (ascorbate-glutathione pathway) for degradation of H_2O_2. In contrast, guaiacol peroxidases, which are not involved in scavenging H_2O_2, are also not inhibited by SA or INA.

Moreover, SA (and INA) appears to induce lipid peroxidation, probably through an SA free radical generated during SA's interaction with catalase and APX (Chen and Klessig, unpubl.). The resulting lipid peroxides activated PR-1 genes. Thus, SA appears to generate two signals: elevated H_2O_2 (ROS) levels and lipid peroxides. Since lipid peroxidation is a chain reaction, a small amount of SA free radical could result in the formation of an effective lipid peroxide signal without a readily perceptible inhibition of catalase or ascorbate peroxidase activities.

In summary, the mechanisms by which SA induces PR gene expression and enhanced disease resistance are not yet firmly established. It seems likely that products of lipid peroxidation, such as lipid peroxides, play a role in development of SAR in uninfected tissue where SA levels may be too low to effectively inhibit catalase and APX. At the site of infection where high levels of SA accumulate, both H_2O_2 and lipid peroxides probably participate. H_2O_2 may act in concert with, or independent of, lipid peroxides to transmit the signal downstream of SA. Alternatively or in addition, it may function to stimulate production of more SA. Finally, SA may have other mechanisms of action yet to be discovered.

VIII. SA Signal Transduction Pathway

Our current level of understanding of the signal transduction pathway from infection to production of defense responses is fragmentary at best. We know that the elevation in endogenous SA levels is a relatively late step in the pathway which is preceded by one, and possibly two, oxidative bursts. The SA signal follows a temperature sensitive step in the TMV–tobacco system and dissemination of the systemic signal in the *P. syringae*–cucumber system. Elevated SA levels inhibit catalase activity possibly leading to yet another increase in ROS. The ROS activate one or more of the defense-related genes, probably indirectly as protein synthesis is required for their transcriptional activation by SA.

One approach to filling in the gaps in our knowledge is to work backwards from an induced gene. Standard practice is to isolate the promoters of induced genes, determine sequences that bind factors involved in induced expression, clone the genes for these factors and determine their expression patterns. This process has been complicated in defense-related gene analysis because in the promoters analyzed so far there are few obviously conserved sequence elements which are necessary and sufficient for induction by SA or pathogens.

Two such elements have been partially characterized. The TCA motif is found in over 30 different stress-induced plant genes (Goldsbrough et al., 1993). The activity of the nuclear factor that binds this element is elevated in extracts from SA-treated plants. Though this element is present in several copies in two of the best studied tobacco PR genes (PR-1a and PR-2d), it is not found in either of the regions of these promoters found to be sufficient to confer SA inducibility. In addition, there is no evidence that this element can, either as a monomer or a concatemer, confer SA inducibility on a heterologous promoter (Ulmasov et al. 1994). Therefore, its role in SA induction of gene expression is unclear.

A second family of elements has been identified which does confer SA inducibility on a heterologous promoter (Qin et al., 1994; Zhang and Singh, 1994). Members of this family include the *ocs* element (originally identified in *Agrobacterium tumefaciens* T-DNA genes, Bouchez et al., 1989) and the *as-1* motif (identified in the CaMV 35S promoter; Lam et al., 1989). Plant genes which contain related sequence elements in their promoters include soybean Gmhsp26-A (Ellis et al., 1993), tobacco GNT35/GNT1 (Y. Takahashi et al., 1990) and par-1 (van der Zaal et al., 1991). These sequences appear to bind a family of related proteins, ASF-1 (Katagiri et al., 1989; Miao et al., 1994), which can form homo- and heterodimers. Promoters containing these elements are induced by a number of substances including auxin, methyl jasmonate, and SA (Kim et al., 1993; Liu and Lam, 1994; Qin et al., 1994; Zhang and Singh, 1994).

The SA-induced expression of genes containing *ocs/as-1* elements in their promoters, however, differs considerably from that seen in the PR genes, which lack these elements. SA induction of promoters containing *ocs/as-1* elements is rapid, with mRNA detectable as early as 30 min and peaking at around 4 h after treatment (Qin et al., 1994; Zhang and Singh, 1994). In addition, induction by SA is not cycloheximide sensitive (Qin et al., 1994). In contrast, PR gene induction is first detectable at 2–4 h post-treatment, peaks much later and is cycloheximide sensitive (Ward et al., 1991; Qin et al., 1994).

While many promoters have been shown to be SA inducible (for review, see Klessig and Malamy, 1994; Somssich, 1994), the best studied are those of the PR genes. Dissection of PR promoters has yielded a complex and still incomplete picture. In general the standard approach has been to fuse the promoter of choice or deletions thereof to a reporter gene (usually *uidA* or GUS). In more detailed studies, various segments of the promoter are joined to a heterologous minimal promoter (usually the 35S CaMV promoter) followed by the reporter gene. These chimeric genes are then used to transform plants. It should be noted that in all cases the promoter constructions behave similarly under SA treatment or TMV infection, providing additional evidence that PR promoters are induced via SA during pathogen attack. Extensive dissection of the tobacco PR-1a promoter suggests that one or more regions between −335 and −661 relative to the transcription start site is necessary for high levels of SA responsiveness (Ohshima et al., 1990; van de Rhee et al., 1990; Beilman et al., 1992; Uknes et al., 1993a). Another study suggests that there are at least three regions (−902 to −691, −659 to −643, and −643 to −287) which provide positive regulation of the tobacco PR-1a gene in response to SA and TMV (van de Rhee and Bol, 1993).

Ohashi and colleagues have identified two sites close to the transcription start site (−37 to −61 and −168 to −179) of PR-1a which bind protein factors present in extracts from untreated, and thus non-PR expressing, resistant tobacco (Hagiwara et al., 1993). These proteins are not detected in extracts from the interspecific hydrid of *N. glutinosa* × *N. debneyi* which constitutively produces PR-1 proteins (Ahl and Gianinazzi, 1982; Ahl-Goy et al., 1992; Ohashi et al., 1992) and has elevated levels of SA (Yalpani et al., 1993b). Though the elements do not contain a common sequence, competition studies suggest that the same factor binds both. Another study demonstrates that at least seven of eight fragments spanning −900 to +29 of the PR-1a promoter bind a nuclear factor(s) with different affinities, with the highest affinity sites between −902 and −656. This factor(s) appears to be reduced in extracts from SA-treated or TMV-infected tobacco (Buchel et al., 1995). Both sets of results imply that there is negative regulation of the PR-1a promoter, in addition to positive regulation.

Characterization of two tobacco PR-2 genes, which encode the acidic β-1,3-glucanases, suggests that regulation of this gene family may also be complex. 5′ and 3′ deletion analysis of the PR-2d promoter and fusion of various segments of it to the core 35S promoter suggest that an SA-responsive element is located between –348 and –324. Southwestern analysis and band shift analysis suggest that multiple proteins interact with this promoter element (Hennig et al., 1993a; Shah and Klessig, unpubl.). Deletion analysis of the PR-2b promoter demonstrates that low level induction by SA requires only ~300 bp 5′ of the transcription start site while an additional 350 bp of 5′ flanking sequence is necessary for high levels of induction (van de Rhee et al., 1993).

The *Arabidopsis* acidic PR-3 gene promoter has also been fairly well characterized. Samac and Shah (1991) demonstrated that only the proximal 193 bp from the start of transcription is necessary to give appropriate basal and induced expression. Two upstream elements are involved in quantitative expression. There is a negative regulatory element between −384 to −590 and a positive element between −590 to −1129. Future work to further define the sequences involved in negative and positive regulation, in combination with in vitro analysis, should define the regulatory motifs and the factors that interact with them.

A genetic approach has been very useful in determining components of signaling pathways in many organisms. This approach has been particularly useful in *Arabidopsis*, which is diploid, has a small, well-mapped genome, is readily transformed and has a relatively short generation time. Though use of this plant in pathogen studies is relatively recent, it has provided a wealth of information. *Arabidopsis* is resistent to a variety of bacterial (Simpson and Johnson, 1990; Davis et al., 1991; Debener et al., 1991; Tsuji et al., 1991; Whalen et al., 1991), fungal (Koch and Slusarenko, 1990; Mauch-Mani and Slusarenko, 1994), and viral pathogens (Melcher, 1989; Ishikawa et al., 1991; Leisner and Howell, 1992; Simon et al., 1992; Dempsey et al., 1993; H. Takahashi et al., 1994) and some of its resistance genes have been cloned and characterized (Dangl, 1995; Staskawicz et al., 1995). As in other plant systems, SA appears to be an integral part of the *Arabidopsis* defense response: (1) SA levels increase in inoculated leaves of resistant plants after infection with a pathogen (Uknes et al., 1993b; Summermatter and Métraux, 1994;

Dempsey and Klessig, unpubl.). (2) Induction of SAR is correlated with increases in SA in the uninoculated leaves (Summermatter et al., 1994). (3) SA treatment induces PR genes (Uknes et al., 1992; Dempsey and Klessig, unpubl.). (4) *Arabidopsis* contains SA-binding activity and a SA-inhibitable catalase activity (Sánchez-Casas and Klessig, 1994). (5) Transgenic plants expressing the *nah G* gene show enhanced susceptibility to pathogens (Delaney et al., 1994).

Dong and colleagues have isolated mutants with increased resistance to the pathogen *P. syringae*. These plants have elevated levels of SA and SAG in the absence of any treatment and constitutively express PR genes. Thus, these mutants appear to be constitutively "on" for at least a subset of defense responses (Bowling et al., 1994). Lawton and coworkers (1993) have also identified mutants with aberrant defense gene expression. These mutants can be divided into two classes—those which constitutively express PR genes and those which fail to activate these genes in response to chemical inducers such as SA or INA. The latter class of mutants, which have been isolated and characterized by several groups (Cao et al., 1994; Delaney et al., 1995; Shah and Klessig, unpubl.), exhibit increased susceptibility to pathogens. Century et al. (1995) have identified a third class of mutants (*ndr*) which are susceptible to strains of *P. syringae* pv. tomato and *Peronospora parasitica* that are avirulent on the wild type host. Characterization of the prototypic ndr1-1 mutant suggests that a common step exists in pathways of resistance to these two unrelated pathogens.

Another class of mutants that have recently been isolated and characterized are lesion mimic mutants. These mutants develop necrotic lesions in the absence of pathogen in a developmentally- and environmentally-influenced manner. Lesion mimic mutants have been previously described in maize (reviewed in Walbot, 1991), barley (reviewed in Jorgensen, 1992), and tomato (Langford, 1945), but their biochemical and molecular characterization has been rather limited. The *Arabidopsis* mutants have been characterized in more detail, particularly with respect to biochemical and molecular changes that are known to occur during a defense reaction to pathogen infection. Greenberg and Ausubel (1993; Greenberg et al., 1994) isolated two sets (complementation groups) of mutants which exhibit accelerated cell death (*acd*) in response to pathogens, as well as formation of lesions in the absence of pathogens. The *acd 1* mutants showed increased susceptibility to pathogens including opportunistic pathogens and pathogens of other plant species that normally do not infect *Arabidopsis*. In contrast, *acd 2* and several similar groups of mutants (four complementation groups), termed lesion simulating disease (*lsd*) mutants (Dietrich et al., 1994), appear to have enhanced resistance to pathogens. In the absence of pathogen, or after inoculation with a normally virulent pathogen, the *acd 2* and several of the *lsd* mutants form lesions that are indistinguishable, by a number of criteria, from necrotic lesions formed after infection by an avirulent pathogen. The lesion-bearing tissue also exhibits other characteristics of plants undergoing an HR, including phytoalexin production, cell wall modification, elevated SA levels and PR gene expression. In addition, the healthy (lesion free) portions of the plant show a number of the characteristics of SAR such as induction of PR genes, elevated SA levels and enhanced resistance to pathogens. In both *acd* and *lsd* mutants, control of necrosis during lesion formation is frequently lost, resulting in spreading lesions that eventually consume the entire leaf.

The large number of complementation groups (at least six; Greenberg et al., 1994; Dietrich et al., 1994) for lesion mimic mutants in *Arabidopsis* suggests that either the pathway for lesion formulation contains many steps or that there are multiple pathways. Epistasis analysis of these complementation groups should facilitate determination of the order of events within a particular pathway. Cloning and characterization of these genes will undoubtedly increase our understanding of the process of lesion formation, development of resistance and SA's role in these events.

IX. Perspectives

It is now clear that SA plays an important signaling function in disease resistance and thermogenesis. In disease resistance it appears to act by inhibiting catalase and APX activities and inducing lipid peroxidation; lipid peroxidation is probably initiated by a salicylate free radical produced during SA's inhibition of these two H_2O_2-scavanging enzymes. But it may have other modes of action in plant defense and/or other processes such as thermogenesis. In animals, SA and its derivative aspirin have a wide range of effects, only some of which appear to be attributable to their inhibition of prostaglandin biosynthesis (for review, see Weissman, 1991). Similarly, SA may have several mechanisms of action in plants and may participate in many other processes.

Even in disease resistance, where SA's role is best understood, most of the components of the signal transduction pathway(s) in which SA participates have yet to be defined. The cloning and characterization of resistance (R) genes (Johal and Briggs, 1992; Martin et al., 1993; Bent et al., 1994; Jones et al., 1994; Mindrinos et al., 1994; Whitham et al., 1994; Grant et al., 1995; Lawrence et al., 1995) hopefully will provide information about the events that initiate these pathways. Recently, mutational analysis in tomato has defined additional genes besides the R genes which are required for full resistance to *Cladosporium fulvum* (Hammond-Kosack et al., 1994) or to *P. syringae* pv. tomato (Salmeron et al., 1994). These genes, which are distinct from their respective R genes, presumably encode other components of the signal transduction cascade. The *Arabidopsis* lesion mimic mutants, *acd* (Greenberg and Ausubel, 1993; Greenberg et al., 1994) and *lsd* (Dietrich et al., 1994), should facilitate dissection of the pathways to HR. Mutants of *Arabidopsis* which constitutively express several of the defense-related PR genes (Lawton et al., 1993; Bowling et al., 1994) are insensitive to inducers of PR genes (Lawton et al., 1993; Cao et al., 1994; Delaney et al., 1995), or are susceptible to a variety of normally avirulent pathogens (Century et al., 1995) will also help define components of the signaling cascade.

Dissection of pathways will be facilitated by identification of different markers for the various steps. The defense induction pathway in TMV-infected tobacco is best characterized in this respect. Production of both SA and ethylene is induced after infection and is preceded by a temperature sensitive step (Van Loon and Antoniw, 1982; Yalpani et al., 1991; Malamy et al., 1992). Three chemical inducers of PR genes and acquired resistance (polyacrylic acid, thiamine-HCl, and INA) have also been shown to enter the pathway at different points (Malamy et al., unpubl.). These markers will not only aid in defining the sequence of events but will provide criteria for identification of mutants defective at different steps in the signal cascade.

Characterization of SA-inducible genes also provides some insights into the likely order of events after production of SA. Since it has been demonstrated that SA can rapidly activate genes with an *ocs/as-1*-type element in their promoter and that this induction does not require protein synthesis (Qin et al., 1994), a gene(s) with an *ocs/as-1*-containing promoter is the likely immediate target for SA activation. The product of this gene, then, may activate (directly or indirectly) PR genes and perhaps other SA-inducible genes whose activation is relatively slow and requires protein synthesis. Given the precedence in animal systems of the induction of defense (immune) genes by H_2O_2-mediated activation of NF-κB, which occurs post-transcriptionally, an NF-κB-like transcription factor may participate (directly or indirectly) in activation of SA-responsive genes.

The signal transduction pathway involving SA is likely to be very complex. There is evidence for extensive interplay between SA and other signals such as ethylene and jasmonic acid. For example, ethylene action inhibitors prevent SA induction of PR-3 genes in tobacco (Raz and Fluhr, 1993). While not required for SA induction of PR genes in *Arabidopsis*, ethylene potentiates their response (Lawton et al., 1994). Thus, ethylene seems to effect the plant's ability to respond to SA. In turn, SA inhibits ethylene synthesis in suspension cells of several species (Roustan et al., 1990; Romani et al., 1989; Leslie and Romani, 1988). The close analog of SA, acetylsalicylic acid (aspirin), blocks jasmonic acid biosynthesis (Peña-Cortés et al., 1993), consistent with the observation that SA and aspirin prevent expression of several wound-induced genes (Doherty et al., 1988; Peña-Cortés et al., 1993), whose expression is activated by jasmonic acid and its methyl ester (Farmer and Ryan, 1990; Farmer et al., 1992; Peña-Cortés et al., 1992).

Perhaps one of the most exciting recent results is the discovery that ROS play roles in plant defense at several points following infection. SA appears to induce defense responses, at least in part, by inhibiting catalase and APX which leads to increases in ROS levels. Interestingly, this process might be initiated by prior ROS production. There is speculation and some correlative evidence to suggest that necrosis during the HR is due, at least in part, to an early oxidative burst. A rapid oxidative burst occurs from 10 min to 4 h after shifting TMV-infected tobacco from 30 °C (restrictive temperature for HR formation) to 20 °C (permissive temperature for HR formation). The amount of ROS produced correlates with the number of lesions formed. Infiltration of the infected leaves with catalase or superoxide dismutase, both of which reduce levels of ROS, depressed lesion formation upon temperature shift (Doke and Ohashi, 1988). Since this early oxidative burst appears to be required for HR, and since the HR has been linked to induction of PAL, a key enzyme in the synthesis of SA, the early oxidative burst may indirectly cause another rise in ROS levels which is mediated by SA.

The association of elevated levels of ROS with HR (e.g., Levine et al., 1994) is also consistent with the view that the HR is a form of programmed cell death (apoptosis) which is being extensively characterized in animal systems. Korsmeyer and colleagues have presented data that suggest ROS may play a role in mammalian apoptosis (Veis et al., 1993, Hockenbery et al., 1993). Mice deficient in *Bcl-2*, an anti-apoptosis gene, suffer a large number of defects which are consistent with elevated levels of ROS. In tissue culture, treatment of cells with H_2O_2 leads to cell death in a manner characteristic of apoptosis. However, a cultured murine cell line over-expressing *Bcl-2* is resistant to normally lethal amounts of added H_2O_2. Further characterization of this resistance suggests that *Bcl-2* does not prevent generation of peroxides but rather acts by blocking ROS-mediated lipid peroxidation. In plants the HR is accompanied by an increase in ROS and lipid peroxidation. In fact, lipid peroxidation appears to be a key signal in induction of phytoalexin biosynthesis (Degousée et al., 1994) and PR protein synthesis (Chen and Klessig, unpubl.), hallmarks of the HR. These similarities suggest that the HR and apoptosis may occur through related mechanisms.

In mammals, apoptosis is mediated by activation of a cysteine protease, termed ICE. ICE also converts pro-interleukin-1β to active IL-1β, which can alert/activate the immune system (Vaux et al., 1994). Given the presence of an analogous gene product (ced-3) in the nematode, *C. elegans*, perhaps plants contain a comparable protease which mediates the HR and produces the mobile signal for activation of SAR.

Interestingly, in animals apoptosis can be induced not only by environmental stresses such as oxidative stress and γ-irradiation, but also by viral infection. During infection, apoptosis appears to be used to eliminate potentially harmful cells and as such can be viewed as a defense response (Vaux et al., 1994), similar to the HR in plants. Plant cells undergoing the HR exhibit some of the other hallmarks of apoptotic cell death, such as DNA fragmentation (Mittler et al., 1995).

One very practical result of this research will be the application to crop protection. A more complete understanding of the role of SA in plant defense responses may allow manipulation of the plant either genetically, by engineering constitutive expression of defense responses, or chemically, by treating the plants with compounds that will induce defense responses. The end result should be substantially decreased crop loss to pathogens. Moreover, further insights into SA and its signal transduction pathway in disease may have implications beyond the plant world given (*i*) the broad physiological effects of SA (and aspirin) in animal systems and (*ii*) the interesting parallels between the HR in plants and apoptosis in animals.

Acknowledgements

The authors would like to thank members of the Klessig laboratory for their reading and criticism of this manuscript. We would also like to extend our heartfelt thanks to the many researchers who contributed their results prior to publication. The authors' studies described in this review were supported in part by grants DCB-9003711 and MCB-9310371 from the National Science Foundation and 92-37301-7599 from the Department of Argiculture to DFK. KKW is a recipient of post-doctoral fellowship BIR-9203814 from the National Science Foundation.

X. References

Abate C, Patel L, Rauscher FJ III, Curran T (1990) Redox regulation of fos and jun DNA-binding activity in vitro. Science 249: 1157–1161

Åberg B (1981) Plant growth regulators XLI. Monosubstituted benzoic acids. Swedish J Agric Res 11: 913–105

Ahl P, Gianinazzi S (1982) b-Protein as a constitutive component of highly (TMV) resistant interspecific hybrid of *Nicotiana glutinosa* × *Nicotiana debneyi*. Plant Sci Lett 26: 173–181

Ahl-Goy P, Felix G, Métraux J-P, Meins F (1992) Resistance to disease in the hybrid *Nicotiana glutanosa* × *Nicotiana debneyi* is associated with high constitutive levels of β-1,3-glucanase, chitinase, peroxidase and polyphenoloxidase. Physiol Mol Plant Pathol 41: 11–21

Alexander D, Goodman RM, Gut-Rella M, Glascock C, Weymann K, Friedrich L, Maddox D, Ahl-Goy P, Luntz T, Ward E, Ryals J (1993) Increased tolerance to two oomycete pathogens in transgenic tobacco expressing pathogenesis-related protein 1a. Proc Natl Acad Sci USA 90: 7327–7331

Antoniw JF, White RF (1980) The effects of aspirin and polyacrylic acid on soluble leaf proteins and resistance to virus infection in five cultivars of tobacco. Phytopathol Z 98: 331–341

Apostol I, Heinstein PF, Low PS (1989) Rapid stimulation of an oxidative burst during elicitation of cultured plant cells. Plant Physiol 90: 109–116

Asselin A, Grenier J, Coté F (1985) Light-influenced extracellular accumulation of b (pathogenesis-related) proteins in Nicotiana green tissue induced by various chemicals on prolonged floating on water. Can J Bot: 1276–1282

Baker CJ, O'Neill NR, Keppler LD, Orlandi EW (1991) Early responses during plant-bacteria interactions in tobacco cell suspensions. Phytopathology 81: 1504–1507

Beilmann A, Albrecht K, Schultze S, Wanner, Pfitzner UM (1992) Activation of a truncated PR-1 promoter by endogenous enhancers in transgenic plants. Plant Mol Biol 18: 65–78

Bent AF, Kunkel BN, Dahlbeck D, Brown KL, Schmidt R, Giraudot J, Leung J, Staskawicz BJ (1994) RPS2 of *Arabidopsis thaliana*: a leucine-rich repeat class of plant disease resistance genes. Science 265: 1856–1860

Ben-Tal Y, Cleland CF (1982) Uptake and metabolism of [^{14}C] salicylic acid in *Lemna gibba* G3. Plant Physiol 70: 291–296

Bi YM, Kenton P, Mur L, Darby R, Draper J (1995) Hydrogen peroxide does not function downstream of salicylic acid in the induction of PR protein expression. Plant J 8: 235–245

Bouchez D, Tokuhisa JG, Llewellyn DJ, Dennis ES, Ellis JG (1989) The ocs element is a component of the promoters of several T-DNA and plant viral genes. EMBO J 8: 4197–4204

Bowles D (1990) Defense-related proteins in higher plants. Annu Rev Biochem 59: 873–907

Bowling SA, Guo A, Cao H, Gordon AS, Klessig DF, Dong X (1994) A mutation in *Arabidopsis* that leads to constitutive expression of systemic acquired resistance. Plant Cell 6: 1845–1857

Bradley DJ, Kjellbom P, Lamb CJ (1992) Elicitor- and wound-induced oxidative cross-linking of a proline-rich plant cell wall protein: a novel, rapid defense response. Cell 70: 21–30

Broglie K, Chet I, Holliday M, Cressman R, Biddle P, Knowlton S, Mauvais CJ, Broglie R (1991) Transgenic plants with enhanced resistance to the fungal pathogen *Rhizoctonia solanni*. Science 254: 1194–1197

Buchel AS, Molenkanp R, Bol JF, Linthorst HJM (1994) The PR-1a promoter contains a number of regulatory elements that bind GT-1 like factors with different affinity. J Cell Biochem 21A: 485

Cao H, Bowling SA, Gordon S, Dong X (1994) Characterization of an *Arabidopsis* mutant that is nonresponsive to inducers of systemic acquired resistance. Plant Cell 6: 1583–1592

Century KS, Holub EB, Staskawicz BJ (1995) NDRI, a locus of *Arabidopsis thaliana* that is required for disease resistance to both a bacterial and a fungal pathogen. Proc Natl Acad Sci USA 92: 6597–6601

Chadha KC, Brown SA (1974) Biosynthesis of phenolic acids in tomato plants infected with *Agrobacterium tumefaciens*. Can J Bot 52: 2041–2046

Chai HB, Doke N (1987) Superoxide anion generation: a response of potato leaves to infection with *Phytophthora infestans*. Phytopathology 77: 645–649

Chen Z, Klessig DF (1991) Identification of a soluble salicylic acid-binding protein that may function in signal transduction in the plant disease resistance response. Proc Natl Acad Sci USA 88: 8179–8183

Chen Z, Ricigliano J, Klessig DF (1993a) Purification and characterization of a soluble salicylic acid-binding protein from tobacco. Proc Natl Acad Sci USA 90: 9533–9537

Chen Z, Silva H, Klessig DF (1993b) Active oxygen species in the induction of plant systemic acquired resistance by salicylic acid. Science 262: 1883–1886

Chen Z, Malamy J, Hennig J, Conrath U, Sánchez-Casas P, Silva H, Ricigliano J, Klessig DF (1995) Induction, modification, and transduction of the salicylic acid signal in plant defense responses. Proc Natl Acad Sci USA 92: 4134–4137

Cohen JD, Bandurski RS (1982) Chemistry and physiology of the bound auxins. Annu Rev Plant Physiol 33: 403–430

Conn EE (1984) Compartmentation of secondary compounds. In: Boudet AM, Alibert G, Marigo G, Lea PJ (eds) Annual Proceeding of the Phytochemical Society of Europe, vol 24, membranes and compartmentation in the regulation of plant functions. Clarendon, Oxford, pp 1–28

Conrath U, Chen Z, Ricigliano J, Klessig DF (1995) Two inducers of plant defense responses, 2,6-dichloroisonicotinic acid and salicylic acid, inhibit catalase activity in tobacco. Proc Natl Acad Sci USA 92: 7143–7141

Cooper-Driver G, Corner-Zamodits JJ, Swain T (1972) The metabolic fate of hydroxybenzoic acids in plants. Z Naturforsch B 27: 943–946

Cutt JR, Klessig DF (1992) Pathogenesis-related proteins. In: Boller T, Meins F Jr (eds) Genes involved in plant defense. Springer, Wien New York, pp 209–243 [Dennis ES et al (eds) Plant gene research. Basic knowledge and application]

Dangl JL (1995) Pièce de résistance: novel classes of plant disease resistance genes. Cell 80: 363–366

Davis KR, Schott E, Ausubel FM (1991) Virulence of selected phytopathogenic *Pseudomonas* in *Arabidopsis thaliana*. Mol Plant Microbe Interact 4: 477–488

Debener T, Lehnackers H, Arnold M, Dangl JL (1991) Identification and molecular mapping of a single *Arabidopsis thaliana* locus determining resistance to a phytopathogenic *Pseudomonas syringae* isolate. Plant J 1: 289–302

Degousée N, Triantaphylidès C, Montillet J-L (1994) Involvement of oxidative processes in the signaling mechanisms leading to the activation of glyceollin synthesis of soybean (*Glycine max*). Plant Physiol 104: 945–952

Delaney T, Ukness S, Vernooij B, Friedrich L, Weymann K, Negrotto D, Gaffney T, Gut-Rella M, Kessmann H, Ward E, Ryals J (1994) A central role of salicylic acid in plant disease resistance. Science 266: 1247–1250

Delaney TP, Friedrick L, Ryals JA (1995) *Arabidopsis* signal transduction mutant defective in chemically and biologically induced disease resistance. Proc Natl Acad Sci USA 92: 6602–6606

Dempsey DA, Klessig DF (1995) Signals in plant disease resistance. Bull Inst Pasteur 93: 167–186

Dempsey DA, Wobbe KK, Klessig DF (1993) Resistance and susceptible responses of *Arabidopsis thaliana* to turnip crinkle virus. Phytopathology 83: 1021–1029

Devary Y, Gottlieb RA, Lau LF, Karin M (1991) Rapid and preferential activation of the c-jun gene during the mammalian UV response. Mol Cell Biol 11: 2804–2811

Dietrich RA, Delaney TP, Uknes SJ, Ward ER, Ryals JA, Dangl JL (1994) *Arabidopsis* mutants simulating disease response. Cell 77: 565–577

Dixon RA, Harrison MJ (1990) Activation, structure and organization of genes involved in microbial defense in plants. Adv Genet 28: 165–234

Doherty HM, Selvendran RR, Bowles DJ (1988) The wound response of plants can be inhibited by aspirin and related hydroxy-benzoic acids. Physiol Mol Plant Pathol 33: 377–384

Doke N (1983a) Generation of superoxide anion by potato tuber protoplasts during the hypersensitive response to hyphal wall components of *Phytophthora infestans* and specific inhibition of the reaction by suppressors of hypersensitivity. Physiol Plant Pathol 23: 359–367

Doke N (1983b) Involvement of superoxide anion generation in the hypersensitive response of potato tuber tissues to infection with an incompatible race of *Phytophthora infestans* and to the hyphal wall components. Physiol Plant Pathol 23: 345–357

Doke N, Ohashi Y (1988) Involvement of O_2^--generating systems in the induction of necrotic lesions on tobacco leaves infected with TMV. Physiol Mol Plant Pathol 32: 163–175

Dong X, Mindrions M, Davis KR, Ausubel FM (1991) Induction of *Arabidopsis* defense genes by virulent and avirulent *Pseudomonas syringae* strains and by a cloned avirulence gene. Plant Cell 3: 61–72

Duchesne M, Fritig B, Hirth L (1977) Phenylalanine ammonia-lyase in tobacco mosaic virus-infected hypersensitive tobacco; density-labelling evidence of de novo synthesis. Biochim Biophys Acta 485: 465–481

Durner J, Klessig DF (1995) Inhibition of ascorbate peroxidase by salicylic acid and 2,6-dichloroisonicotinic acid, two inducers of plant defense responses. Proc Natl Acad Sci USA 92: 11312–11316

El-Basyouni SZ, Chen D, Ibrahim RK, Neish AC, Towers GHN (1964) The biosynthesis of hydroxybenzioc acids in higher plants. Phytochemistry 3: 485–492

Ellis JG, Tokuhisa JG, Llewellyn DJ, Bouchez D, Singh K, Dennis ES, Peacock WJ (1993) Does the ocs-element occur as a functional component of the promoters of plant genes? Plant J 4: 433–443

Enyedi AJ, Raskin I (1993) Induction of UDP-glucose: salicylic acid glucosyltransferase activity in tobacco mosaic virus-inoculated tobacco (*Nicotiana tabacum*) leaves. Plant Physiol 101: 1375–1380

Enyedi AJ, Yalpani N, Silverman P, Raskin I (1992a) Signal molecules in systemic plant resistance to pathogens and pests. Cell 70: 879–886

Enyedi A, Yalpani N, Silverman P, Raskin I (1992b) Localization, conjugation and function of salicylic acid in tobacco during the hypersensitive reaction to tobacco mosaic virus. Proc Natl Acad Sci USA 89: 2480–2484

Farmer EE, Ryan CA (1990) Interplant communication: airborne methyl jasmonate induces synthesis of proteinase inhibitor in plant leaves. Proc Natl Acad Sci USA 87: 7713–7716

Farmer EE, Johnson RR, Ryan CA (1992) Regulation of expression of proteinase inhibitor genes by methyl jasmonate and jasmonic acid. Plant Physiol 98: 995–1002

Fritig B, Gosse J, Legrande M, Hirth L (1973) Changes in phenylalanine ammonia-lyase during the hypersensitive reaction of tobacco to TMV. Virology 55: 371–379

Gaffney T, Friedrich L, Vernooij B, Negrotto D, Nye G, Uknes S, Ward E, Kessmann H, Ryals J (1993) Requirement of salicylic acid for the induction of systemic acquired resistance. Science 261: 754–756

Gianinazzi S (1970) Hypersensibilite aux virus, temperatures et proteines solubles chez le *Nicotiana tabacum* cv. Xanthi-nc. C R Acad Sci Paris D 270: 2382–2386

Gianinazzi S, Ahl P (1993) The genetic and molecular basis of b-proteins in the genus *Nicotiana*. Neth J Plant Pathol 89: 275–281

Gianinazzi S, Kassanis B (1974) Virus resistance induced in plants by polyacrylic acid. C R Acad Sci Paris D 270: 2382–2386

Goldsbrough AP, Albrecht H, Stratford R (1993) Salicylic acid-inducible binding of a tobacco nuclear protein to a 10 bp sequence which is highly conserved among stress-inducible genes. Plant J 3: 563–571

Grant MR, Godiard L, Straube E, Ashfield T, Lewald J, Sattler A, Innes RW, Dangl JL (1995) Structure of the *Arabidopsis* RPM1 gene enabling dual specificity disease resistance. Science 269: 843–846

Greenberg JT, Ausubel FM (1993) *Arabidopsis* mutants compromised for the control of cellular damage during pathogenesis and aging. Plant J 4: 27–341

Greenberg JT, Guo A, Klessig DF, Ausubel FM (1994) Programmed cell death in plants: a pathogen-triggered response activated coordinately with multiple defense functions. Cell 77: 551–563

Hagiwara H, Matsuoka M, Ohshima M, Watanabe M, Hosokawa D, Ohashi Y (1993) Sequence-specific binding of protein factors to two independent promoter regions of the acidic tobacco pathogenesis-related-1 protein gene (PR-1). Mol Gen Genet 240: 197–205

Hammond-Kosack KE, Jones DA, Jones JDG (1994) Identification of two genes required in tomato for full Cf-9-dependent resistance to *Cladosporium fulvum*. Plant Cell 6: 361–374

Harborne JB (1964) Phenolic glycosides and their natural distribution. In: Harborne JB (ed) Biochemistry of phenolic compounds. Academic Press, London, pp 129–169

Hennig J, Dewey RE, Cutt JR, Klessig DF (1993a) Pathogen, salicylic acid and developmental dependent expression of a β-1,3-glucanase/GUS gene fusion in transgenic tobacco plants. Plant J 4: 481–493

Hennig J, Malamy J, Grynkiewicz G, Indulski J, Klessig DF (1993b) Interconversion of the salicylic acid signal and its glucoside in tobacco. Plant J 4: 593–600

Hockenbery DM, Oltvai ZN, Yin X-M, Milliman CL, Korsmeyer SJ (1993) Bcl-2 functions in an antioxidant pathway to prevent apoptosis. Cell 75: 241–251

Ishikawa M, Obata F, Kumagai T, Ohno T (1991) Isolation of mutants of *Arabidopsis thaliana* in which accumulation of tobacco mosaic virus coat protein is reduced to low levels. Mol Gen Genet 230: 33–38

Johal G, Briggs S (1992) Reductase activity encoded by the MH1 disease resistance gene in maize. Science 258: 985–987

Jones DA, Thomas CM, Hammond-Kosack KE, Balint-Kurti PJ, Jones JDG (1994) Isolation of the tomato Cf-9 gene for resistance to *Cladosporium fulvum* by transposon tagging. Science 266: 789–793

Jorgensen JH (1992) Discovery, characterization and exploitation of Mlo powdery mildew resistance in barley. Euphytica 63: 141–152

Kapulnik Y, Yalpani N, Raskin I (1992) Salicylic acid induces cyanide-resistant respiration in tobacco cell-suspension cultures. Plant Physiol 100: 1921–1926

Kassanis B (1952) Some effects of high temperature on the susceptibility of plant to infection with viruses. Ann Appl Biol 39: 358–369

Katagiri F, Lam E, Chua N-H (1989) Two tobacco DNA-binding proteins with homology to the nuclear factor CREB. Nature 340: 727–730

Keppler LD, Baker CJ (1989) O_2^--initiated lipid peroxidation in a bacteria-induced hypersensitive reaction in tobacco cell suspensions. Phytopathology 79: 555–562

Keppler LD, Novacky A (1986) Involvement of membrane lipid peroxidation in the development of a bacterially induced hypersensitive reaction. Phytopathology 76: 104–108

Keppler LD, Baker CJ, Atkinson MM (1989) Active oxygen production during a bacteria induced hypersensitive reaction in tobacco suspension cells. Phytopathology 79: 974–978

Kirkman HN, Gaetani GF (1984) Catalase: a tetrameric enzyme with four tightly bound molecules of NADPH. Proc Natl Acad Sci USA 81: 4343–4347

Kim SR, Yim, Y, An G (1993) Identificaton of methyl jasmonate and salicylic acid response elements from the nopaline synthase (nos) promoter. Plant Physiol 103: 97–103

Klessig DF, Malamy J (1994) The salicylic acid signal in plants. Plant Mol Biol 26: 1439–1458

Koch E, Slusarenko A (1990) *Arabidopsis* is susceptible to infection by a downy mildew fungus. Plant Cell 2: 437–445

Lam E, Benfey P, Gilmarten PM, Fang R-X, Chua N-H (1989) Site-specific mutations alter in vitro factor binding and change promoter expression patterns in transgenic plants. Proc Natl Acad Sci USA 86: 7890–7894

Langford AN (1948) Autogenous necrosis in tomatoes immune from *Cladosporium fulvum* Cooke. Can J Res 26: 35–64

Lawrence GJ, Finnegan EJ, Ayliffe MA, Ellis JG (1995) The L6 gene for flax rust resistance is related to the *Arabidopsis* bacterial resistance gene RPS2 and the tobacco viral resistance gene N. Plant Cell 7: 1195–1206

Lawton KA, Uknes S, Friedrich L, Gaffney T, Alexander D, Goodman R, Métraux JP, Kessmann H, Ahl-Goy P, Gut-Rella M, Ward D, Ryals J (1993) The molecular biology of systemic acquired resistance. In: Fritig B, Legrande M (eds) Developments in plant biology, mechanisms of plant defense responses. Kluwer, Dordrecht, pp 422–432

Lawton KA, Potter SL, Uknes S, Ryals J (1994) Acquired resistance signal transduction in *Arabidopsis* is ethylene independent. Plant Cell 6: 581–588

Legendre L, Rueter S, Heinstein PF, Low PS (1993) Characterization of the oligogalacturonide-induced oxidative burst in cultured soybean (*Glycine max*) cells. Plant Physiol 102: 233–240

Legrande M, Fritig B, Hirth L (1976) Enzymes of the phenylpropanoid pathway and the necrotic reaction of hypersensitive tobacco to TMV. Phyochemistry 15: 1353–1359

Leisner SM, Howell SH (1992) Symptom variation in different *Arabidopsis thaliana* ecotypes produced by cauliflower mosaic virus. Phytopathology 82: 1042–1046

León J, Yalpani N, Raskin I, Lawton MA (1993) Induction of benzoic acid 2-hydroxylase in virus-inoculated tobacco. Plant Physiol 103: 323–328

León J, Lawton MA, Raskin I (1995) Hydrogen peroxide stimulates salicylic acid biosynthesis in tobacco. Plant Physiol 108: 1673–1678

Leslie CA, Romani RJ (1988) Inhibition of ethylene biosynthesis by salicylic acid. Plant Physiol 88: 833–837

Letham DS, Palni LMS (1983) The biosynthesis and metabolism of cytokinins. Annu Rev Plant Physiol 34: 163–197

Levine A, Tenhaken R, Dixon R, Lamb C (1994) H_2O_2 from the oxidative burst orchestrates the plant hypersensitive disease resistance response. Cell 79: 583–593

Linthorst HJM (1991) Pathogenesis-related proteins of plants. Crit Rev Plant Sci 10: 123–150

Liu D, Raghothama KG, Hasegawa PM, Bressan RA (1994) Osmotin overexpression in potato delays development of disease symptoms. Proc Natl Acad Sci USA 91: 1888–1892

Liu X, Lam E (1994) Two binding sites for the plant transcription factor ASF-1 can respond to auxin treatments in transgenic tobacco. J Biol Chem 269: 668–675

McGurl B, Pearce G, Orizco-Cardensa M, Ryan C (1992) Structure, expression and antisense inhibition of the systemin precursor gene. Science 255: 1570–1573.

Madamanchi NR, Kuć J (1991) Induced systemic resistance in plants. In: Cole GT, Hoch HC (eds) The fungal spore and disease initiation in plants and animals. Plenum, New York, pp 347–362

Malamy J, Klessig DF (1992) Salicylic acid and plant disease resistance. Plant J 2: 643–654

Malamy J, Carr JP, Klessig DF, Raskin I (1990) Salicylic acid—a likely endogenous signal in the resistance response of tobacco to viral infection. Science 250: 1001–1004

Malamy J, Hennig J, Klessig DF (1992) Temperature-dependent induction of salicylic acid and its conjugates during the resistance response to tobacco mosaic virus infection. Plant Cell 4: 359–366

Martin GB, Brommonschenkel SH, Chunwongse J, Frary S, Ganal MW, Spivey R, Wu T, Earle ED, Tanksley SD (1993) Map-based cloning of a protein kinase gene conferring disease resistance in tomato. Science 262: 1432–1436

Matthews REF (1991) Plant virology, 3rd edn. Harcourt Brace Jovanovich, San Diego

Mauch F, Mauch-Mani B, Boller T (1988) Antifungal hydrolases in pea tissue. II. Inhibition of fungal growth by combinations of β-1,3 glucanase and chitinase. Plant Physiol 88: 936–942

Mauch-Mani B, Slusarenko, A (1994) Systemic acquired resistance in *Arabidopsis thaliana* induced by a predisposing infection with a pathogenic isolate of *Fusarium oxysporum*. Mol Plant Microbe Interact 7: 378–383

May JM, deHaven C (1979) Insulin stimulated intracellular hydrogen peroxide production in rat epididyamal fat cells. J Biol Chem 254: 2214–2220

Meeuse BJD (1975) Thermogenic respiration in aroids. Annu Rev Plant Physiol 26: 117–126

Melcher U (1989) Symptoms of cauliflower mosaic virus infection in *Arabidopsis thaliana* and turnip. Bot Gaz 150: 139–147

Métraux JP, Signer H, Ryals J, Ward E, Wyss-Benz M, Gaudin J, Raschdorf K, Schmid E, Blum W, Inverardi B (1990) Increase in salicylic acid at the onset of systemic acquired resistance in cucumber. Science 250: 1004–1006

Meuwly P, Mölders W, Summermatter K, Sticher L, Métraux JP (1994) Salicylic acid and chitinase in infected cucumber plants. Acta Hort 1: 371–374

Miao Z-H, Liu X, Lam E (1994) TGA3 is a distinct member of the TGA family of bZIP transcription factors in *Arabidopsis thaliana*. Plant Mol Biol 9: 1–11

Mills PR, Wood RKS (1984) The effects of polyacrylic acid, aspirin and salicylic acid on resistance of cucumber to *Colletotrichum lagenarium*. Phytopathol Z 111: 209–216

Minagawa N, Koga S, Nakano M, Sukajo S, Yoshimoto A (1992) Possible involvement of superoxide anion in the induction of cyanide-resistant respiration in *Hansenula anomala*. FEBS 302: 217–219

Mindrinos M, Katagiri F, Yu G-L, Ausubel FM (1994) The *A. thaliana* disease resistance gene RPS2 encodes a protein containing a nucleotide-binding site and leucine-rich repeats. Cell 78: 1089–1099

Mittler R, Shulaev V, Lam E (1995) Coordinated activation of programmed cell death and defense mechanism in transgenic plants expressing a bacterial proton pump. Plant Cell 7: 39–42

Murphy TM, Raskin I, Enyedi AJ (1993) Plasma membrane effects of salicylic acid treatment on cultured rose cells. Environ Exp Bot 33: 267–272

Neuenschwander U, Vernooij B, Friedrich L, Uknes S, Kessmann H, Ryals J (1995) Is hydrogen peroxide a second messenger of salicylic acid in systemic acquired resistance? Plant J 8: 221–233

Ohashi Y, Ohshima M, Itoh H, Matsuoka M, Watanabe S, Murakami T, Hosokawa D (1992) Constitutive expression of stress-inducible genes, including pathogenesis-related 1 protein gene in a transgenic interspecific hybrid of *Nicotiana glutinosa* × *Nicotiana debneyi*. Plant Cell Physiol 33: 177–18

Ohshima M, Itoh H, Matsuoka M, Murakami T, Ohashi Y (1990) Analysis of stress-induced or salicylic acid-induced expression of the pathogenesis-related 1a protein gene in transgenic tobacco. Plant Cell 2: 95–106

Orlandi EW, Hutcheson SW, Baker CJ (1992) Early physiological responses associated with race-specific recognition in soybean leaf tissue and cell suspensions treated with *Pseudomonas syringae* pv. *glycinea*. Physiol Mol Plant Pathol 40: 173–180

Palva TK, Hurtig M, Saindrenan P, Palva ET (1994) Salicylic acid induced resistance to *Erwinia carotovora* subsp. *carotovora* in tobacco. Mol Plant Microbe Interact 7: 356–363

Pegg GF (1976) The involvement of ethylene in plant pathogenesis. In: Heitefuss R, Williams PH (eds) Physiological plant pathology. Springer, Berlin Heidelberg New York, pp 582–591 [Pirson A, Zimmermann MH (eds) Encyclopedia of plant physiology, NS, vol 4]

Peña-Cortés H, Liu X, Sanchez-Serrano JJ, Schmid R, Willmitzer L (1992) Factor affecting gene expression of patatin and proteinase-inhibitor-II gene families in detached potato leaves: implications for their co-expression in developing tubers. Planta 186: 495–502

Peña-Cortés H, Albrecht T, Prat S, Water EW, Willmitzer L (1993) Aspirin prevents wound-induced gene expression in tomato leaves by blocking jasmonic acid biosynthesis. Planta 191: 123–128

Ponstein AS, Bres-Vloemans SA, Sela-Buurlage MB, van den Elzen PJM, Melchers LS, Cornelissen BJC (1994) A novel pathogen- and wound-inducible tobacco (*Nicotiana tabacum*) protein with antifungal activity. Plant Physiol 104: 109–118

Prasad TK, Anderson MD, Martin BA, Stewart CR (1994) Evidence for chilling-induced oxidative stress in maize seedlings and a regulatory role for hydrogen peroxide. Plant Cell 6: 65–74

Qin XF, Holuigue L, Horvath DM, Chua N-H (1994) Immediate early transcription activation by salicylic acid via the cauliflower mosaic virus *as-1* element. Plant Cell 6: 863–874

Rainsford KD (1984) Aspirin and the salicylates. Butterworth, London

Raskin I (1992) Role of salicylic acid in plants. Annu Rev Plant Physiol Plant Mol Biol 43, 439–463.

Raskin I, Ehmann A, Melander WR, Meeuse BJD (1987) Salicylic acid—a natural inducer of heat production in Arum lilies. Science 237: 1601–1602

Raskin I, Turner IM, Melander WR (1989) Regulation of heat production in the inflorescences of an arum lily by endogenous salicylic acid. Proc Natl Acad Sci USA 86: 2214–2218

Raskin I, Skubatz H, Tang W, Meeuse, BJD (1990) Salicylic acid levels in thermogenic and non-thermogenic plants. Ann Bot 66: 369–373

Rasmussen JB, Hammerschmidt R, Zook M (1991) Systemic induction of salicylic acid accumulation in cucumber after inoculation with *Pseudomonas syringae* pv. *syringae*. Plant Physiol 97: 1342–1347

Raz V, Fluhr R (1992) Calcium requirement for ethylene-dependent responses. Plant Cell 4: 1123–1130.

Reinecke DM, Bandurski RS (1988) Auxin biosynthesis and metabolism. In: Davis PJ (ed) Plant hormones and their role in plant growth and development. Martinus Nijhoff, Dordrecht, pp 24–42

Rhoades DM, McIntosh L (1992) Salicylic acid regulation of respiration in higher plants: alternative oxidase expression. Plant Cell 4: 1131–1139

Rhoades DM, McIntosh L (1993) Cytochrome and alternative pathway respiration in tobacco; effects of salicylic acid. Plant Physiol 103: 877–883

Roggero P, Pennazio S (1988) Effects of salicylate on systemic invasion of tobacco plants by various viruses. J Phytopathol 123: 207–216

Roggero P, Pennazio S (1991) Salicylate does not induce resistance to plant viruses or stimulate pathogenesis-related protein production in soybean. Microbiologica, 14: 65–69

Romani RJ, Hess BM, Leslie CA (1989) Salicylic acid inhibition of ethylene production by apple discs and other plant tissues. J Plant Growth Regul 8: 63–70

Roustan JP, Latche A, Fallot J (1990) Inhibition of ethylene production and stimulation of carrot somatic embryogenesis by salicylic acid. Biol Plant 32: 273–276

Ryals J, Ward E, Ahl-Goy P, Métraux JP (1992) Systemic acquired resistance: an inducible defense mechanism in plants. In: Wray JL (ed) Inducible plant proteins: their biochemistry and molecular biology. Cambridge University Press, New York, pp 205–229 (Society for Experimental Biology seminar series, vol 49)

Saint-Pierre B, Miville L, Dion P (1984) The effects of salicylates on phenomena related to crown gall. Can J Bot 62: 729–734

Salmeron JM, Barker SJ, Carland FM, Mehta AY, Staskawicz BJ (1994) Tomato mutants altered in bacterial disease resistance provide evidence for a new locus controlling pathogen recognition. Plant Cell 6: 511–520

Samac DA, Shah DM (1991) Developmental and pathogen-induced activation of the *Arabidopsis* acidic chitinase promoter. Plant Cell 3: 1063–1072

Sánchez-Casas P, Klessig DF (1994) Salicylic acid-binding activity correlates with inhibition by salicylic acid of catalase acivity in various plants. Plant Physiol 106: 1675–1679

Schenk H, Klein M, Erdbrügger W, Dröge W, Schulze-Osthoff K (1994) Distinct effects of thioredoxin and antioxidants on the activation of transcription factors NF-κB and AP-1. Proc Natl Acad Sci USA 91: 1672–1676

Schreck R, Baeuerle PA (1991) A role for oxygen radicals as second messengers. Trends Cell Biol 1: 39–42

Schreck R, Rieber P, Baeuerle PA (1991) Reactive oxygen intermediates as apparently widely used messengers in the activation of the NF-κB transcription factor and HIV-1. EMBO J 10: 2247–2258

Shulaev V, León J, Raskin I (1995) Is salicylic acid a translocated signal of systemic acquired resistance in tobacco? Plant Cell 7: 1691–1701

Silverman P, Nuckles E, Ye XS, Kuć J, Raskin I (1993) Salicylic acid, ethylene, and pathogen resistance in tobacco. Mol Plant Microbe Interact 6: 775–781

Simon AE, Li XH, Lew JE, Stange R, Zhang C, Polacco M, Carpenter CD (1992) Susceptibility and resistance of *Arabidopsis thaliana* to turnip crinkle virus. Mol Plant Microbe Interact 5: 496–503

Simpson RB, Johnson LJ (1990) *Arabidopsis thaliana* as a host for *Xanthomonas campestris* pv. *campestris*. Mol Plant Microbe Interact 3: 233–237

Singh L (1978) In vitro screening of some chemicals against three phytopathogenic fungi. J Ind Bot Soc 57: 191–195

Smith JA, Hammerschmidt R, Fulbright DW (1991) Rapid induction of systemic resistance in cucumber by *Pseudomonas syringae* pv. *syringae*. Physiol Mol Plant Pathol 38: 223–235

Somssich IE (1994) Several regulatory elements governing pathogenesis-related gene expression. In: Nover L (ed) Plant promoters and transcription factors. Springer, Berlin Heidelberg New York Tokyo, pp 163–179 [Henning W et al (eds) Results and problems in cell differentiation, vol 20]

Staskawicz BJ, Ausubel FM, Baker BJ, Ellis JG, Jones JDG (1995) Molecular genetics of plant disease resistance. Science 268: 661–667

Stein B, Rahmsdorf HJ, Steffen A, Litfin M, Herrlich P (1989) Uv-induced DNA damage is an intermediate step in UV-induced expression of human immunodeficiency virus type 1, collagenase, c-fos and metallothionein. Mol Cell Biol 9: 5169–5181

Sutherland MW (1991) The generation of oxygen radicals during host plant responses to infection. Physiol Mol Plant Pathol 39: 79–93

Summermatter K, Meuwly P, Mölders W, Métraux JP (1994) Salicylic acid levels in *Arabidopsis thaliana* after treatments with *Pseudomonas syringae* or synthetic inducers. Acta Hort 1: 367–370

Takahashi H, Goto N, Ehara Y (1994) Hypersensitive response in cucumber mosaic virus-inoculated *Arabidopsis thaliana*. Plant J 6: 369–377

Takahashi Y, Niwa Y, Machida Y, Nagata T (1990) Location of the cis-acting auxin responsive region in the promoter of the par gene from tobacco mesophyll protoplasts. Proc Natl Acad Sci USA 87: 8013–8016

Tanaka S, Hayakawa K, Umetani Y, Tabata M (1990) Glucosylation of isomeric hydroxybenzoic acids by cell suspension cultures of *Mallotus japonicus*. Phytochemistry 29: 1555–1558

Towers GHN (1964) Metabolism of phenolics in higher plants and microorganisms. In: Harborne JB (ed) Biochemistry of phenolic compounds. Academic Press, London, pp 249–294

Tsuji J, Somerville, SC, Hammerschmidt R (1991) Identification of a gene in *Arabidopis thaliana* that controls resistance to *Xanthomonas campestris* pv. *campestris*. Physiol Mol Plant Pathol 38: 57–65

Uknes S, Mauch-Mani B, Moyer M, Potter S, Williams S, Dincher S, Chandler D, Slusarenko A, Ward E, Ryals J (1992) Acquired resistance in *Arabidopsis*. Plant Cell 4: 645–656

Uknes S, Dincher S, Friedrich L, Negrotto D, Williams S, Thompson-Taylor H, Potter S, Ward E, Ryals J (1993a) Regulation of pathogenesis-related protein-1a gene expression in tobacco. Plant Cell 5: 159–169

Uknes S, Winter AM, Delaney T, Vernooij B, Morse A, Friedrich L, Nye G, Potter S, Ward E, Ryals J (1993b) Biological induction of systemic acquired resistance in *Arabidopsis*. Mol Plant Microbe Interact 6: 692–698

Ulmasov T, Hagen G, Guilfoyle T (1994) The ocs element in the soybean GH2/4 promoter is activated by both active and inactive auxin and salicylic acid analogues. Plant Mol Biol 26: 1055–1064

Umetani Y, Kodakari E, Yamamura T, Tanaka S, Tabata M (1990) Glucosylation of salicylic acid by cell suspension cultures of *Mallotus japonicus*. Plant Cell Rep 9: 325–327

van de Rhee MD, Bol JF (1993) Induction of the tobacco PR-1a gene by virus infection and salicylate treatment involves an interaction between multiple regulatory elements. Plant J 3: 71–82

van de Rhee MD, Van Kan JAL, Gonzalez-Jaen MT, Bol JF (1990) Analysis of regulatory elements involved in the induction of two tobacco genes by salicylate treatment and virus infection. Plant Cell 2: 357–366

van de Rhee MD, Lemmers R, Bol JF (1993) Analysis of regulatory elements involved in stress-induced and organ-specific expression of tobacco acidic and basic β-1,3-glucanase genes. Plant Mol Biol 21: 451–461

van der Zaal EJ, Droog FNJ, Boot CJM, Hensgens LAM, Hoge JHC, Schilperoort RA, Libbenga KR (1991) Promoters of auxin-induced genes from tobacco can lead to auxin-inducible and root tip-specific gene expression. Plant Mol Biol 16: 983–998

Van Loon LC, Antoniw JF (1982) Comparison of the effects of salicylic acid and ethephon with virus-induced hypersensitivity and acquired resistance in tobacco. Neth J Plant Pathol 88: 237–256

Vaux DL, Haecker G, Strasser A (1994) An evolutionary perspective on apoptosis. Cell 76: 777–779

Veis DJ, Sorenson CM, Shutter JR, Korsmeyer SJ (1993) Bcl-2-deficient mice demonstrate fulminant lymphoid apoptosis, polycystic kidneys and hypopigmented hair. Cell 75: 229–240

Vernooij B, Friedrich L, Morse A, Reist R, Kolditz-Jawhar R, Ward E, Uknes S, Kessmann H, Ryals J (1994) Salicylic acid is not the translocated signal responsible for inducing systemic acquired resistance but is required in signal transduction. Plant Cell 6: 959–965

Vernooij B, Friedrich L, Ahl Goy P, Staub T, Kessmann H, Ryals J (1995) 2,6-Dichloroisonicotinic acid-induced resistance to pathogens without the accumulation of salicylic acid. Mol Plant Microbe Interact 8: 228–234

Vigers AJ, Roberts WK, Selitrennikoff CP (1991) A new family of plant antifungal proteins. Mol Plant Microbe Interact 4: 315–323.

Wagner AM (1995) A role for active oxygen species as second messenger in the induction of alternative oxidase gene expression in *Petunia hybrida* cells. FEBS 368: 339–342

Walbot V (1991) Maize mutants for the 21st century. Plant Cell 3: 851–856

Ward ER, Uknes SJ, Williams SC, Dincher SS, Wiederhold DL, Alexander DC, Ahl Goy P, Métraux JP, Ryals JA (1991) Coordinate gene activity in responses to agents that induce systemic acquired resistance. Plant Cell 3: 1085–1094

Weete JD (1992) Induced systemic resistance to *Alternaria cassiae* in sicklepod. Physiol Mol Plant Pathol 40: 437–445

Weissman G (1991) Aspirin. Sci Am 264: 84–90

Whalen MC, Innes RW, Bent AF, Staskawicz BJ (1991) Identification of *Pseudomonas syringae* pathogens of *Arabidopsis* and a bacterial locus determining avirulence on both *Arabidopsis* and soybean. Plant Cell 3: 49–59

White RF (1979) Acetylsalicylic acid (aspirin) induces resistance to tobacco mosaic virus in tobacco. Virology 99: 410–412

White RF, Antoniw JF (1991) Virus-induced resistance responses in plants. Crit Rev Plant Sci 9: 443–455

White RF, Dumas E, Shaw P, Antoniw JF (1986) The chemical induction of PR (b) proteins and resistance to TMV infection in tobacco. Antiviral Res 6: 177–185

Whitham S, Dinesh-Kumar SP, Choi D, Hehl R, Corr C, Baker B (1994) The product of the tobacco mosaic virus resistance gene N: similarity to toll and interleukin-1 receptor. Cell 18: 1101–1115

Wilson DC, Thain JF, Minchin PEH, Gubb IR, Reilly AJ, Skipper YD, Doherty HM, O' Donnell PJ, Bowles DJ (1992) Electrical signalling and systemic proteinase inhibitor induction in the wounded plant. Nature 360: 62–65

Woloshuk CP, Meulenhoff JS, Sela-Buurlage M, van den Elzen PJM, Cornelissen BJC (1991) Pathogen-induced proteins with inhibitory activity toward *Phytophthora infestans*. Plant Cell 3: 619–628

Xanthoudakis S, Miao G, Wang F, Pan Y-CE, Curran T (1992) Redox activation of fos-jun DNA binding activity is mediated by a DNA repair enzyme. EMBO J 11: 3323–3335

Yalpani N, Silverman P, Wilson TMA, Kleier DA, Raskin I (1991) Salicylic acid is a systemic signal and an inducer of pathogenesis-related proteins in virus-infected tobacco. Plant Cell 3: 809–818

Yalpani N, Schulz M, Davis MP, Balke NE (1992) Partial purification and properties of an inducible uridine 5-diphosphate-glucose:salicylic acid glucosyltransferase from oat roots. Plant Physiol 100: 457–463

Yalpani N, León J, Lawton MA, Raskin I (1993a) Pathway of salicylic acid biosynthesis in healthy and virus-inoculated tobacco. Plant Physiol 103: 315–321

Yalpani N, Shulaev V, Raskin I (1993b) Endogenous salicylic acid levels correlate with accumulation of pathogenesis-related proteins and virus resistance in tobacco. Phytopathology 83: 702–708

Yang SF, Pratt HK (1978) The physiology of ethylene in wounded plant tissue. In: Wahl G (ed) Biochemistry of wounded plant tissues. Walter de Gruyter, Berlin, pp 595–622

Ye XS, Pan SQ, Kuć J (1989) Pathogenesis-related proteins and systemic resistance to blue mould and tobacco mosaic virus induced by tobacco mosaic virus, *Peronospora tabacina* and aspirin. Physiol Mol Plant Pathol 35: 161–175

Yoshikawa M, Tsuda M, Takeuchi Y (1993) Resistance to fungal diseases in transgenic tobacco plants expressing the phytoalexin elicitor-releasing factor, β-1,3-endoglucanase from soybean. Naturwissenschaften 80: 417–420

Zhang B, Singh KB (1994) ocs element promoter sequences are activated by auxin and salicylic acid in *Arabidopsis*. Proc Natl Acad Sci USA 91: 2507–2511

Biologically Active Lipids and Lipid-modulated Protein Kinase in Plants

Günther F. E. Scherer

Contents

I. Introduction

The career of lipids in activating protein kinase began with a brain protein kinase M which differed from cAMP- or cGMP-activated protein kinase from the same source (Inoue et al., 1977; Takai et al., 1977). Somewhat later it was found by Nishizuka and coworkers that this protein kinase was activated by calcium and by a heat-stable factor as coactivator which was identified as membrane lipid (Takai et al., 1979). The new protein kinase was renamed as protein kinase C (C standing for calcium). The heat stable factor was identified as the lipid activators diacylglycerol and phosphatidylserine. After protein kinase C was found, Nishizuka and coworkers soon realized that receptor-activated phospholipase C (PLC) provided the lipid activator diacylglycerol in animal cells (Kishimoto et al., 1980). By now PKC not only is known to be a central enzyme in animal signal transduction but it also turned out to be a small gene family, instead of a single enzyme, the subgroups of which interact with different lipid mediators, thus making the initial picture much more complex (Asaoka et al., 1992; Nishizuka, 1992; Dekker and Parker, 1994; Exton, 1994; Liscovitch and Cantley, 1994). PKC also provides the major paradigm in signal transduction how lipid breakdown products in the (plasma) membrane can act as lipid second messengers in between hormones, the first messengers, and their receptors and the typical target enzymes for second messengers, protein kinases and phosphatases. These latter enzymes are destined to regulate by phosphorylation/dephosphorylation the activity of other enzymes. Besides PKC, at least one additional lipid-activated protein kinase was discovered in animal cells (Pushkareva et al., 1992; Joseph et al., 1993; Kolesnik and Golde 1994) and a lipid-activated protein phosphatase (Dobrowsky et al., 1993; Hannun, 1994).

Why can the animal lipid-modulated protein kinase(s) be relevant to plant signal transduction? In principle, an animal PKC is neither positive nor negative evidence that plants would possess a similar lipid-modulated protein kinase. However, in many laboratories knowledge is being gathered about more and more enzymes in plants which are regulated by protein phosphorylation (Huber et al., 1993), and we know about homologous protein kinases in plants and animals (Lawton et al., 1989; Ferreira et al., 1991; Hirt et al., 1991; Duerr et al., 1993; Kieber et al., 1993) but also apparently new and different ones (Harper et al., 1991; Roberts and Harmon, 1992; Kieber et al., 1993). Secondly, clear proof of receptors to which hormones or elicitors bind at the extracytoplasmic side of the plasma membrane (Barbier-Brygoo et al., 1989, 1991; Cosio et al., 1990; Cheong and Hahn, 1991; Hooley et al., 1991; Denarie and Cullimore, 1993; B. Anderson et al., 1994; Gilroy and Jones, 1994) should necessitate biochemical mechanisms for signal transduction into the cytoplasm which is at least analogous—if not homologous—to the function of lipid second messengers in animals. Thirdly, plants possess agonist-activated phospholipases A and C which provide lipid breakdown products in a completely analogous (homologous?) way as in animal cells (Morse et al., 1987; Scherer and André, 1989; Einspahr and Thompson, 1990; Droeback, 1992; S.-S. Lee et al., 1992; Cho et al., 1993; Legendre et al., 1993) so that lipid-modulated enzymes (protein kinases?) should be very likely and, last but not least, there is direct experimental evidence for lipid-modulated protein kinase in plants. In order to put such evidence into the right background, we should first have a look at enzymes in animals and in plants that provide the lipid breakdown products having a function as second messengers.

II. What Enzymes Do Generate Biologically Active Lipids in Animals and in Plants?

By looking at the rather long list of biologically active lipids in animal systems one can hardly escape the impression that this list could be fairly complete by now. Quite simply, of almost every possible type of phospholipid or sphingomyelin hydrolysing hydrolase (Fig. 1), every one seems to have a function in signal transduction in animal cells, each generating a lipid breakdown product that is an activator or inhibitor for a certain target enzyme (Fig. 2). The corresponding figure for potential plant lipid-derived second messengers looks far less crowded (Fig. 3). So, either plants are really very different in this aspect or we should expect the discovery of more details in the coming years.

The long-known example, phospholipase C (PLC) (Michell, 1985), is now known to be a small gene family of enzymes and each subtype is unique in its mode of activation by different G protein-subunits or tyrosine phosphorylation. Moreover, not only phosphatidylinositol-4,5-bisphosphate is hydrolysed, yield-

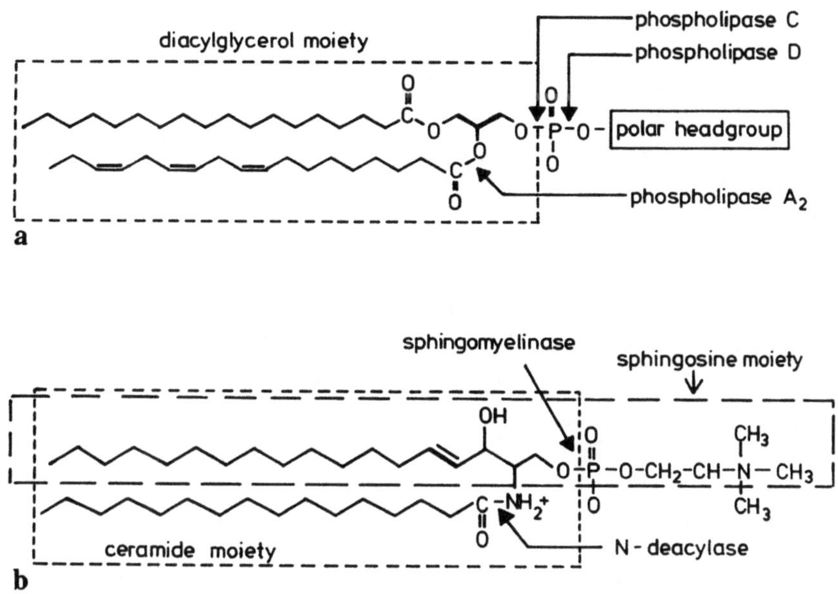

Fig. 1. Phospholipases and sphingolipases. a, The three major phospholipases, phospholipase A_2, phospholipase C, and phospholipase D and the bonds hydrolyzed by them in glycerophospholipids are shown. b, Sphingomyelin as an example for (animal) sphingolipids is chosen to demonstrate which bonds are hydrolyzed by sphingomyelinase and by N-deacylase

ing two biologically active hydrolysis products, diacylglycerol and inositol-1,4,5-trisphosphate, but also a phosphatidylcholine-specific PLC was found the function of which is to provide a lasting burst of diacylglycerol (Liscovitch, 1992). Diacylglycerol is the biologically active lipid generated by all PLC subtypes and it is an activator for PKC (Bell and Burns, 1991; Nishizuka, 1992). Inositol-1,4,5-trisphosphate, the second breakdown product of phosphatidylinositol-4,5-bisphosphate hydrolysis, activates a calcium channel in the ER in animal cells whereas for phosphorylcholine no second messenger function is known (Berridge, 1993).

PLC specific for phosphatidylinositol-4,5-bisphosphate was partially purified from plant sources and one of the breakdown products, inositol-1,4,5-trisphosphate, activates a calcium channel in plant tonoplast. Also, more and more evidence for an in vivo function of PLC in plant systems is accumulating (for reviews, see Einspahr and Thompson, 1990; Droeback, 1992), so that only the potential function for the diacylglycerol breakdown product in plant signal transduction remains mysterious in plants since no evidence for a true plant PKC (genetically related to the animal PKC) can be found in the literature as yet (see below).

Phospholipase A_2 (PLA2) was the second enzyme in animals to be found to generate biologically active lipid breakdown products (Burch et al., 1986). Besides the Ca^{2+}-activated cytosolic PLA2 (Clark et al., 1991; Kramer et al.,

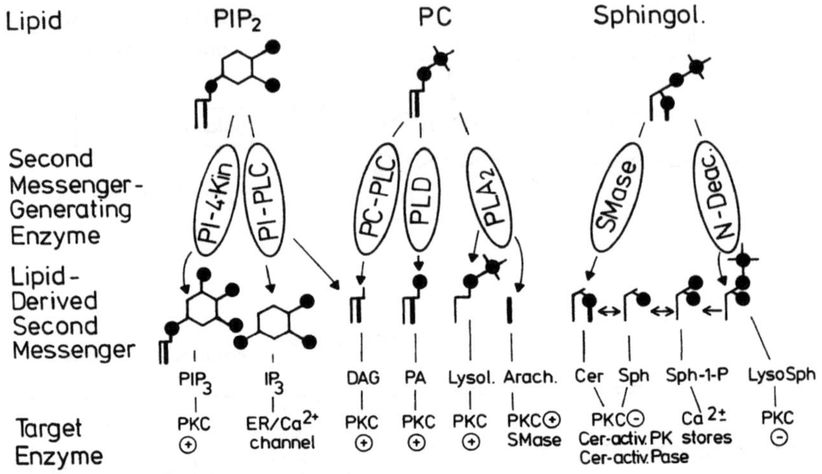

Fig. 2. Lipid-derived second messengers in animals. Major lipids as precursors to lipid-derived second messengers are shown schematically at the top: PIP_2, phosphatidylinositol-4,5-bisphosphate; PC, phosphatidylcholine; Sphingol, sphingomyelin. The second messenger-generating enzymes are shown below: PI-3-kin, phosphatidylinositol 3′-kinase; PI-PLC, phosphatidylinositol-specific phospholipase C; PC-PLC, phosphatidylcholine-specific phospholipase C; PLD, phospholipase D; SMase, sphingomyelinase; N-Deac, N-deacylase. Second messengers: PIP_3, phosphatidylinositol-3,4,5-trisphosphate; IP_3, *myo*-inositol-1,4,5-trisphosphate; DAG, diacylglycerol; PA, phosphatidic acid; Lysol, lysophospholipid; Arach, arachidonic acid; Cer, ceramide; Sph, sphingosine; Sph-1-P, sphingosine-1-phosphate; LysoSph, lysosphingolipid. Target enzymes: PKC, protein kinase C; Cer-activ. PK/Pase, ceramide-activated protein kinase or phosphatase

1991; Sharp et al., 1991; Lin et al., 1992) a Ca^{2+}-independent cytosolic PLA2 was found (Ackermann et al., 1994) and both enzymes are specific for arachidonic acid in the C2 position of the glycerophospholipid backbone but are less specific for a certain headgroup. Phosphatidylcholine and phosphatidylethanolamine, however, seem to be preferred substrates. Animal signal transduction-related PLA2 enzymes have been reviewed recently (Dennis, 1994). Arachidonic acid is the precursor to leukotrienes and prostaglandins, hormones in their own right (Samuelson et al., 1987), but arachidonic acid also activates a subtype of PKC (Asaoka et al., 1992; Dekker and Parker, 1994). Similarly, the second lipid breakdown product of PLA2 hydrolysis, lysophospholipid, activates another subtype of PKC as lipid second messenger.

A plant phospholipase A, possibly of an A_2 type, was described in plants (Scherer and André, 1989) and both fatty acids and lysophospholipids modulate protein kinase in plants (see below). A further similarity to animal systems could exist because linolenic acid is the precursor to jasmonic acid (J. Anderson, 1989), a plant hormone in its own right (Parthier, 1991).

The lipid breakdown product of phospholipids by phospholipase D (PLD) is phosphatidic acid (PA) which may either be a source for the second messenger diacylglycerol (generated by phosphatidic acid phosphatase) in animal cells or it may be a biologically active lipid in its own right (Moritz et al., 1992). No PLD having a function in signal transduction in plants has been described as yet, only enzymes having the properties of vacuolar enzymes (Dyer et al., 1994).

In animals, sphingomyelinase-mediated signal transduction has emerged as an important pathway (Exton, 1994; Hannun, 1994; Kolesnik and Golde, 1994; Liscovitch and Cantley, 1994). Ceramide, the lipid second messenger generated by sphingomyelinase in animals, would have to be generated by an analogous enzyme in plants from related sphingolipids, since they lack sphingomyelin. In animal systems, a ceramide-activated protein kinase was described as well as a ceramide-activated protein phosphatase and, additionally, this lipid inhibits PKC. Lysosphingolipids, generated by N-deacylase, inhibit PKC but the properties of plant N-deacylase are unknown. There is only one report for a possible function of sphingosine in plants as an activator for the tonoplast H^+-pyrophosphatase (Bille et al., 1992).

What enzymes interact with biologically active lipids in animal cells? Suspiciously, most of them have a function in protein phosphorylation, namely the different isoforms of PKC, ceramide-activated protein kinase and ceramide-activated protein phosphatase (Fig. 2). This fits into the general pattern for most other animal second messengers, e.g., CAMP, CGMP or Ca^{2+}/(calmodulin) all of which can activate protein kinases.

For comparison, our current knowledge about (potentially) biologically active lipids in plants is summarized in Fig. 3. It is in all respects a picture of a much more scant knowledge. But again, it is protein kinase activity which is modulated by several lipids, fatty acids, lysophospholipids, and phosphorylated phosphatidylinositols. The details of this regulation will be described in Sect. III. It is possible, however, that plants may directly regulate membrane-bound enzymes without protein kinase as a mediator. Such examples of lipid regulation in plants could be the effects of lysophosphatidylcholine on a plasma membrane redox system (Brightman et al., 1991), on plasma membrane H^+-ATPase (Serrano et al., 1988; Palmgren and Sommarin, 1989; Martiny-Baron and Scherer, 1989; Palmgren, 1991; Scherer, et al., 1993b) or of acidic lipids on the plasma membrane H^+-ATPase (Memon et al., 1989; Chen and Boss, 1991), of the lysophospholipid-like platelet-activating factor (PAF) on tonoplast H^+-ATPase (Martiny-Baron et al., 1992), of fatty acids on a potassium channel regulating stomatal aperture (Y. Lee et al., 1994), and of sphingosine on the tonoplast H^+-pyrophosphatase (Bille et al., 1992). Either, these enzymes are all target proteins for potential lipid-derived second messengers or, conceivably, the action of these lipid breakdown products might be also mediated by protein kinase or protein phosphatase as was suggested for the two major plant H^+-ATPases (Palmgren, 1991; Martiny-Baron et al., 1992). The function(s) for these lipid-enzyme interactions not involving protein kinase are even more speculative than those for lipid-modulated protein kinase from plants since very little is known about lipid specificity in these examples. Therefore, these are not included into the scheme in Fig. 3.

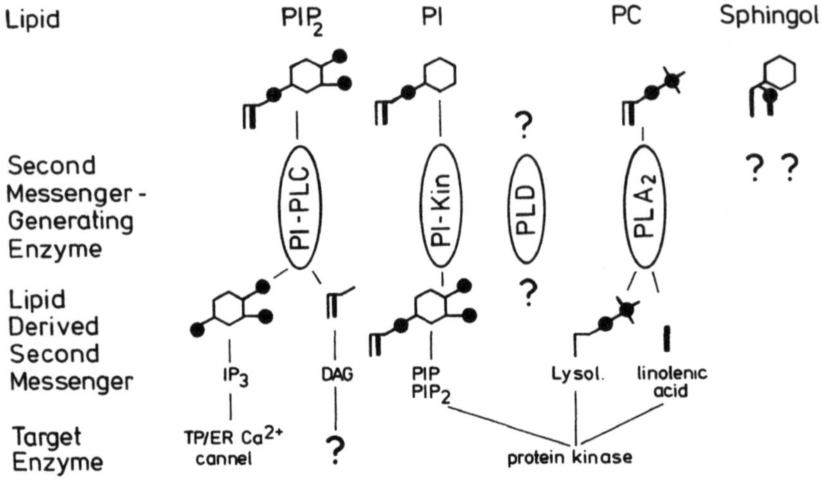

Fig. 3. Potential lipid-derived second messengers in plants. Lipid precursors to potential plant lipid-derived second messengers are shown schematically at the top: PIP_2, phosphatidylinositol-4,5-bisphosphate; PI, phosphatidylinositol; PC, phosphatidylcholine; Sphingol, (glyco)-sphingolipid. The second-messenger-generating enzymes are shown below: PI-PLC, phosphatidylinositol-specific phospholipase C; PI-Kin, phosphatidylinositol kinase(s); PLD, phospholipase D; PLA_2, phospholipase A_2. Potential second messengers: IP_3, *myo*-inositol-1,4,5-trisphosphate; DAG, diacylglycerol; Lysol, lysophospholipid; PIP, phosphatidylinositol-4-phosphate

III. Lipid-modulated Protein Kinase in Plants

A. *Specificity of Lipid–Protein Interactions and Assay Types for Lipid-modulated Protein Kinases*

Lipids present several unique problems to their biochemistry. Most are very hydrophobic and then immersed in an existing native membrane or, in vitro, in a substitute membrane such as pure lipid vesicles but most lipids are not water-soluble as single molecules. Others, such as lysophospholipids and fatty acids, are more or less water-soluble but are also membrane active as detergents so that a distinction between lipid effects and detergent effects must be made clear. This means that at least some kind of specificity of the lipid effect in a particular reaction must be demonstrated. Criteria for specificity are low concentrations of the lipid applied, indicating the possibility of selective binding to the effector protein, and comparison of the obtained effects with effects of molecules of similar structure.

The types of assays used to demonstrate lipid-modulated protein kinase in plant sources are shown in Fig. 4 a and b, for comparison, the principle of a mixed-micelle assay is shown in Fig. 4 c. Several workers preferred to use soluble or solubilized enzyme and added lipids either as micelles or as vesicles

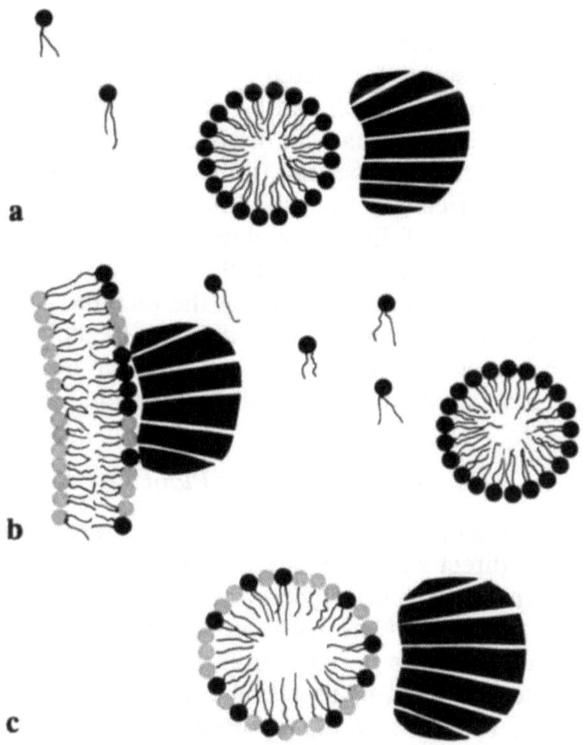

Fig. 4. Schematic illustrations for assays of lipid-interacting enzymes, i.e., protein kinases. a, The enzyme is in soluble or solubilized form and lipids are added as micelles or single molecules, depending on their solubility (shown), or liposomes (not shown). b, Assay of membrane vesicles with adhering protein kinase and lipids added exogenously. Some of the lipid, depending on the partitioning coefficient, will intersperse into the vesicle (presumably only at the outer surface), and some of the lipid will remain as micelles, liposomes or single molecules, depending on solubility. c, Mixed-micelle assay: A detergent micelle (gray) is drawn into which added lipid is interspersed. The soluble or solubilized enzyme interacts with a mixed surface of detergent and lipid of defined molar concentrations and ratios

or both, depending on the lipid solubility (Fig. 4a). The advantage of this assay is that assay conditions are well defined and exogenous substrate can be added. In a second type of assay, both as an enzyme source and substrate source, isolated vesicles were used (Fig. 4b). Here, the advantage could be that only membrane-associated protein kinase activity enters the test system, thereby discarding other sources of enzyme. The substrates are endogenous membrane proteins and, therefore, perhaps the natural ones to these membrane-associated protein kinases. Hence, the substrates can be identified which helps to characterize the function of the respective kinase. Added lipids will, partially, intersperse into the membrane vesicles and, partially, be free lipids, micelles, or pure lipid vesicles.

The third type of assay is the mixed-micelle assay (Fig. 4c). The micelles are provided by a detergent which can be mixed with lipids in defined molar ratios or at defined concentrations with the lipids to be tested. It is necessary to demonstrate that the detergent does not influence the enzyme to be tested. This type of assay is not restricted to lipid-modulated protein kinase but applicable, in principle, to all kinds of lipid-modulated enzymes. It was often used in assaying animal PKC (Bell and Burns, 1991). The advantage of this assay is that a soluble enzyme can be tested so that the enzyme source, the lipid concentrations and the substrates are all well-defined. Especially for an attempt to purify a lipid-modulated protein kinase such an assay seems to be indispensable even though it was not yet used for a plant protein kinase.

B. The Search for a "Plant PKC"

In several early reports on lipid-activated protein kinase in plants (1985–1988) animal PKC was the direct and only paradigm for a search of the plant counterpart. Up to that time only diacylglycerol and phosphatidylserine were known as lipid activators for PKC so that, in consequence, few other lipids were investigated as potential activators in plant protein kinase assays. One report about stimulation by diacylglycerol of membrane-associated protein kinase in soybeans (Morré et al., 1984) was not supported by reports on a lack of effect of diolein by others (Martiny-Baron and Scherer, 1989; Scherer et al., 1990; Scherer, 1990; Schaller et al., 1992). Two reports on effects of diacylglycerol on physiological responses, one on mitosis (Larsen et al., 1991) and one on guard cell swelling via channel regulation (Y. Lee and Assman, 1991), might support the notion of an as yet elusive homologue plant PKC. However, involvement of a protein kinase in these responses and in the effect of diacylglycerol was not directly demonstrated.

Other early experiments on activation of plant protein kinase activity by phosphatidylserine as the second potential lipid activator were done with partially purified and soluble Ca^{2+}-activated protein kinase (Schäfer et al., 1985; Elliott and Skinner, 1986; Olah and Kiss, 1986; Elliott and Kokke, 1987; Favre and Turian, 1987; Elliott et al., 1988). Common to all these experiments were no clear demonstration of specificity or preference for phosphatidylserine as an activator, i.e., other phospholipids tested were equally effective as activating substances. These results cannot, therefore, be regarded as evidence of a plant PKC. When phosphatidylserine or a mixture of phosphatidylserine and diacylglycerol was added to plant membrane vesicles endogenous membrane-associated protein kinase was not affected by these lipids (Martiny-Baron and Scherer, 1989; Scherer et al., 1990). A similar negative result was obtained with soluble enzyme and diolein as activator (Schaller et al., 1992). In conclusion, the search for a diacylglycerol-activated protein kinase from a plant source failed so far, though certainly many more attempts in various laboratories to find it went unreported.

C. Identification of a Lipid-modulated Protein Kinase in Plants

Other lipids as activators for plant protein kinase were discovered, namely lysophospholipids and platelet-activating factor (Scherer et al., 1988; Martiny-Baron and Scherer, 1988, 1989). Platelet-activating factor (PAF), an ether lipid (1-O-alkyl-*sn*-glycero-2-acetyl-3-phosphorylcholine), is not found in plants but, due to its chemical similarity to lysophosphatidylcholine, it might be a kind of artificial substitute for lysophospholipid in plants. The other groups of lipid activators are fatty acids (Klucis and Polya, 1987; Lucantoni and Polya, 1987) and inositol phospholipids (Schaller et al., 1992; Binder et al., 1994).

Two different approaches and assay systems were used in the description and characterization of lipid-activated protein kinase from plant sources. In microsomes isolated from zucchini hypocotyls, initially, it was found that platelet-activating factor activated ATP-dependent H^+ transport (Scherer, 1985). A protein factor could be removed from the membranes which was necessary as a cofactor for lipid activation of H^+ transport (Scherer and Martiny-Baron, 1987). Moreover, a plant lipid or, more precisely, at least one plant lysophospholipid could replace the animal platelet-activating factor as an activator of ATP-dependent H^+ transport (Scherer and Stoffel, 1987). Then it became clear that the protein factor for transport activation copurified with a protein kinase and that this protein kinase was activated by Ca^{2+} ions and platelet-activating factor (Scherer et al., 1988). The activation by lipid or by Ca^{2+} was only two-fold in these initial experiments. Since the membrane-associated enzyme responded stronger to added lipids and allowed approaches to identify the endogenous substrates (Martiny-Baron and Scherer, 1988) we chose this type of assay to investigate this Ca^{2+}/lysophospholipid-activated protein kinase and its substrates.

A completely different approach was chosen by Polya and coworkers and also by Sussman's and Harmon's group. As a starting point, they all took partially or completely purified Ca^{2+}-dependent (and calmodulin-independent) protein kinase (CDPK) from various sources (soybean: Harmon et al., 1987; wheat germ, silver beet leaves, oat leaves: Klucis and Polya, 1987; Lucantoni and Polya, 1987; Minichello et al., 1991; oats roots: Schaller and Sussman, 1988). Such a purified protein kinase could also be activated by fatty acids (Klucis and Polya, 1987; Lucantoni and Polya, 1987) and lysophospholipids and inositol phospholipids (Schaller et al., 1992). From the combined work of Sussman's and Harmon's group it became clear that this Ca^{2+}-activated protein kinase was a novel protein kinase with a kinase and calmodulin-like domain in one and the same molecule (Harper et al., 1991; Roberts and Harmon, 1992). It seems now quite possible that a number of laboratories previously also characterized this enzyme which, typically, can be identified as an autophosphorylating kinase with an apparent molecular mass of 45–69 kDa and, sometimes higher, up to 85 kDa (Olah and Kiss, 1986; Polya et al., 1987; Battey and Venis, 1988; Klucis and Polya, 1988; Verhey et al., 1993). The use of monoclonal antibodies developed in Harmon's laboratory further made clear that this CDPK actually comprises a small group of genetically and serologically related enzymes (Harper et al., 1991; Suen and Choi, 1991; Roberts and Harmon, 1992; Schaller et al., 1992; Verhey et al., 1993; Zhao et al., 1993).

It is not yet clear, however, whether a single or a few isoforms of CDPK is modulated by lipids (Schaller et al., 1992; Harper et al., 1993; Binder et al., 1994). Binder et al. (1994) definitely identified a single isoform, expressed in *E. coli*, as lipid-modulated and Polya and coworkers chromatographically separated two isoforms from wheat germ which responded differentially to added lipids (Klucis and Polya, 1987, 1988; Lucantoni and Polya, 1987; Minichello et al., 1989). The latter results indicate that at least not all CDPK isoforms respond in the same way to lipids and some perhaps not at all. The enzyme investigated in the author's laboratory is a membrane-associated Ca^{2+}/lysophospholipid-activated protein kinase. Since in our assays always several proteins in a range of 53–62 kDa are strongly stimulated in phosphorylation by Ca^{2+} and by lysophospholipids and, since three proteins in this molecular size range were found in an autophosphorylation assay (Scherer et al., 1993a), it can be tentatively concluded that among this group of autophosphorylating proteins is a Ca^{2+}/lysophospholipid-activated protein kinase which is most likely a member of the CDPK family. This conclusion is further supported by the very similar lipid specificity of the activation of the enzyme investigated in our and in Sussman's laboratory (Martiny-Baron and Scherer, 1989; Scherer, 1990; Scherer et al., 1990, 1993a; Schaller et al., 1992; Binder et al., 1994).

In conclusion about this still somewhat fuzzy picture, it seems likely that the soluble CDPK, described by Harmon, by Polya, and by Sussman and their co-workers, and the membrane-associated Ca^{2+}/lysophospholipid-activated protein kinase in our work, all belong to the same genetically related group, now called CDPK. However, we know very little about the number and the properties of isoforms, especially with respect to differences in lipid modulation.

D. Lipid Specificity of Plant Lipid-modulated Protein Kinase

The lipid specificity of modulation of CDPK activity was investigated by Polya's group mostly for fatty acids. The enzyme sources were silver beet (*Beta vulgaris*) leaves, containing an isoform I (51–56 kDa) and an isoform II (53–57 kDa), both as soluble and as a membrane-associated form (Klucis and Polya, 1987, 1988; Polya et al., 1987), oat (*Avena sativa*) leaves, containing a 79 kDa enzyme (Minichello et al., 1989), and wheat (*Triticum vulgare*) embryo, containing a 85 kDa enzyme (Lucantoni and Polya, 1987; Jinsart et al., 1991). The property common to all these enzymes is that they are activated by fatty acids in the absence of Ca^{2+} ions. Maximal activation is found at about 200 μM fatty acid (Klucis and Polya, 1987; Lucantoni and Polya, 1987). If, however, the level of activation by Ca^{2+} ions alone is compared to this level of activation by fatty acids, activation by fatty acids is never higher than that by Ca^{2+} alone, more often it is lower. The isoform II from silver beet leaves exhibits much stronger activation by fatty acids as does isoform I, indicating clear differences between CDPK isoforms. The spectrum of activating fatty acids may be also different for enzymes from different sources since desaturated fatty acids (oleic, linoleic, arachidonic acid) are good activators for the silver beet leaf and wheat germ enzymes but not for the oat leaf enzyme. This latter one is activated best by long chain fatty acids (behenic acid and lignoceric acid) but also by palmitic and by stearic acid to an appreciable extent. The level of activation was 2–3-fold in activity for long chain fatty acids (Minichello et al., 1989) and up to ten-fold by arachidonic and linoleic acid (Lucantoni and Polya, 1987; Klucis and Polya, 1987). In the presence of Ca^{2+} ions, fatty acid either do not stimulate (at low concentrations) or they are inhibitory at high fatty acid concentrations. This is a different property found by Polya and coworkers for fatty acids as lipid modulators and distinguishes them from lysophospholipids and other phospholipids which always act synergistic to Ca^{2+} in their mode of activation (Martiny-Baron and Scherer, 1989; Schaller et al., 1992). In our assay system, which is different from Polya's, no influence of oleic acid (without Ca^{2+}) on phosphorylation was found (Martiny-Baron and Scherer, 1989). At present, these results with fatty acids are still difficult to interpret in terms of function even though, conceivably, fatty acids could act as second messengers and their level in plant membranes might be regulated without a rise in cytoplasmic Ca^{2+} levels.

For the wheat germ enzyme the influence of other amphiphilic compounds on CDPK activity was investigated (Polya et al., 1990a). Interestingly, in the absence of Ca^{2+}, very low dihydrosphingosine concentrations ($<20\,\mu M$) inhibited CDPK but concentrations higher than $200\,\mu M$ were activating. A number of artificial amphiphilic substances also modulated this CDPK, amongst these were acyl carnitins (activation up to two-fold), zwittergents (activation up to two-fold), alkyl sulfates (weakly activating), and long chain trimethylammonium alkanes (inhibitory). Among the above-mentioned compounds, only sphingosines can be expected to function as plant lipid second messengers at all.

CDPK was investigated with respect to phospholipid specificity by Sussman's group and in the author's laboratory. Sussman and coworkers first characterized the CDPK from oat root plasma membranes (Schaller et al., 1992) and later of an isoform from *Arabidopsis* expressed as a fusion protein in *E. coli* (Harper et al., 1993; Binder et al., 1994), using histone IIIs as an exogenous substrate (see Fig. 4 a). We used zucchini and the microsome assay (Fig. 4 b). Despite the different assays and enzyme sources quite similar results with respect to lipid specificity were obtained. Also, in both laboratories it was found that Ca^{2+} and lipids always acted synergistic (Martiny-Baron and Scherer, 1989; Scherer et al., 1990; Scherer, 1990; Schaller et al., 1992; Harper et al., 1993; Binder et al., 1994). Most likely, we also used CDPK. This can be concluded from a comparison of our autophosphorylation experiments where we found several protein kinases having a molecular weight from 55–62 kDa (Scherer et al. 1993b) which is similar to results from Sussman and coworkers using well-defined CDPK and autophosphorylation (Binder et al., 1994) and to similar immunodetection experiments with zucchini membranes (Verhey et al., 1993).

Two groups of lipids are activators, one of which is comprised by the zwitterionic lysophosphatidylcholine and the chemically similar PAF. Intermediate activator properties were found for *N,N'*-dimethyl-lysophosphatidylethanolamine and none for lysophosphatidylethanolamine, lysophosphatidylglycerol, or lysophosphatidylserine (Scherer et al., 1990; Binder et al., 1993; Scherer and Führ, unpubl.). A second group of activating lipids are acidic lysophospholipids, lysophosphatidylinositol and lysophosphatidic acid (Binder et al., 1994; Scherer and Führ, unpubl.), and the phospholipids phosphatidylinositol, phosphatidylinositol-4-phosphate, and phosphatidyl-4,5-bisphosphate (Schaller et al., 1992; Binder et al., 1994). In our assay, phosphatidylinositol was inactive as well as all other common phospholipids, namely phosphatidylcholine, phosphatidylethanolamine, phosphatidylserine, phosphatidylglycerol, and phosphatidic acid (Martiny-Baron and Scherer, 1989) so that the only divergent result was obtained for phosphatidylinositol, being active in Sussman's assay but not in ours. Hence, a fairly large group of acidic lysophospholipids and phosphorylated inositol phospholipids and the zwitterionic lysophosphatidylcholine are activators. The non-plant lipid PAF having the same head group as lysophosphatidylcholine must be regarded as very effective but artificial activator for plant CDPK. With most of these lipids activation was found at about 200 µM, for phosphatidylinositol-4-phosphate maximal activation was found already at a concentration as low as 50 µM.

Since lysophospholipids, especially lysophosphatidylcholine, have also detergent properties, activation of protein kinase by detergents was also studied. Binder et al. (1994) found no activation of CDPK by CHAPS and obtained half of the level of maximal activation by the zwittergent SB-16 as that obtained by lysophosphatidylcholine. We also found no activation by CHAPS up to 1 mM concentrations. At lower concentrations only zwittergent and Triton X-100 were activating membrane-associated enzyme but clearly weaker than lysophosphatidylcholine or than PAF. At higher concentrations (>1 mM), in our assays eventually all detergents became activating (Scherer et al., 1993a; Scherer and Führ, unpubl.). With an assay using membranes it is difficult to distinguish between the solubilizing activity and the activating property of a detergent. Therefore, we also compared the solubilizing properties and the activator properties of lysophospholipids and detergents. We found that detergents were always the better solubilizing reagents and the weaker activating reagents than the lysophospholipids. Hence, it seems reasonable to assume that the activation of CDPK by certain lipids is specific.

Only circumstantial evidence exists that might provide an explanation for the activating properties of lipids on CDPK. It is thought that a basic domain of the CDPK could exert an autoinhibitory effect (Roberts and Harmon, 1992; Binder et al., 1994; Harper et al., 1994). Polya et al. (1990) found that CDPK is inhibited by basic polypeptides and Binder et al. (1994) showed that acidic lipids attenuated this inhibition so that this could be a mode of action for these acidic lipid activators. This, however, provides no explanation for the activation of CDPK by the zwitterionic lysophosphatidylcholine or platelet-activating factor.

The substrates of the lipid-activated protein kinase in plants are mostly not clearly identified. Several unknown polypeptides were identified in tonoplast and plasma membranes as substrates of the lipid-activated protein kinase and two of which could be identified, the B-subunit of the tonoplast H^+-ATPase (Martiny-Baron et al., 1992) and the plasma membrane H^+-ATPase (Martiny-Baron and Scherer, 1989; Nickel et al., 1991; Scherer et al., 1993b). The plasma membrane H^+-ATPase had been identified as a phosphoprotein before (Schaller and Sussman, 1988) and regulation of this enzyme in plants by phosphorylation seems possible (Palmgren, 1991) whereas in yeast this is strongly suggested (Chang et al., 1991; Na et al., 1993; Eraso and Portillo, 1994). Another substrate for CDPK is nodulin, a channel-forming protein in the peribacteroid membrane, derived from the host plasma membrane after *Rhizobium* infection (Weaver and Roberts, 1992). However, whether lipid-activated protein kinase or other protein kinase(s) do not only phosphorylate these enzymes in plant extracts but also exert regulation by phosphorylation remains still an open question for the plant plasma membrane H^+-ATPase and the other enzymes mentioned. For the plant tonoplast H^+-ATPase, regulation by phosphorylation is purely speculative even though one report about animal endosomal H^+-ATPase being regulated by PKC lends support to this idea (Nanda et al., 1992).

The function of lipid activation of CDPK in plants must be regarded as hypothetical at present even though changes of lysophospholipid content due to agonists have been demonstrated (Scherer and André, 1989; Scherer, 1990, 1995; Scherer et al., 1990; S.-S. Lee et al., 1992). Moreover, changes in phosphorylated phosphatidylinositols due to agonists seem likely (Memon et al., 1989; Memon and Boss, 1990; Chen and Boss, 1991; Yang et al., 1993; Yang and Boss, 1994), which could provide regulation of CDPK by light, osmotic and pathogen stress. This in vivo modulation of potential second messengers in plants has been linked to the regulation of plasma membrane H^+-ATPase only. Similarly, addition of biologically active lipids was shown to increase proton secretion in cultured plant cells (Scherer and Nickel, 1988; Nickel et al., 1991) and growth (Scherer, 1994). Both effects might or might not be mediated by a regulatory effect on the plasma membrane H^+-ATPase and, additionally, the involvement of the CDPK is not proven by these effects. Hence, all the above-mentioned effects remain purely speculative in their relationship to CDPK (or any other protein kinase) but seem to provide a tantalizingly coherent picture. However, it has not yet been tried to directly demonstrate a change in CDPK activity in vivo (or in vitro) due to agonist treatment. There are more open questions: Is the CDPK the only lipid-modulated protein kinase (or is the plant PKC still to be discovered)? What are the (additional) substrates and is their activity really regulated by lipid-dependent phosphorylation/dephosphorylation? What distinguishes the lipid-modulated CDPK isoforms from the other isoforms? These open questions show that we are still at the very beginning of a research chapter termed lipid-modulated protein kinase in plants.

IV. References

Ackermann EJ, Kempner ES, Dennis EA (1994) Ca^{2+}-independent cytosolic phospholipase A_2 from macrophage-like P388D_1 cells. Isolation and characterization. J Biol Chem 269: 9227–9233

Anderson B, Ward JM, Schroeder JI (1994) Evidence for an extracellular reception site for abscisic acid in *Commelina* guard cells. Plant Physiol 104: 1177–1183

Anderson JM (1989) Membrane-derived fatty acids as precursors to second messengers. In: Boss WF, Morré DJ (eds) Second messengers in plant growth and development. AR Liss, New York, pp 181–212

André B, Scherer GFC (1991) Stimulation by auxin of phospholipase A in membrane vesicles from an auxin-sensitive tissue is mediated by an auxin receptor. Planta 185: 209–214

Asaoka Y, Nakamura S-I, Yoshida K, Nishizuka Y (1992) Protein kinase C, calcium and phospholipid degradation. Trends Biochem Sci 17: 414–417

Barbier-Brygoo H, Ephretikine G, Klämbt D, Ghislain M, Guern J (1989) Functional evidence for an auxin receptor at the plasmalemma of tobacco mesophyll protoplasts. Proc Natl Acad Sci USA 86: 891–895

Barbier-Brygoo H, Ephretekine G, Klämbt D, Maurel C, Palme K, Schell J, Guern J (1991) Perception of the auxin signal at the plasma membrane of tobacco mesophyll protoplasts. Plant J 1: 83–93

Battey NH, Venis MA (1988) Calcium-dependent protein kinase from apple fruit membranes is calmodulin-independent but has calmodulin-like properties. Planta 176: 91–97

Bell RM, Burns DJ (1991) Lipid activation of protein kinase C. J Biol Chem 266: 4661–4664

Berridge MJ (1993) Inositol trisphosphate and calcium signalling. Nature 361: 315–325

Bille J, Weiser T, Bentrup F-W (1992) The lysolipid sphingosine modulates pyrophosphatase activity in tonoplast vesicles and isolated vacuoles from a heterotrophic cell suspension culture of *Chenopodium rubrum*. Physiol Plant 84: 250–254

Binder BM, Harper JF, Sussman MR (1994) Characterization of an *Arabidopsis* calmodulin-like domain protein kinase purified from *Escherichia coli* using an affinity sandwich technique. Biochemistry 33: 2033–2041

Brightman A, Zhu X-Z, Morré DJ (1991) Activation of plasma membrane NADH oxidase activity by products of phospholipase A. Plant Physiol 96: 1314–1320

Burch RM, Luini A, Axelrod J (1986) Phospholipase A_2 and phospholipase C are activated by distinct GTP-binding proteins in response to A-adrenergic stimulation of FRTL5 thyroid cells. Proc Natl Acad Sci USA 83: 7201–7205

Chang A, Slayman CW (1991) Maturation of the yeast plasma membrane H^+-ATPase involves phosphorylation during intracellular transport. J Cell Biol 115: 289–295

Chen Q, Boss WF (1991) Neomycin inhibits the PIP and PIP_2 stimulation of plasma membrane ATPase activity. Plant Physiol 96: 340–343

Cheong J-J, Hahn MG (1991) A specific, high-affinity binding site for the hepta-β-glucoside elicitor that exists in soybean membranes. Plant Cell 3: 137–147

Cho M-H, Shears SB, Boss WF (1993) Changes in phosphatidylinositol metabolism in response to hyperosmotic stress in *Daucus carota* L. cells grown in suspension culture. Plant Physiol 103: 637–647

Clark JD, Lin L-L, Kriz RW, Ramesha CS, Sultzman LA, Lin AY, Milona N, Knopf JL (1991) A novel arachidonic acid-selective cytosolic PLA_2 contains a Ca^{2+}-dependent translocation domain with homology to PKC and GAP. Cell 65: 1043–1051

Cosio EG, Frey T, Verduyn R, van Boom J, Ebel J (1990) High-affinity binding of a synthetic heptaglucoside and fungal glucan phytoalexin elicitors to soybean membranes. FEBS Lett 271: 223–226

Dekker LV, Parker PJ (1994) Protein kinase C—a question of specificity. Trends Biochem Sci 19: 73–77

Dénarié J, Cullimore J (1993) Lipo-oligosaccharide nodulation factors: a new class of signalling molecules mediating recognition and morphogenesis. Cell 74: 951–954

Dennis E (1994) Diversity of group types, regulation, and function of phospholipase A_2. J Biol Chem 269: 13507–13060

Dobrowsky RT, Kamiyabasi C, Mumby M, Hannun YA (1993) Ceramide activates heterotrimeric protein phosphatase 2A. J Biol Chem 268: 15523–15530

Droeback BK (1992) The plant phosphoinositide system. Biochem J 288: 697–712

Duerr B, Gawienowski M, Ropp T, Jacobs T (1993) MsERK1: a mitogen-activated protein kinase from a flowering plant. Plant Cell 5: 87–96

Dyer JH, Ryu SB, Wang X (1994) Multiple forms of phospholipase D following germination and during leaf development of castor bean. Plant Physiol 105: 715–724

Einspahr KJ, Thompson GA Jr (1990) Transmembrane signaling via phosphatidylinositol 4,5-bisphosphate hydrolysis in plants. Plant Physiol 93: 361–366

Elliott DC, Kokke YS (1986) Calcium-dependent, phospholipid-activated protein kinase in plants. Phytochemistry 25: 39–44

Elliot DC, Kokke YS (1987) Partial purification and properties of a protein kinase C type enzyme from plants. Phytochemistry 26: 2929–2935

Elliot DC, Fourier A, Kokke YS (1988) Phosphatidylserine activation of plant protein kinase C. Phytochemistry 27: 3725–3730

Eraso P, Portillo F (1994) Molecular mechanism of regulation of yeast plasma membrane H^+-ATPase by glucose. Interaction between domains and identification of new regulatory sites. J Biol Chem 269: 10393–10399

Exton JH (1994) Messenger molecules derived from membrane lipids. Curr Opin Cell Biol 6: 226–229

Favre B, Turian G (1987) Identification of a calcium- and phospholipid-dependent protein kinase (protein kinase C) in *Neurospora crassa*. Plant Sci 49: 15–21

Ferreira PCG, Hemerly AS, Villaroel R, Van Montagu M, Inzé D (1991) The *Arabidopsis* functional homolog of the $p34^{cdc2}$ protein kinase. Plant Cell 3: 531–540

Gilroy S, Jones RL (1994) Perception of gibberellin and abscisic acid at the external face of the plasma membrane of barley (*Hordeum vulgare* L.) aleurone protoplasts. Plant Physiol 104: 1185–1192

Hannun YA (1994) The sphingomyelin cycle and the second messenger function of ceramide. J Biol Chem 269: 3125–328

Harmon AC, Putnam-Evans C, Cormier JM (1987) A calcium-dependent but calmodulin-independent protein kinase. Plant Physiol 83: 830–837

Harper JF, Sussman MR, Schaller GE, Putnam-Evans C, Charbonneau H, Harmon A (1991) A calcium-dependent protein kinase with a regulatory domain similar to calmodulin. Science 252: 951–954

Harper JF, Binder BM, Sussman MR (1993) Calcium and lipid regulation of an *Arabidopsis* protein kinase expressed in *Escherichia coli*. Biochemistry 32: 3282–3290

Hirt H, Pay A, Gyorgyey J, Baco L. Nemeth K, Bogre L, Schweyen RJ, Heberle-Bors E, Dudits D (1991) Complementation of a yeast cell cycle mutant by an alfalfa cDNA encoding a protein kinase homologous to $p34^{cdc2}$. Proc Natl Acad Sci USA 82: 1636–1640

Hooley R, Beale MH, Smith SJ (1991) Gibberellin perception at the plasma membrane of *Avena fatua* aleurone protoplasts. Planta 183: 274–280

Huber SC, Huber JL, McMichael RW Jr (1994) Control of plant enzyme activity by reversible protein phosphorylation. Int Rev Cytol 149: 47–98

Inoue M, Kishimoto A, Takai Y, Nishizuka Y (1977) Studies on a cyclic nucleotide-independent protein kinase and its proenzyme in mammalian tissues. II. Proenzyme and its activation by calcium-dependent protease from rat brain. J Biol Chem 252: 7610–7616

Jinsart W, Ternai B, Polya GM (1991) Inhibition and activation of wheat embryo calcium-dependent protein kinase and inhibition of avian myosin light chain kinase by long chain aliphatic amphiphiles. Plant Sci 78: 165–175

Joseph CK, Byun HS, Bittman R, Kolesnik RN (1993) Substrate recognition by ceramide-activated protein kinase. Evidence that kinase activity is proline-directed. J Biol Chem 268: 20002–20006

Kieber JJ, Rothenberg M, Roman G, Feldmann KA, Ecker JR (1993) *CTR1*, a negative regulator of the ethylene response pathway in *Arabidopsis*, encodes a member of the raf family of protein kinases. Cell 72: 427–441

Kishimoto A, Takai Y, Mori T, Kikkawa U, Nishizuka Y (1980) Activation of calcium and phospholipid-dependent protein kinase by diacylglycerol, its possible relation to phosphatidylinositol turnover. J Biol Chem 255: 2273–2276

Klucis E, Polya GM (1987) Calcium-independent activation of two plant leaf calcium-regulated protein kinases by fatty acids. Biochem Biophys Res Commun 147: 1041–1047

Klucis E, Polya GM (1988) Localization, solubilization and characterization of plant membrane-associated calcium-dependent protein kinases. Plant Physiol 88: 164–171

Kolesnik R, Golde DW (1994) The sphingomyelin pathway in tumor necrosis factor and interleukin-1 signaling. Cell 77: 325–328

Kramer RM, Roberts EF, Manetta J, Putnam JE (1991) The Ca^{2+}-sensitive cytosolic phospholipase A_2 is a 100-kDa protein in human monoblast U937 cells. J Biol Chem 266: 5268–5272

Larsen PM, Chen T-L, Wolniak SM (1991) Neomycin reversibly disrupts mitotic spindle progression in stamen hair cells of *Tradescantia.* J Cell Sci 98: 159–168

Lawton MA, Yamamoto RT, Hanks SK, Lamb CJ (1989) Molecular cloning of plant transcripts encoding protein kinase homologs. Proc Natl Acad Sci USA 86: 340–3144

Lee S-S, Kawakita K, Tsuge T, Doke N (1992) Stimulation of phospholipase A_2 in strawberry cells treated with AF-toxin 1 produced by *Alternaria alternata* strawberry phenotype. Physiol Mol Plant Pathol 41: 283–294

Lee Y, Assmann SM (1991) Diaclyglycerols induce both ion pumping in patch-clamp guard cells and opening of intact stomata. Proc Natl Acad Sci USA 88: 2127–2131

Lee Y, Lee, HJ, Crain RC, Lee A, Korn SJ (1994) Polyunsaturated fatty acids modulate stomatal aperture and two distinct K^+ channel currents in guard cells. Cell Signal 6: 181–186

Legendre L, Yueh YG, Crain R, Haddock N, Heinstein PF, Low PS (1993) Phospholipase C activation during elicitation of the oxidative burst in cultured plant cells. J Biol Chem 268: 24559–24563

Lin L-L, Lin Y, Knopf JL (1992) Cytosolic phospholipase A_2 is coupled to hormonally regulated release of arachidonic acid. Proc Natl Acad Sci USA 89: 6147–6151

Liscovitch M (1992) Crosstalk among multiple signal-activated phospholipases. Trends Biochem Sci 17: 393–399

Liscovitch M, Cantley LC (1994) Lipid second messengers. Cell 77: 329–334

Lucantoni A, Polya GM (1987) Activation of wheat embryo Ca^{2+}-regulated protein kinase by unsaturated fatty acids in the presence and absence of calcium. FEBS Lett 221: 33–36

Martiny-Baron G, Scherer GFE (1988) A plant protein kinase and plant microsomal H^+ transport are stimulated by the ether lipid platelet-activating factor. Plant Cell Rep 7: 579–582

Martiny-Baron G, Scherer GFE (1989) Phospholipid-stimulated protein kinase in plants. J Biol Chem 264: 18052–18059

Martiny-Baron G, Hecker D, Manolson MF, Poole RJ, Scherer GFE (1992) Proton transport and phosphorylation of tonoplast polypeptides from zucchini are stimulated by the phospholipid platelet-activating factor. Plant Physiol 99: 1635–1641

Memon AR, Boss WF (1990) Rapid light-induced changes in phosphoinositide kinases and H^+-ATPase in plasma membrane of sunflower hypocotyls. J Biol Chem 265: 14817–14821

Memon AR, Chen Q, Boss WF (1989) Inositol phospholipids activate plasma membrane ATPase in plants. Biochem Biophys Res Commun 162: 1295–1301

Mitchell, RH (1985) Inositol phospholipids and cell surface receptor function. Biochim Biophys Acta 415: 81–147

Minichello J, Polya GM, Keane PJ (1989) Inhibition and activation of oat leaf calcium-dependent protein kinase by fatty acids. Plant Sci 65: 143–152

Moritz A, De Graan PNE, Gispen WH, Wirtz KWA (1992) Phosphatidic acid is a specific activator of phosphatidylinositol-4-phosphate kinase. J Biol Chem 267: 7207–7210

Morré DJ, Morré JT, Varnold R (1984) Phosphorylation of membrane-located proteins of soybean in vitro and response to auxin. Plant Physiol 75: 265–268

Morse MJ, Crain R, Satter RL (1987) Light-stimulated inositolphospholipid turnover in *Samanea saman* leaf pulvini. Proc Natl Acad Sci USA 84: 7075–7078

Na S, Perlin, S, Seto-Young D, Wang GF, Haber JE (1993) Characterization of yeast plasma membrane H^+-ATPase mutant *pmal-A135V* and its revertants. J Biol Chem 268: 11792–11797

Nanda A, Gukovskaya A, Tseng J, Grinstein S (1992) Activation of vacuolar-type proton pumps by protein kinase C—role in neutrophil pH regulation. J Biol Chem 267: 22740–22746

Nickel R, Schütte M, Hecker D, Scherer GFE (1991) The phospholipid platelet-activating factor stimulates proton extrusion in cultured soybean cells and protein phosphorylation and ATPase activity in plasma membranes. J Plant Physiol 139: 205–211

Nishizuka Y (1992) Intracellular signaling by hydrolysis of phospholipids and activation of protein kinase C. Science 258: 607–614

Olah Z, Kiss Z (1986) Occurrence of lipid and phorbol ester activated protein kinase in wheat cells. FEBS Lett 195: 33–37

Palmgren MG (1991) Regulation of plant plasma membrane H^+-ATPase activity. Physiol Plant 83: 314–323

Palmgren MG, Sommarin M (1989) Lysophosphatidylcholine stimulates ATP-dependent proton accumulation in isolated oat root plasma membrane vesicles. Plant Physiol 90: 1009–1014

Parthier B (1991) Jasmonates, new regulators of plant growth and development: many facts and few hypotheses on their actions. Bot Acta 6: 446–454

Polya GM, Klucis E, Haritou M (1987) Resolution and characterization of two soluble calcium-dependent protein kinases from silver beet leaves. Biochim Biophys Acta 931: 68–77

Polya GM, Nott R, Klucis E, Minichiello J, Chandra S (1989) Inhibition of plant calcium dependent protein kinases by basic polypeptides. Biochim Biophys Acta 1037: 259–262

Polya GM, Minichiello J, Nott R, Klucis E, Keane PJ (1990) Differential inhibition of plant calcium-dependent protein kinases by long-chain fatty acids and other amphiphiles. Plant Sci 71: 45–54

Pushkareva MY, Kahn Wa, Allessenko AV, Shayoun N (1992) Sphingosine activation of protein kinase in Jurkat T cells. In vitro phosphorylation of endogenous protein substrates and specificity of action. J Biol Chem 267: 15246–15251

Roberts DM (1993) Protein kinase with calmodulin-like domains: novel targets of calcium signals in plants. Curr Opin Cell Biol 3: 242–246

Roberts DM, Harmon AC (1992) Calcium-modulated proteins: targets of intracellular calcium signals in plants. Annu Rev Plant Physiol Mol Biol 43: 375–414

Samuelson B, Dahlen S, Lindgren JA, Rouzer CA, Serhan CN (1987) Leukotriens and lipoxins: structures, biosynthesis, and biological effects. Science 237: 1171–1175

Schäfer A, Bygrave F, Matzenauer F, Marmé D (1985) Identification of a calcium- and phospholipid-dependent protein kinase in plant tissue. FEBS Lett 201: 282–286

Schaller GE, Sussman MG (1988) Phosphorylation of the plasma membrane H-ATPase of oat roots by calcium-stimulated protein kinase. Planta 173: 509–518

Schaller GE, Harmon A, Sussman MA (1992) Characterization of a calcium- and lipid- dependent protein kinase associated with the plasma membrane of oat. Biochemistry 31: 1721–1727

Scherer GFE (1985) 1-O-alkyl-2-acetyl-*sn*-glycero-3-phosphocholine (platelet-activating factor) stimulates plant H^+ transport and growth. Biochem Biophys Res Commun 133: 1160–1167

Scherer GFE (1990) Phospholipid-activated protein kinase in plants coupled to phospholipase A_2? In: Ranjeva R, Boudet AM (eds) Signal perception and transduction in higher plants. Springer, Berlin Heidelberg New York Tokyo, pp 69–82 (NATO-ASI series, series H, vol 47)

Scherer GFE (1994) Phospholipid signalling by phospholipase A_2 in plants. The role of mastoparan and lysophospholipids as weak "auxin-like" agonists. In: Blatt MR, Leigh RA, Sanders D (eds) Membrane transport in plants and fungi: molecular mechanisms and control. Cambridge University Press, Cambridge, pp 229–242 (Society for Experimental Biology seminar series, vol 58)

Scherer GFE (1995) Activation of phospholipase A by auxin and mastoparan in hypocotyl segments from zucchini and sunflower. J Plant Physiol 145: 483–490

Scherer GFE, André B (1989) A rapid response to a plant hormone: auxin stimulates phospholipase A_2 in vivo and in vitro. Biochem Biophys Res Commun 163, 111–117

Scherer GFE, André B (1993) Stimulation of phospholipase A_2 by auxin in microsomes from suspension-cultured soybean cells is receptor-mediated and influenced by nucleotides. Planta 191: 515–523

Scherer GFE, Martiny-Baron G (1987) Stimulation of in vitro H^+ transport in zucchini microsomes by the ether lipid platelet activating factor and a soluble protein. In: Klämbt D (ed) Plant hormone receptors. Springer, Berlin Heidelberg New York Tokyo, pp 163–176 (NATO ASI series, series H, vol 10)

Scherer GFE, Nickel R (1988) The animal ether phospholipid platelet-activating factor stimulates acidification of the incubation medium by cultured soybean cells. Plant Cell Rep 7: 575–578

Scherer GFE, Stoffel B (1987) A plant lipid and the platelet-activating factor stimulate ATP-dependent H^+ transport in isolated plant membrane vesicles. Planta 172: 127–130

Scherer GFE, Martiny-Baron G, Stoffel B (1988) A new set of regulatory molecules in plants: a plant phospholipid similar to platelet-activating factor stimulates protein kinase and proton-translocating ATPase in membrane vesicles. Planta 175: 241–253

Scherer GFE, André B, Martiny-Baron G (1990) Hormone-activated phospholipase A_2 and lysophospholipid-activated protein kinase: a new signal transduction chain and a new second messenger system in plants? Curr Top Plant Biochem Physiol 9: 190–218

Scherer GFE, Führ A, Schütte M (1993a) Activation of membrane-associated protein kinase by lipids, its substrates, and its function in signal transduction. In: Battey NH, Dickinson HG, Hetherington AM (eds) Post-translational modifications in plants. Cambridge University Press, Cambridge, pp 109–121 (Society for Experimental Biology series, vol 53)

Scherer GFE, Hecker D, Müller J (1993b) Ca^{2+} ions and lysophospholipids activate phosphorylation of different proteins in plasma membranes and tonoplast purified by free-flow electrophoresis. J Plant Physiol 142: 432–437

Serrano R, Montesinos C, Sanchez J (1988) Lipid requirements of the plasma membrane ATPase from oat roots and yeast. Plant Sci 56: 117–122

Sharp JD, White DL, Chiou XG, Goodson T, Gamboa GC, McClure D, Burgett S, Hoskins J, Skatrud PL, Sportsman JR, Becker GW, Kang LH, Roberts EF, Kramer RM (1991) Molecular cloning and expression of human Ca^{2+}-sensitive cytosolic phospholipase A_2. J Biol Chem 266: 14850–14853

Suen K-L, Choi JH (1991) Isolation and sequence analysis of a cDNA clone for a carrot calcium-dependent protein kinase: homology to calcium/calmodulin-dependent protein kinases and to calmodulin. Plant Mol Biol 17: 581–590
Takai Y, Kishimoto A, Inoue M, Nishizuka Y (1977) Studies on a cyclic nucleotide-independent protein kinase and its proenzyme in mammalian tissue. I. Purification and characterization of an active enzyme from bovine cerebellum. J Biol Chem 252: 7603–7609
Takai Y, Kishimoto A, Iwasa Y, Kawahara Y, Mori T, Nishizuka Y (1979) Calcium-dependent activation of a multifunctional protein kinase by membrane lipids. J Biol Chem 254, 3692–3695
Verhey SD, Gaiser C, Lomax TL (1993) Protein kinases in zucchini. Characterization of calcium-requiring plasma membrane kinases. Plant Physiol 103: 413–419
Weaver CD, Roberts DM (1992) Determination of the site of phosphorylation of nodulin 26 by the calcium-dependent protein kinase from soybean nodules. Biochemistry 31: 8954–8959
Yang W, Boss WF (1994) Regulation of phosphatidylinositol 4-kinase by the protein activator PIK-A49. Activation requires phosphorylation of PIK-A49. J Biol Chem 269: 3852–3857
Yang W, Burckhardt W, Cavallius J, Merrick WC, Boss WF (1993) Purification and characterization of a phosphatidyl-inositol 4-kinase activator in carrot cells. J Biol Chem 268: 392–398
Zhao Y, Kappes B, Franklin RM (1993) Gene structure and expression of an unusual protein kinase from *Plasmodium falciparum* homologous at its carboxyl terminus with the EF hand calcium-binding proteins. J Biol Chem 268: 4347–4354

S-Locus Receptor Kinase Genes and Self-incompatibility in *Brassica napus*

Daphne R. Goring and Steven J. Rothstein

Contents

I. Introduction

In some species of *Brassica*, there exists a sporophytic self-incompatibility system to control fertilization. The diploid species, *B. campestris* (2n = 20, aa) and *B. oleracea* (2n = 18, cc) generally are found as self-incompatible lines. However, *B. napus* (2n = 38, aacc), an amphidiploid consisting of the genomes from these two diploid species, generally is found to be self-compatible (Downey and Rakow, 1987). There are some naturally occurring self-incompatible *B. napus* (Gowers, 1989), and in addition, self-incompatibility alleles from *B. campestris* have been crossed into self-compatible *B. napus* (Mackay, 1977). There has been particular interest in transferring self-incompatibility traits into the oil seed crop *B. napus* ssp. *oleifera* as a method for producing hybrid lines. Given that seed is required for production of canola oil and meal, yield is extremely important and can be increased through their use (Thompson, 1983; Grant and Beversdorf, 1985). The prerequisite for developing these lines is the use of a pollen-control system to prevent self-pollination of the female line.

The effect of the self-incompatibility system is to block "self" pollen from fertilizing the *Brassica* plant. This block usually occurs at the initial stages of pollination by preventing pollen hydration, pollen germination, or pollen tube penetration of the stigma papillae (Heslop-Harrison, 1975). This is thought to occur as a result of a recognition event between the pollen and the stigma papillae. The recognition event must be very specific since pollen derived from self-compatible lines or from parental plants carrying other S-alleles is not affected. The outcome of this system is to force outcrossing in the self-incompatible lines.

II. Genetics of the Self-incompatibility System

Genetic analysis has shown that *Brassica* self-incompatibility is controlled by a single genetic locus called the S-locus (Bateman, 1955). A systematic characterization of the *B. oleracea* S-loci through genetic crosses has shown that there are nearly 50 different alleles present (Ockendon, 1974, 1982). *Brassica* self-incompatibility has been described as sporophytic, because while pollen grains are haploid, the phenotype of the pollen reflects that of the diploid pollen parent. Thus, if two different S-alleles are present in the pollen parent, the phenotype of the haploid pollen grain is generally derived from both alleles (de Nettancourt, 1977). This genetic analysis therefore demonstrated that there must be an S-product present in pollen that when recognized by the stigma prevents fertilization. Further, this S-product must be either synthesized by the diploid pollen mother cell before mitosis, or more likely is secreted from the surrounding diploid tapetal cells during pollen development for deposition in the exine of the pollen grain.

Many of the characterized S-alleles behave as co-dominant alleles when in a heterozygous state (Thompson and Taylor, 1966). In these cases, the action of the S-locus is to block fertilization by pollen grains originating from a pollen parent which has one or two common S-alleles with the pistil (Fig. 1). For example, an S6/S10 pistil will prevent germination of any pollen originating from the S6/S7 anther regardless of the pollen genotype. Conversely, the same pollen landing on a S10/S12 pistil will successfully fertilize this plant. However, for certain *B. oleracea* S-alleles, Thompson and Taylor (1966) found that certain combinations of S-alleles did not display co-dominance, but showed dominant and recessive interactions. These interactions were not linear and changed for different allele combinations as well as for the site of the interaction (pollen or pistil). For example, when the pollen recessive S2 allele was present in combination with the S1 allele, the S1/S2 pollen did not

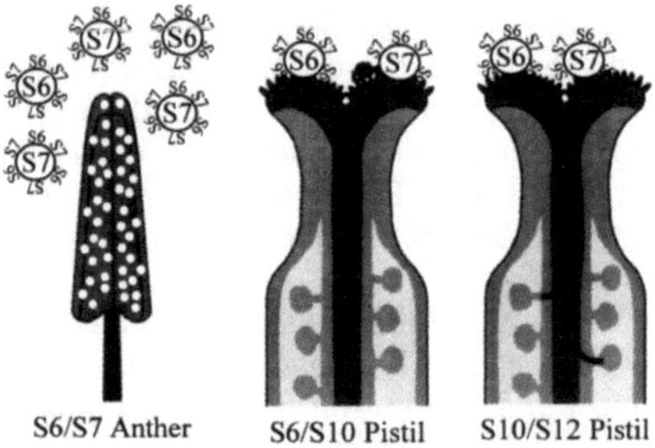

Fig. 1. Sporophytic self-incompatibility in *Brassica*. The phenotype of the haploid pollen grain is determined by the genotype of the diploid pollen parent. Thus, pollen grains from an S6/S7 anther which are genetically either S6 or S7 will not germinate on an S6/S10 pistil because the S6 allele is present in both parents. These pollen grains will fertilize the S10/S12 pistil because no S-alleles are shared by the parents

germinate on the S1/S1 pistil, but unexpectedly fertilized S2/S2 pistils. Conversely, S1/S18 pistils prevented S18/S18 pollen from germinating, but allowed fertilization by S1/S1 pollen indicating that the S1 allele was recessive to S18 in the pistil. Thus, in certain combinations, one allele was able to block the function of another allele.

III. Molecular Biology of the Self-incompatibility System

Recent molecular analyses have revealed that the S-locus is a complex locus and so far consists of two different genes which appear to be required for the self-incompatibility trait. The first gene to be characterized is the S-locus glycoprotein (*SLG*) gene (J. Nasrallah et al., 1985). This gene encodes for an abundant, allele specific, glycosylated protein which appears to be secreted and localized to the stigma papillae cell wall (M. Nasrallah et al., 1970; Hinata and Nishio, 1978; Kandasamy et al.,1989). *SLG* cDNAs for several alleles from different *Brassica* species have been isolated and sequence analyses show that they have several features in common (J. Nasrallah et al., 1985, 1987; Trick and Flavell, 1989; Chen and Nasrallah, 1990; Dwyer et al., 1991; Goring et al., 1992a, b; Scutt and Croy, 1992). All of these genes have putative signal peptides and several potential N-glycosylation sites as predicted for a secreted glycoprotein. In addition, they contain 11–12 conserved cysteine residues towards the C-terminal end of the predicted coding region (Fig. 2).

A second gene at the S-locus was subsequently isolated based on sequence similarity to the *SLG* gene (Stein et al., 1991; Goring and Rothstein, 1992). This gene, called the S-locus receptor kinase (*SRK*), is also predominantly expressed in the pistil. Analysis of the predicted amino acid sequence suggested a structure composed of an N-terminal domain resembling the *SLG* coding region, followed by a putative transmembrane domain and a C-terminal region showing homologies to serine/theronine kinases (Fig. 2). As seen for the *SLG* genes, the putative SRK extracellular domain contains the 11–12 conserved cysteine residues. Conserved cysteine rich domains are found in the extracellular domains of a variety of animal receptor tyrosine kinases (Aaronson, 1991) and TGF-β receptor serine/theronine kinases (Massague et al., 1994). In the TGF-β receptor family, the type II TGF-β receptors show variable patterns of cysteines which appear to be involved in ligand binding specificity, while the type I TGF-β receptors have a conserved pattern of cys-

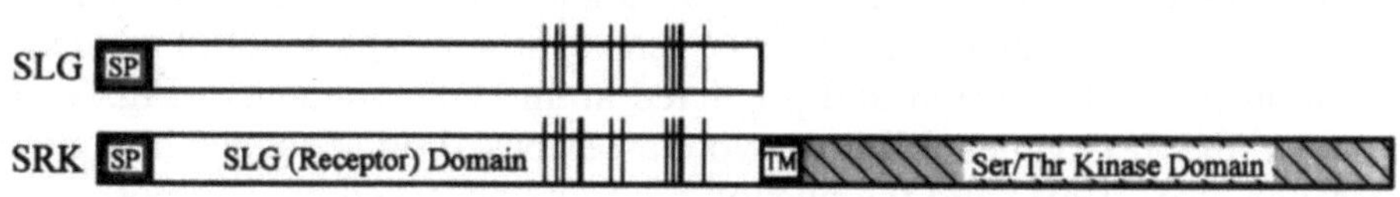

Fig. 2. Predicted structure of the *Brassica* S-locus proteins. The putative receptor region of the SRK shows high sequence homology to the SLG. Vertical bars in this region represent conserved cysteine residues. SLG, S-locus glycoprotein; SRK, S-receptor kinase; SP, signal peptide; TM, transmembrane domain; Ser/Thr, serine/threonine

teines which do not influence the ligand binding pattern (Massague et al., 1994). Since the cysteines are invariable in the S-locus proteins, they probably do not play a role in the specificity of ligand binding, but may be involved in the secondary structure of this SLG region. Another conserved region present in the extracellular domain of many animal receptor kinases is the immunoglobulin-like repeat (Aaronson, 1991). A comparison of the putative SRK extracellular domain to the consensus immunoglobulin motif revealed significant homology to this repeat in the cysteine rich region (Glavin et al., 1994).

IV. Self-incompatibility in *Brassica napus*

We have characterized the *SLG* and *SRK* genes in two different lines of self-incompatible *B. napus*. In both cases, the original cultivar was self-compatible and was converted into a self-incompatible line by crossing in a functional S-allele. The first line, W1, was generated by crossing a self-incompatible *B. campestris* plant to the self-compatible *B. napus* ssp. *oleifera* cv. Westar. The second line, T2, was generated by crossing a naturally occurring *B. napus* ssp. *rapifera* line to the self-compatible *B. napus* ssp. *oleifera* cv. Topas. Using a variety of library screening and PCR techniques, the *SLG* and *SRK* genes for these two S-loci were isolated (Goring et al., 1992a, b; Goring and Rothstein, 1992; Glavin et al., 1994). The functional W1 S-locus was called the 910-locus, while the functional T2 S-locus was called the A14-locus. RNA blot analysis showed that the genes were all expressed primarily in the pistils from maturing buds, and the *SLG* genes showing much higher levels of steady-state mRNA compared to the *SRK* genes. In all cases, these genes were found to segregate with self-incompatibility in a F_2 population derived by crossing the self-incompatible lines to the original self-compatible lines. Using PCR, we have screened for recombination events between the *SLG* and *SRK* genes from the 910 and A14 loci in almost 300 double haploid plants derived from a W1/T2 heterozygous plant. The absence of any SLG/SRK recombinants suggested that these genes are tightly linked (Goring et al., unpubl.). For the *B. oleracea* S6 locus, the *SLG* and *SRK* genes have been mapped to be approximately within 200 kb of each other using pulsed-field gel electrophoresis (Boyes and Nasrallah, 1993). Pulse-field gel analysis of the 910 locus has revealed that the *SLG-910* and *SRK-910* genes are within approximately 25 kb of each other (Yu and Rothstein, unpubl.).

Alignments of the *SLG-A14* (*B. napus* allele) and *SLG-910* (originally a *B. campestris* allele) genes to other *SLG* genes show a high level of sequence conservation for both the DNA and predicted amino acid sequence (Fig. 3). In *B. oleracea*, the *SLG* genes fall into two classes based on sequence homology (J. Nasrallah et al., 1991) and the phenotypic traits observed by Thompson and Taylor (1966). The class II alleles belong to the group of pollen recessive alleles which tend to have weak self-incompatibility phenotypes, while the class I alleles show stronger self-incompatibility phenotypes. *SLG* cDNA clones isolated from *B. napus* and *B. campestris* are most similar to the class I

AMINO ACID \ DNA	*B. napus* SLG-A14	*B. campestris* SLG-910 SLG-8	*B. oleracea* Class I SLG-6 SLG-13 SLG-29	*B. oleracea* Class II SLG-2 SLG-5 *
B. napus SLG-A14		76, 78	76 - 82	65, 66
B. campestris SLG-910 SLG-8	84, 87	85 \ 79	78 - 90	66 - 68
B. oleracea Class I SLG-6 SLG-13 SLG-29	87 - 90	86 - 94	88 - 90 \ 82 - 84	65 - 68
B. oleracea Class II SLG-2 SLG-5 *	73	73 - 75	73 - 74	93 \ 89

Fig. 3. Comparison of the *SLG* gene sequences within and between *Brassica* spp. DNA and amino acid comparisons were made using the predicted coding regions for each gene. Shaded boxes represent DNA/amino acid homologies within a group. Ranges were listed for alignments involving several genes. *Only a partial sequence was available for the *SLG-5* gene. References for the sequences are the following: *SLG-A14*, Goring et al. (1992b); *SLG-910*, Goring et al. (1992a); *SLG-8* and *SLG-13*, Dwyer et al. (1991); *SLG-6*, J. Nasrallah et al. (1985); *SLG-29*, Trick and Flavell (1989); *SLG-2*, Chen and Nasrallah (1990); *SLG-5*, Scutt and Croy (1992)

B. oleracea alleles. Comparison of all the class I-like *SLG* genes also shows that similar levels of homology exist both within and between species (Fig. 3). The putative receptor portion of the *SRK* genes which resemble the *SLG* gene similarly show close homology between class I-like alleles (Fig. 4). An examination of the SLG/SRK pairs from the 910-locus and the S6 locus initially suggested that highest levels of homology were present within a gene pair (Stein et al., 1991; Goring and Rothstein, 1992). However, characterization of the A14 locus had revealed that this is not always the case. The SLG domain of the *SRK-A14* gene is more closely related to the *B. oleracea SLG-6* gene (81%, amino acid level) than it is to the *SLG-A14* gene (78%, amino acid level) (Fig. 4).

The SRK domain was originally identified as a serine/threonine kinase based on homology to other kinases and the presence of two conserved regions which specify serine/threonine catalytic activity (Stein et al., 1991). A comparison of the SRK-910 domain in the data banks shows several interesting features. In plants, outside of the *SRK* related genes, the kinase domain shows only 31–38% homology to other putative serine/threonine kinases. In animals, the highest homologies are seen to tyrosine kinases of the trk (31%) and insulin receptor (29%) families and to the *Drosophila* pelle serine/threonine

AMINO ACID / DNA	SLG-910	SRK-910	SLG-A14	SRK-A14	SLG-6	SRK-6
SLG-910		84	76	75	80	76
SRK-910	90		73	75	79	75
SLG-A14	84	83		78	82	75
SRK-A14	83	83	87		81	80
SLG-6	86	86	90	87		86
SRK-6	85	84	86	88	91	

Fig. 4. Sequence comparisons of the SLG domains in the *SLG* and *SRK* genes. DNA and amino acid comparisons were made using the predicted SLG domain from each gene. Shaded boxes represent sequence identities between the *SLG* gene and *SRK* gene from the same S-locus. References for the sequences are the following: *SLG-910*, Goring et al. (1992a); *SRK-910*, Goring and Rothstein (1992); *SLG-A14*, Goring et al. (1992b); *SRK-A14*, Glavin et al. (1994); SLG-6, J. Nasrallah et al. (1985); *SRK-6*, Stein et al. (1991)

kinase (29%). Confirmation that the *SRK-910* gene did encode a functional serine/threonine kinase was obtained by expressing the kinase domain as a glutathione S-transferase (GST) fusion protein in *E. coli*, testing the fusion protein for kinase activity, and identifying the phosphorylated amino acids (Goring and Rothstein, 1992). As a negative control, a conserved lysine which had been shown to be required for activity in other kinases (Hanks et al., 1988) was mutated to an alanine (Fig. 5 a). Only the wild-type kinase was found to have autophosphorylation activity (Fig. 5 b). When the phosphorylated kinase domain was subjected to 2-dimensional phosphoamino acid analysis, it was found to contain only phosphoserines and phosphothreonines (Fig. 5 c). Similar results were found for the kinase domain encoded by the *B. oleracea* *SRK*-6 gene (Stein and Nasrallah, 1993).

V. Self-compatibility in *Brassica napus*

Most cultivars of *B. napus* are self-compatible despite their origins from two self-incompatible species. Molecular hybridization of various cultivars has shown that *SLG/SRK*-like genes are present in these lines, but presumably are no longer functional. We have analyzed one of these loci in the cultivar Westar (Goring et al., 1993). The *SLG* and *SRK* genes from this locus, called the A10 locus, showed highest levels of homology to the class I S-alleles based on DNA

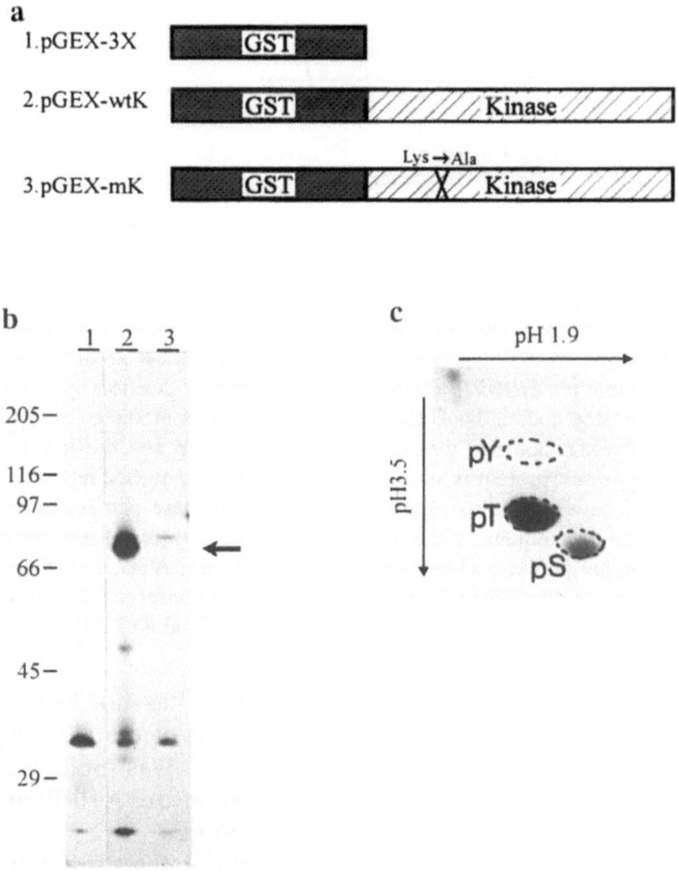

Fig. 5. Detection of serine/threonine kinase activity in the SRK-910 kinase domain. a, The constructs used to produce fusion proteins in *E. coli* are (1) the glutathione-S-transferase (GST) vector alone, (2) the SRK-910 kinase domain starting at the end of the transmembrane domain fused to the *GST* gene, and (3) a mutant version of the SRK-910 kinase domain (one amino acid change, lysine to alanine) fused to the *GST* gene. b, Autoradiography of fusion proteins tested for kinase activity (autophosphorylation) in the presence of [γ-^{32}P]ATP. Lane 1, GST; lane 2, GST-wtkinase; lane 3, GST-mkinase. The positions of the molecular weight markers are indicated on the left. The GST kinase fusion protein has an approximate molecular weight of 72 kDa as marked by the arrow. A signal representing kinase activity is only detected in lane 2 (GST-wtkinase). c, Phosphoamino acid analysis of the GST-wtkinase protein. The labelled GST-wtkinase protein was hydrolysed and ran with phosphoamino-acid markers on a two-dimensional thin-layer electrophoresis system. The positions of phosphoamino acid markers (visualized with ninhydrin) are marked by dotted circles. Only labelled phosphoserines and phosphothreonines were detected. Figure adapted from Goring and Rothstein (1992)

blot analysis. Cloning of the *SLG* and *SRK* cDNAs from this locus revealed that both of these genes are expressed in the spatial and temporal pattern observed for functional *SLG* and *SRK* genes. However, sequence analysis of the cDNAs revealed two interesting features. One feature is a region of 100% homology

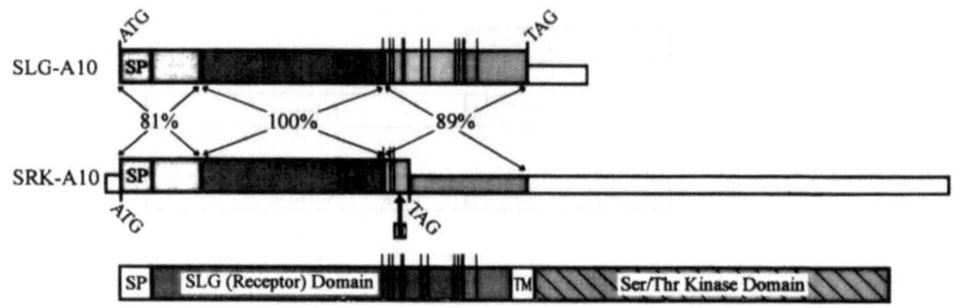

Fig. 6. The A10 S-locus genes. The structures of the predicted coding regions for the *SLG-A10* and *SRK-A10* genes are shown by large boxes. The *SLG-A10* CDNA is predicted to encode a full length SLG protein, while the *SRK-910* CDNA has a one-base-pair deletion (marked by D) which results in a premature stop codon. Thus, the SRK-A10 protein is predicted to be lacking the C-terminal region of the SLG domain, the transmembrane domain and the kinase domain. The structure of a wild type SRK protein is shown underneath. DNA sequence homologies in the SLG domain are indicated between the two CDNAS. There is a 590 base pair region showing 100% homology between the two sequences and is thought to have occurred as a result of a gene conversion event. SP, signal peptide; TM, transmembrane domain; ATG, initiation codon; TAG, stop codon; Ser/Thr, serine/threonine. Vertical bars represent conserved cysteines found in the SLG domain. Figure adapted from Goring et al. (1993)

between the *SLG-A10* and *SRK-A10* genes which has not been detected in other SLG/SRK pairs (Fig. 6). Secondly, near the end of the SLG domain in the *SRK* gene, there was a one base pair deletion which was predicted to cause a frame-shift resulting in a truncated protein. Thus, at the A10 locus, the *SRK-A10* gene was inactivated by a one-base-pair deletion.

Analysis of the A10 locus has raised some interesting questions. For example, will a correction of the one-base-pair deletion in the *SRK-A10* gene result in a functional S-locus? The predicted sequence of the kinase domain did not reveal any obvious alterations potentially affecting kinase activity. Analysis of the *SLG* gene also suggested that it should be functional. However, if there is a third component to the S-locus, a pollen ligand (see below), then this component also must be intact. While the A10 locus suggests that self-incompatibility was lost in *B. napus*, it is not known if self-compatibility occurred simultaneously with the origin of *B. napus* or resulted from centuries of domestication of *B. napus*. Selection for self-compatibility could occur with the origins of *B. napus* if it conferred an advantage to the plant. For example, self-compatibility would be a desired trait if there were cross-pollination difficulties due to low density, and if the polyploid nature of *B. napus* circumvented severe inbreeding depression (Jain, 1976).

VI. Molecular Model for *Brassica* Self-incompatibility

While direct proof has not been obtained to show that the *SLG/SRK* genes are required for *Brassica* self-incompatibility, there is growing evidence for their involvement. First of all, in several cases these genes have been genetically linked to the self-incompatibility response (J. Nasrallah et al., 1985; Chen and Nasrallah, 1990; Stein et al., 1991; Goring et al., 1992a, b; Glavin et al., 1994). Secondly, analysis of S-locus genes in self-compatible lines in *B. napus* and *B. oleracea* have revealed mutations in the *SRK* genes (Goring et al., 1993; J. Nasrallah et al., 1994). Lastly, loss of *SLG* expression in *B. oleracea* has been associated with loss of self-incompatibility (Toriyama et al., 1991; M. Nasrallah et al., 1992). Thus, from these studies, we assume that both the *SLG* and *SRK* genes are required to produce a *Brassica* self-incompatibility response.

In order to understand the possible role of the S-receptor kinase, it is interesting to look at what has been discovered about analogous proteins in other species. One of the best characterized systems is the receptor tyrosine kinase and the cascade of events that are triggered by receptor activation in animal cells (Fantl et al., 1993). In this system, a ligand binds to the receptor leading to receptor dimerization, activation of the kinase domain, and phosphorylation of key tyrosine residues in the cytoplasmic domain. These phosphotyrosines represent binding sites for cytosolic proteins which upon binding and phosphorylation become activated. These proteins subsequently initiate a cascade of events which may lead to elevation of cytosolic calcium (Ca^{2+}) levels, changes in cytosolic pH and potassium levels, cytoskeletal rearrangements, or gene regulation (Fig. 7). Receptor serine/threonine kinases also exist in animal systems as represented by the TGF-β receptor family. While the ligands are known for these receptors, the pathways which are activated are relatively uncharacterized.

Two generalizations can be made from what is known about animal receptor kinases. One is that generally a protein ligand binds to the extracellular domain of the receptor kinase resulting in kinase activation. Secondly, receptor

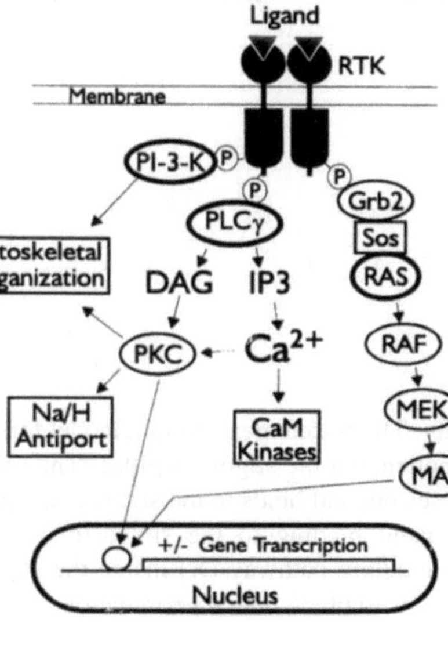

Fig. 7. Signalling pathways activated by receptor tyrosine kinases. RTK, receptor tyrosine kinase; PI-3-K, phosphatidylinositol-3-kinase; PLCγ, phospholipase Cγ; DAG, diacyglycerol; IP3, inositol-1,4,5-trisphosphate; PKC, protein kinase C; Ras is a guanosine trisphosphatase; Raf, Mek, and MAP are protein kinases. Figure adapted from Crew and Erikson (1993); Juliano and Haskill (1993); Posada and Cooper (1992)

activation leads to a cascade of events inside the cell in response to the external signal. By analogy, the self-incompatibility reaction might involve the following events, although other scenarios are certainly possible. The gene encoding a putative ligand must be linked to the S-locus, since genetically this system is inherited as a single locus. In addition, the ligand gene must have evolved along with the *SLG* and *SRK* genes to be specific for the particular SLG/SRK pair, and represent the pollen component of the self-incompatibility system. When the pollen lands on the pistil surface, the ligand (in the pollen exine) diffuses out and if the corresponding receptor is present in the stigma papillae cell wall binds to the receptor and stimulates pathways leading to a block in pollen germination (Fig. 8). While a specific ligand for the SRK protein has not been identified, small molecular weight proteins from the pollen surface have been shown to bind non-specifically to SLG proteins and may represent related proteins (Doughty et al., 1993).

Examination of other receptor systems does not offer an explanation for the role of the SLG protein in this response. In effect, the SLG represents a secreted form of the SRK receptor domain. These secreted receptor forms have

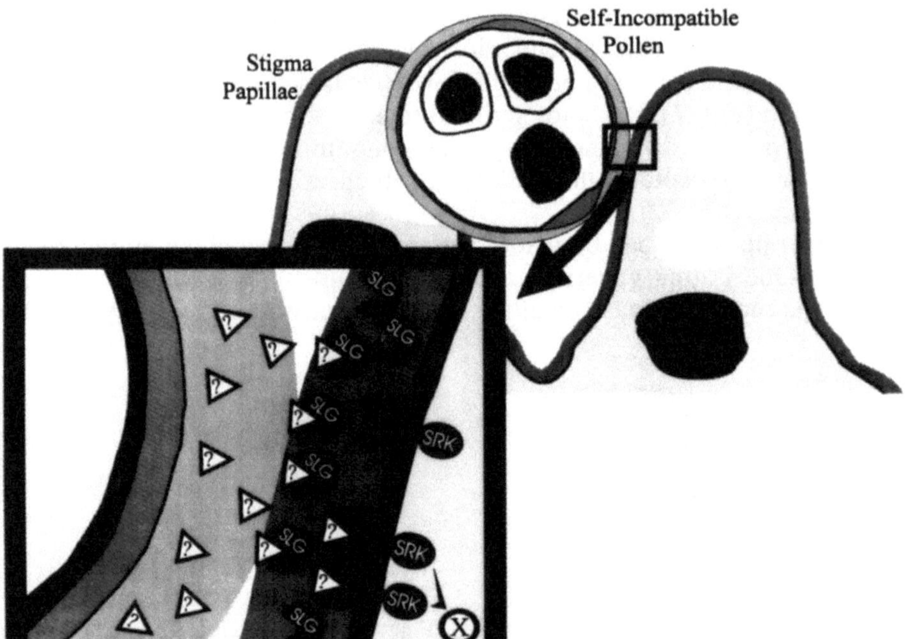

Fig. 8. Proposed model for *Brassica* self-incompatibility. The boxed area shows a model for the molecular interactions between the self-incompatible pollen and the stigma papillae. The triangles represent an unidentified pollen ligand which diffuses out and binds to the SLG and SRK proteins, all encoded by genes at the same S-locus. Ligand binding to the SRK activates the cytoplasmic kinase domain which in turn initiates a signalling pathway (X) inside the stigma papillae. The end result of the signalling pathway is to block pollen germination

been detected in other systems, but it is unknown what role they play (Johnson et al., 1990; Petch et al.,1990). They have been suggested to play an inhibitory role by preventing the ligand from reaching the receptor. It is not clear how this could be the role of the SLG in self-incompatibility and if it were, it would imply a quite different mechanism for self-recognition to that described earlier. An alternative role for the SLG protein is that it helps the ligand cross the cell wall thus allowing it to come into contact with the SRK protein. While the SLG/SRK proteins from the same locus are not necessarily most similar to each other, it is difficult to know whether the important amino acids for ligand binding are conserved. For example, the sequence differences between the SLG and SRK may result in different ligand affinities. The ligand would therefore first bind the soluble, abundant SLGs which appear to be associated with the cell wall. The ligand–SLG complex would then come into contact with the SRK extracellular domain, and the ligand transferred to the SRK due to its higher affinity (Fig. 8).

There are several conditions which repress the self-incompatibility response and possibly these affect the ligand/receptor interaction. For example, both high temperature and the addition of an NaCl solution can overcome self-incompatibility and these conditions would influence the ability of a ligand to bind to its receptor. Further, the dominance of certain alleles may also be due to the differential binding affinities between different ligands and receptors. For example, the recessive nature of the S1 allele to the S18 allele in the pistil may result from a disruption of S1 ligand binding to the S1 receptor. If the SLG-18 protein had a higher affinity for the S1 ligand compared with the SLG-1 or SRK-1 proteins, then the SLG-18 protein may prevent the S1 ligand from binding to the S1 receptor. In the case of the pollen recessive S2 allele, Tantikanjana et al. (1993) have shown that the *SLG-2* gene in addition to producing a secreted glycoprotein also produces an anomalous membrane bound SLG which may interfere with the function of the SRK-2 protein.

Little is known about the pathway activated by the receptor kinase. One of the visible markers of self-incompatibility is the deposition of callose (Kerhoas et al., 1993). However, callose has not been shown to be necessary for this reponse. Interestingly, calcium, an important second messenger in receptor tyrosine kinase signalling and other signalling pathways, has been shown to activate callose synthase in soybean cultures (Kauss, 1985). As mentioned for the receptor tyrosine kinases, receptor activation ultimately leads to the binding of cytosolic proteins to the kinase domain. A number of techniques have been developed which exploit the physical association of two proteins to isolate unknown components (Guarente, 1993). If the SRK protein functions in a similar manner to the receptor tyrosine kinases, these approaches can be used to isolate the target of the activated SRK protein. One such approach is the yeast two-hybrid system (Fields and Song, 1989). It is based on the use of yeast transcription factors which can be separated into two domains, a DNA binding domain and a transcription activation domain. These two domains once separated cannot associate with each other and must be fused to two interacting proteins (e.g., kinase domain and its substrate) to bring the two domains together. This interaction can be detected by the transcriptional activation of reporter genes (e.g., *LacZ, HIS3*) in the yeast genome.

VII. Conclusion

Self-incompatibility involves the recognition of self-pollen, the subsequent prevention of fertilization and enforces outcrossing in self-incompatible species. The discovery of the *SRK* gene will undoubtedly lead to a much greater understanding of the biochemical basis of self-incompatibility than is presently available. However, at this time there are many questions that remain unanswered. Amongst these are the following: What is the pollen factor recognized by the *SRK*? How are the pollen factors from 50 different S-alleles distinguished from each other, thus giving the specificity of the self-incompatible reaction? Finally, once the SRK protein is activated, how is fertilization prevented? There are a variety of biochemical and genetic techniques that can be brought to bear to address these questions that are largely derived from studies on animal receptor systems. Consequently, it is likely that in the next few years answers to these questions will be elucidated.

VIII. References

Aaronson SA (1991) Growth factors and cancer. Science 254: 1146–1153

Bateman AJ (1955) Self-incompatibility systems in angiosperms. III. Cruciferae. Heredity 9: 53–68

Boyes, DC, Nasrallah JB (1993) Physical linkage of the *SLG* and *SRK* genes at the self-incompatibility locus of *Brassica oleracea*. Mol Gen Genet 236: 369–373

Chen CH, Nasrallah JB (1990) A new class of S sequences defined by a pollen recessive self-incompatibility allele of *Brassica oleracea*. Mol Gen Genet 222: 241–248

Crews CM, Erikson RL (1993) Extracellular signals and reversible protein phosphorylation: what to mek of it all. Cell 74: 215–217

Doughty J, Hedderson F, McCubbin A, Dickinson H (1993) Interaction between a coating-borne peptide of the *Brassica* pollen grain and stigmatic S (self-incompatibility)-locus-specific glycoproteins. Proc Natl Acad Sci USA 90: 467–471

Downey RK, Rakow GFW (1987) Rapeseed and mustard. In: Fehr WR (ed) Principles of cultivar development. Macmillan, New York, pp 437–486

Dwyer KG, Balent MA, Nasrallah JB, Nasrallah ME (1991) DNA sequences of self-incompatibility genes from *Brassica campestris* and *B. oleracea*: polymorphism predating speciation. Plant Mol Biol 16: 481–486

Frantl WJ, Johnson DE, Williams LT (1993) Signalling by receptor tyrosine kinases. Annu Rev Biochem 62: 453–481

Fields S, Song O-K (1989) A novel genetic system to detect protein-protein interactions. Nature 340: 245–246

Glavin TL, Goring DR, Shafer U, Rothstein SJ (1994) Features of the extracelular domain of the S-locus receptor kinase from *Brassica*. Mol Gen Genet 244: 630–637

Goring DR, Rothstein SJ (1992) The S-locus receptor kinase gene in a self-incompatible *Brassica napus* line encodes a functional serine/threonine kinase. Plant Cell 4: 1273–1281

Goring DR, Banks P, Beversdorf WD, Rothstein SJ (1992a) Use of the polymerase chain reaction to isolate an S-locus glycoprotein cDNA introgressed from *Brassica campestris* into *B. napus* ssp. *oleifera*. Mol Gen Genet 234: 185–192

Goring DR, Banks P, Fallis L, Baszczynski CL, Beversdorf WD, Rothstein SJ (1992b) Identification of an S-locus glycoprotein allele introgressed from *B. napus* ssp. *rapifera* to *B. napus* ssp. *oleifera*. Plant J 2: 983–989

Goring DR, Glavin TL, Schafer U, Rothstein SJ (1993) The S-locus receptor kinase gene in self-compatible *Brassica napus* has a frame-shift mutation. Plant Cell 5: 531–539

Gower S (1989) Self-incompatibility interactions in *Brassica napus*. Euphytica 42: 99–103

Grant I, Beversdorf WD (1985) Heterosis and combining ability estimates in spring-planted oilseed rape (*Brassica napus* L.). Can J Genet Cytol 27: 472–478

Guarente L (1993) Strategies for the identification of interacting proteins. Proc Natl Acad Sci USA 90: 1639–1641

Hanks S, Quinn AM, Hunter T (1988) The protein kinase family: conserved features and deduced phylogeny of the catalyic domains. Science 241: 42–52

Heslop-Harrison J (1975) Incompatibility and the pollen-stigma interaction. Annu Rev Plant Physiol 26: 403–425

Hinata K, Nishio T (1978) S allele specificity of stigma proteins in *Brassica oleracea* and *Brassica campestris*. Heredity 41: 93–100

Jain SK (1976) The evolution of inbreeding plants. Annu Rev Ecol Syst 7: 469–495

Johnson DE, Lee PL, Lu J, Williams LT (1990) Diverse forms of a receptor for acidic and basic fibroblast growth factors. Mol Cell Biol 1: 4728–4736

Juliano RL, Haskill S (1993) Signal transduction from the extracellular matrix. J Cell Biol 120: 577–585

Kandasamy MK, Paolillo DJ, Faraday CD, Nasrallah JB, Nasrallah ME (1989) The S-locus specific glycoproteins of *Brassica* accumulate in the cell wall of developing stigma papillae. Dev Biol 134: 462–472

Kauss H (1985) Callose biosynthesis as a Ca^{2+}-regulated process and possible relations to the induction of other metabolic changes. J Cell Sci [Suppl] 2: 89–103

Kerhoas C, Knox RB, Dumas C (1983) Specificity of the callose response in stigmas of *Brassica*. Ann Bot 52: 597–602

Mackay GR (1977) The introgression of S alleles into forage rape *Brassica napus* L. from turnip *Brassica campestris* ssp. *rapifera*. Euphytica 26, 511–519

Massague J, Attisano L, Wrana JL (1994) The TGF-β family and its composite receptors. Trends Cell Biol 4: 172–178

Nasrallah JB, Goldberg ML, Nasrallah ME (1985) A cDNA clone encoding an S-locus-specific glycoprotein from *Brassica oleracea*. Nature 318: 263–267

Nasrallah JB, Kao TH, Chen CH, Goldberg ML, Nasrallah ME (1987) Amino-acid sequence of glycoprotein encoded by three alleles of the S-locus of *Brassica oleracea*. Nature 326: 617–619

Nasrallah JB, Nishio T, Nasrallah ME (1991) The self-incompatibility genes of *Brassica*. Annu Rev Plant Physiol Plant Mol Biol 42: 393–422

Nasrallah JB, Rundle SJ, Nasrallah ME (1994) Genetic evidence for the requirement of the *Brassica* S-locus receptor kinase gene in the self-incompatibility response. Plant J 5: 373–384

Nasrallah ME, Barber JT, Walllace DH (1970) Self incompatibility proteins in plants: detection, genetics and possible mode of action. Heredity 25: 23–27

Nasrallah ME, Kandasamy MK, Nasrallah JB (1992) A genetically defined trans-acting locus regulates S-locus function in *Brassica*. Plant J 2: 497–506

de Nettancourt D (1977) Incompatibility in angiosperms. Springer, Berlin Heidelberg New York

Ockendon DJ (1974) Distribution of self-incompatibility alleles and breeding structure of open-pollinated cultivars of Brussel sprouts. Heredity 33: 159–171

Ockendon DJ (1982) An S-allele survey of cabbage (*Brassica oleracea* var. capitata). Euphytica 31: 325–331

Petch LA, Harris J, Raymond VW, Blasband A, Lee DC, Earp HS (1990) A truncated, secreted form of the epidermal growth factor receptor is encoded by an alternatively spliced transcript in normal rat issue. Mol Cell Biol 10: 2973–2989

Scutt C, Croy RRD (1992) An S5 self-incompatibility allele-specific cDNA sequence from *Brassica oleracea* shows high homology to the *SLR2* gene. Mol Gen Genet 232: 240–246

Stein JC, Nasrallah JB (1993) A plant receptor-like gene, the S-locus receptor kinase of *Brassica oleracea* L., encodes a functional serine/threonine kinase. Plant Physiol 101: 1103–1106

Stein JC, Howlett B, Boyes DC, Nasrallah ME, Nasrallah JB (1991) Molecular cloning of a putative receptor protein kinase gene encoded at the self-incompatibility locus of *Brassica oleracea*. Proc Natl Acad Sci USA 88: 8816–8820

Thompson KF (1983) Breeding winter oilseed rape, *Brassica napus*. Adv Appl Biol 7: 1–104

Thompson KF, Taylor JP (1966) Non-linear dominance relationships between S-alleles. Heredity 21: 345–362

Toriyama K, Stein JC, Nasrallah ME, Nasrallah JB (1991) Transformation of *Brassica oleracea* with an S-locus gene from *B. campestris* changes the self-incompatibility phenotype. Theor Appl Genet 81: 769–776

Trick M, Flavell RB (1989) A homozygous S genotype of *Brassica oleracea* expresses two S-like genes. Mol Gen Genet 218: 112–117

Mechanisms of Input and Output in Circadian Transduction Pathways

Isabelle Carré and Steve A. Kay

Contents

I. Introduction

A. Circadian Rhythms

Many important cellular processes are observed to occur rhythmically with peak activity occurring once every twenty-four hours. These rhythms have been demonstrated to be under endogenous control, as they occur under constant environmental conditions (for review, see Edmunds, 1988). The cellular machinery that generates circadian rhythms is known collectively as the biological clock. An understanding of the molecular components that constitute these ubiquitous pacemakers will have broad significance for many organisms, including humans. Recent progress in systems such as *Drosophila* (Dunlap, 1993; Hall, 1990) and *Neurospora* (Dunlap, 1990, 1993) as well as vertebrate species (Takahashi, 1991, 1993; Ralph et al., 1990) has demonstrated that clocks are accessible to genetic and molecular dissection. In the case of *Drosophila* and *Neurospora*, genetic approaches have identified, respectively, the *period* (*per*) and *frequency* (*frq*) loci (amongst others), mutations of which lead to a variety of abnormal clock functions (Dunlap, 1990, 1993; Hall, 1990; see below). Physiological studies in the invertebrates, *Bulla* (McMahon and Block, 1987), *Aplysia* (Jacklet and Lotshaw, 1986), the avian pineal gland (Takahashi, 1991), and the mammalian suprachiasmatic nucleus (Takahashi,

1993; Ralph et al., 1990) have demonstrated the existence of pacemaker organs and identified some of the intracellular messengers that are involved in generating rhythms. In the marine unicellular alga *Gonyaulax*, biochemical studies have revealed that translation of the luciferin binding protein mRNA is circadian-regulated (Morse et al., 1989a; Mittag et al., 1994). The cycling of mRNA levels in several species has also been shown to be under clock control (reviewed in Hall and Rosbash, 1993; Chalmers and Kyriacou, 1993). In the case of higher plants, cis-acting DNA elements have been characterized that mediate transcription of specific genes by a circadian clock (Nagy et al., 1988; Fejes et al., 1990; Kay, 1993; McClung and Kay, 1994; Anderson et al., 1994; Anderson and Kay, 1995) and we have recently identified mutants with aberrant clock function (Millar et al., 1995a; see below).

B. Oscillator Mechanisms

Circadian systems are considered to be composed of three general components (Fig. 1).

1. A "central oscillator" that is responsible for the generation of the cyclic activity comprising phase, period length, and amplitude information. Oscillators have been modeled in many forms, but the currently accepted idea is to consider a negative feedback loop in which state variables oscillate due to feedback autoregulation. The feedback loop is controlled by parameters which could be biochemical processes such as transcription, translation, or nuclear transport.

2. Input pathways must exist so that environmental information, most importantly light and temperature, can be used to synchronize (i.e., *entrain,* which is effected through *phase-shifting*) the endogenous clock. This allows the organism to use its endogenous circadian periodicity to *anticipate* the onset of dawn and dusk as daylength changes throughout the solar year. In the case

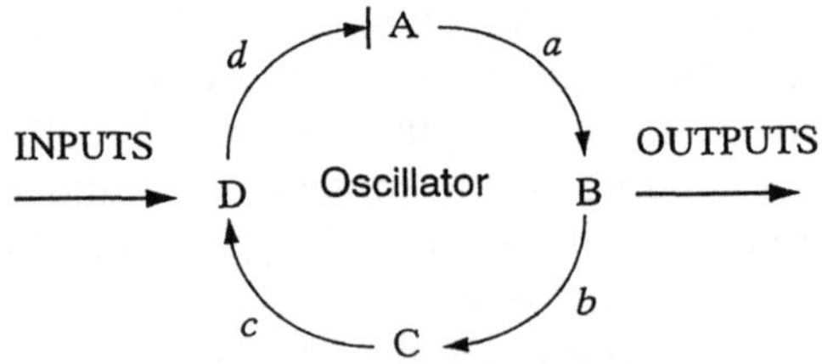

Fig. 1. A hypothetical model for a circadian regulatory system (after Page, 1994). The central oscillator is proposed to consist of a negative feedback loop (A, B, C, D, state variables; a, b, c, d, parameters) that incorporates enough time delay to generate an approximate 24 h periodicity. The clock is entrained under natural conditions by input pathways, most commonly light and temperature. The input pathways can be studied as either responses to light and dark pulses, or by altering ambient light in constant conditions. The clock regulates many different cellular events via output pathways, which may also interact with cellular networks to give complex patterns of response

of higher plants, both phytochromes and blue-light photoreceptors have been shown to affect phase and period length (see below).

3. Output pathways are used to communicate temporal information from the circadian clock to a wide range of biological processes. These include regulation of translation and transcription, membrane conductance, cell division, and behavior. In the case of higher plants, specific molecular output processes that have been characterized include the regulation of K^+ channels and the definition of cis-acting DNA elements that mediate circadian-regulated transcription.

C. Genetic Analysis of Circadian Clocks

The majority of genes that affect circadian rhythmicity have been isolated in brute force screens for mutant individuals with altered cycling patterns, such as period length, arrhythmicity or phase. The identification of the *period* locus in *Drosophila* (Dunlap, 1993; Hall and Rosbash, 1993), followed later by the identification of the *frequency* locus in *Neurospora* (Dunlap, 1993; Hall and Rosbash, 1993) demonstrated that genetics was a powerful approach to identifying potential components of the timing mechanism. Allelic series at each locus include long- and short-period alleles, whereas the null alleles are largely arrhythmic. Furthermore, this work has established the rationale that loci such as *per* and *frq* are likely to encode a fundamental component of the timing mechanism itself, as it is difficult to envision how else independent mutations at a locus could exhibit all three mutant phenotypes (Dunlap, 1993). Recent studies following the cloning of *per* and *frq* have borne out this hypothesis, and both *per* and *frq* appear to be analogous components of the circadian oscillator in their respective organisms (Aronson et al., 1994; Huang et al., 1993; Hardin et al., 1990). Both gene products are part of a negative feedback loop in which they act to repress the transcription of their own genes, thereby causing the gene products themselves to cycle in a circadian fashion. Any autonomous cellular clock would need to be composed of a series of regulated processes (check points) that result in a negative feedback loop. In addition, *per* and *frq* mutant alleles affect several circadian-regulated processes in the same manner (Dunlap, 1990, 1993; Hall and Rosbash, 1993). Any component of the circadian oscillator would also be expected to fulfill two further criteria: sudden changes in the level of the component should cause stable phase shifts in the overt rhythm being studied; and clamping one component to a constant level should cause arrhythmicity (Page, 1994; Aronson et al., 1994; Zeng et al., 1994). Both of these criteria have been met by Per and Frq gene products using either constitutively expressed or inducible constructs in either the mutant or wild-type backgrounds (Aronson et al., 1994; Kyriacou, 1994; Zeng et al., 1994). Studies in both *Drosophila* and *Neurospora* are now focusing on biochemical characterization of function, as well as identifying other genes involved in timekeeping. One such gene in *Drosophila* is *timeless*, which was identified on its basis to cause arrhythmic behavior in adult flies (Sehgal et al., 1994; Vosshall et al., 1994). Interestingly, this *tim* allele causes the *per* RNA feedback loop to also be arrhythmic, which places *tim* conceptually upstream of *per* in regulating circadian rhythmicity. Furthermore *tim* appears to be required for the nuclear localization of *per*, and may therefore represent a protein "partner" of *per* that forms a complex in vivo.

Despite several operational similarities between *per* and *frq*, these genes do not appear to be homologous. This implies that there may be several different ways of building circadian clocks amongst diverse organisms, rather than a single, universal mechanism built from homologous components. Alternatively, *per* and *frq* may occupy separate places within a conserved regulatory loop. In mammals, one circadian rhythm mutant was fortuitously identified in hamster (Ralph et al., 1990), whereas a concerted effort in mouse has identified one rhythm mutant to date (Vitaterna et al., 1994). Given the large size of the mammalian genome it is unlikely that all clock components can be identified and characterized in a reasonable timespan (Takahashi et al., 1994). Recently, an array of circadian rhythm mutants have been identified in cyanobacteria, which may well provide a powerful prokaryotic model system for identifying all clock components (Kondo et al., 1994). Given the central role of circadian rhythms to environmental adaptation, plants provide an extremely interesting system for studying circadian regulation. In particular, higher plants are probably the best model system for studying the interaction of circadian clocks with light signals, which are used to set the phase of the clock. Due to their sessile lifestyle, higher plants need to be particularly attuned to the day/night changes in their environment and as such are dependent upon internal circadian clock(s) to anticipate dawn and dusk. Thus despite the progress in *Drosophila* and *Neurospora*, higher plants offer several important and unique insights into circadian function.

Our laboratory has recently taken advantage of *Arabidopsis* as a powerful genetic system for dissection of circadian function in higher plants. Transgenic *Arabidopsis* carrying *cab2*-luciferase (*cab2::luc*) fusion constructs made possible a new genetic approach to study the regulation of *cab* (chlorophyll *a/b*-binding protein) gene expression by the circadian clock. Luciferase activity can be assayed in vivo after spraying the plants with the substrate luciferin, using a photon counting camera (Millar et al., 1992a, b; Kay et al., 1994). A screen based on this non-invasive assay was initiated to isolate mutants that show aberrant *cab* cycling. Mutants, in which *cab* cycles with a low amplitude, as well as short and long period mutants were recovered from this screen (Millar et al., 1995a). One of the short period mutants, *toc1*, also shortens the period of the leaf movement rhythm (this work), indicating that the clock(s) governing these two different output rhythms share at least one common element. This is particularly interesting, as the leaf movement rhythm, which occurs in petiole cells, has a slightly longer periodicity of 25.2 h (versus 24.7 h for *cab2::luc* transcription in mesophyll cells), and could therefore be driven by a different biological clock. Alternatively, one of the parameters of the same clock could be modified in a cell-specific manner to give a distinct period length.

II. Rhythmic Outputs

A. *Morphology*

Many circadian rhythms have been described in leaf and leaflet movements, flower opening, and stomatal conductance (Sweeney, 1987). Leaf movements have been mostly studied in nyctinastic legumes, some of which exhibit rapid movements in response to mechanical stimuli, as well as slower responses to light and circadian rhythms (Satter, 1990). Leaf movements are mediated by asymmetric changes in turgor pressure within a specialized organ called the pulvinus, located at the basis of petioles, leaves or leaflets. Since pulvini from different species move in different directions, the different types of motor cells within the pulvinus are defined by their function, rather than by their position. Cells whose turgor increases during leaflet opening and decreases during leaflet closure are called *extensor* cells, while those that exhibit the opposite changes are called *flexor* cells (Satter et al., 1974). This swelling and shrinking of motor cells depends on massive fluxes of ions (mainly K^+ and Cl^-) across the plasma membranes and on their redistribution across the pulvinus. These ion movements have been demonstrated by techniques as diverse as flame photometry assays, scanning electron microscope/X-ray analysis, and uptake of $^{42}K^+$ and $^{45}Ca^{2+}$ (Satter and Galston, 1981; Satter and Morse, 1990). Depolarization-activated (K_d) and hyperpolarization-activated (K_h) potassium channels have been described in *Samanea* pulvini motor cells, as well as Cl^-, Ca^{2+} and stretch-activated (SA) channels (Moran, 1990), but little is known about the role of these different channel types in leaf movements. Electrogenic H^+ pumping may energize K^+ uptake into the motor cells (Satter and Galston, 1981). Since calcium is known to regulate ion channels in animal cells, it has been proposed to act downstream of the clock to regulate the opening and closing of K^+ channels in flexor and extensor cells (Satter and Galston, 1981). There is no direct evidence, however, in support of this hypothesis. How K^+ channels in flexor and extensor cells are regulated in opposite manners by the circadian clock is still an outstanding question.

B. *Stomatal Conductance*

Stomatal function is tightly regulated by light, hormones and CO_2 concentration. In addition, stomatal opening is under the control of a circadian clock. Circadian rhythms in stomatal conductance persist in isolated guard cells from *Vicia faba* epidermal peels (Gorton et al., 1989), indicating that the stomatal rhythm can be uncoupled from the circadian rhythm in photosynthetic activity in epidermal and mesophyll cells. Opening and closing of stomata occur by a mechanism similar to that of leaf movements (Gorton, 1990). Increases in hydrostatic pressure cause the swelling of guard cells and the opening of stomatal pores, while decreases in hydrostatic pressure cause guard cells to shrink and stomata to close. Once again K^+ and Cl^- are the chief ions involved in these huge osmotic changes (Gorton, 1990; Assmann, 1993). There is increasing evidence that K^+ uptake during the swelling of guard cells occurs by a passive mechanism. Proton extrusion mediated by a plasma membrane P-type ATPase hyperpolarizes the plasma membrane and creates an electrochemical gradient for K^+ uptake (Lohse and Hedrich, 1992; Thiel et al., 1992; Assmann, 1993). Voltage-dependent inward-rectifying K^+ channels have been described, that may mediate K^+ uptake during stomatal opening (Schroeder and Fang, 1991). In contrast, the loss of K^+ during the closing of guard cells requires the depolarization of the plasma membrane. This depolarization could be mediated by the inhibition of the plasma membrane H^+-ATPase, by the opening of plasma membrane anion channels, or by Ca^{2+} entry (Assmann, 1993). Calcium and pH have been shown to trigger these changes, and thus may participate in the circadian regulation of stomatal opening. In patch-clamp experiments, Ca^{2+} has been shown to inhibit voltage-dependent inward rectifying K^+ channels, while permeabilizing the plasma membrane to anions, thus presumably driving K^+ efflux through outwards K^+ channels and closing stomatal pores (Schroeder and Hagiwara, 1989). The inhibitory effect of Ca^{2+} on K^+ channels was prevented by the action of the immunosuppressants Cyclosporin A, FK506, and FKBP12 (Luan et al., 1993). This result indicated the role of a Ca^{2+}-calmodulin-dependent phosphatase (PP2B, calcineurin) in the regulation of K^+ channels by Ca^{2+}. Another role of calcium may be to induce the phosphorylation of the H^+ pump in the plasma membrane by a Ca^{2+}-dependent kinase, thus causing a change in H^+ pumping activity (Schaller and Sussman, 1988). There is also evidence, however, for a role of Ca^{2+} and calmodulin in K^+ extrusion and stomatal opening. Thus, a variety of calmodulin antagonists inhibit H^+ extrusion by *Vicia faba* guard cells, and block stomatal opening in *C. benghalensis* epidermal peels (Shimazaki et al., 1992). Calcium may mediate both the opening and the closing of stomata, depending on the magnitude, location, and temporal pattern of Ca^{2+} elevation (Assmann, 1993).

In addition to the rhythm in stomatal conductance, *Vicia faba* guard cells exhibit a rhythmic sensitivity to both red and blue light (Gorton et al., 1993). Light signals induced stomatal opening throughout the circadian cycle. Signals given in the middle of the subjective day were more effective, however, since they resulted in a greater stomatal conductance. This result suggests that in addition to regulating the basal level of stomatal conductance in DD, the clock acts to gate the opening of stomata in response to red or blue light.

Despite the similarity of mechanisms that mediate leaf movements and stomatal opening, there is evidence that these two circadian rhythms may be under the control of different circadian clocks. The circadian rhythm in stomatal opening in *Phaseolus vulgaris* has a free-running period that is similar to that (24 h) of the rhythm in CO_2 assimilation, but that is different from that (27 h) of the leaflet movement rhythm, measured in parallel (Hennessey and Field, 1992). Evidence for at least two oscillators within a single cell has been obtained in the unicellular alga *Gonyaulax polyedra*, in which the aggregation and the glow rhythms exhibit different free-running periods under certain experimental conditions (Roenneberg and Morse, 1993). Another possibility is that the same circadian oscillator may operate in pulvini motor cells and in guard cells, but that the free-running period may be regulated in a cell-type-specific manner.

C. Gene Expression

Mechanism for the regulation of gene expression by the circadian clock have been emerging recently from mammalian systems. Thus, the liver-enriched transcriptional activator DBP exhibits free-running, circadian variations in its cellular levels in most tissues examined, as detected by Western blot analysis (Wuarin and Schibler, 1990). The circadian regulation of DBP expression occurs at the transcriptional level, and may be dependent on the rhythmic secretion of glucocorticoid hormones by the adrenal gland (Wuarin and Schibler, 1990; Wuarin et al., 1992). The transcription of one of DBP's target genes, albumin, cycles with a phase consistent with that of DBP acting as an activator. In contrast, the cytochrome P_{450} gene CYP2C6, which contains a DBP-binding site within its promoter, does not appear to be regulated by the clock. This discrepancy may be explained by the large difference in affinity of DBP for the two promoters, that is 17-fold greater for albumin than for CYP2C6 (Wuarin et al., 1992). More recently, a novel isoform of the transcription factor CREM (cAMP-responsive element modulator) was also shown to exhibit circadian rhythmicity in the pineal gland of rats (Stehle et al., 1993). This isoform, named ICER (inducible cAMP early regulator), is generated from an alternative, intronic promoter (P2). The ICER protein is unusually small (13.4 kDa) compared to other CREM products. It contains an intact DNA-binding domain, that enables it to bind CREs (cAMP-responsive elements). Coexpression of cAMP-inducible reporter genes with ICER completely abolished cAMP-mediated induction, indicating that ICER functions as a repressor of cAMP-induced transcription (Stehle et al., 1993). The P2 promoter contains four CRE-like elements, and is transiently induced by cAMP (Molina et al., 1993). Other CREM products, generated by the promoter P1 do not fluctuate between night and day and are regulated by cAMP-induced phosphorylation. Thus, the rhythmic stimulation of β-adrenergic receptors by noradrenaline, secreted at night by post ganglionic fibers originating from the superior cervical ganglia, causes a surge of cAMP level in pineal cells. cAMP activates transcription by inducing the phosphorylation of CREM transcriptional activators. The subsequent elevation of ICER expression causes the repression of cAMP-dependent transcription. Finally, ICER downregulates its own transcription, constituting a negative autoregulatory loop, that is driven by β-adrenergic input (Molina et al., 1993).

Although many genes have now been described in higher plants, that are under the control of the circadian clock (McClung and Kay, 1994), the cis-analysis of most circadian-regulated promoters is still at an early stage, and no transcription factor has yet been cloned that mediates such regulation. In tomato nuclear extracts, the DNA-binding activity IBF2a has been reported to exhibit circadian variations in an UV cross-linking assay (Borello et al., 1993). IBF2a binds to a 27-mer oligonucleotide that contains a conserved GATAAG sequence, found in the circadian-regulated promoters of *cab* and nitrate-reductase genes. Mutations within this conserved motif, do not abolish IBF2a binding, however (Borello et al., 1993). In transgenic tobacco, circadian-regulated elements have been mapped to the –357/–90 region of the wheat *cab1* enhancer (Fejes et al., 1990). More recently, the –111/–38 region of the *Arabidopsis cab2* promoter was shown to confer circadian-regulated expression to a luciferase reporter gene (Anderson et al., 1994). A complex named CGF-1 (*cab* GATA factor-1) was identified in tobacco extracts, that binds a conserved triple GATA motif within the –111/–38 region. Mutation of all three GAT repeats to CCC prevented CGF-1 from binding in vitro, and reduced the level and the amplitude of the luminescense rhythm in transgenic plants carrying a *cab2::luc* gene fusion. The mutation strongly reduced the rapid induction of luciferase by red light in etiolated plants, but did not abolish rhythmicity of luciferase expression (Anderson and Kay, 1995). Furthermore, a trimer of the CGF-1 binding site conferred high expression levels, but not rhythmicity to the luciferase reporter gene (Anderson and Kay, 1995). These results indicate that CGF-1 plays a role in the induction of *cab2* expression by phytochrome and contributes to the amplitude of *cab2* oscillations, but is not sufficient for the perception of rhythmic signal(s) from the circadian clock. Competition assays with the BoxII sequence (Green et al., 1987; Lam and Chua, 1990), as well as supershift experiments with antibodies to *Arabidopsis* GT-1 indicated that CGF-1 is closely related to a transcription factor that plays a role in the regulation of the pea *RBCS3A* gene by light (Anderson et al., 1994; G. Teakle and S. Kay, unpubl. results). Thus the CGF-1 complex may correspond to the GT-1 protein, binding with low affinity to the triple GATA motif in the *cab2* promoter (Hiratsuka et al., 1994). Five additional DNA-binding activities (CUF-2, CUF-3, Tic, Tac, and Toe) have been identified in *Arabidopsis* whole cell extracts, that bind within the –123/–55 region (I. Carré and S. Kay, unpubl.). A 37 base-pair sequence from –111 to –74, containing binding sites for CUF-2, CUF-3, and Tac conferred circadian regulation to the luciferase reporter gene in transgenic *Arabidopsis*. Tac is a particularly interesting activity, since its binding site overlaps a CCAAT box that is conserved amongst *cab* promoters. In addition, the effects of promoter mutations indicated that its binding precludes binding of an adjacent complex named Tic. Thus Tac may act to prevent binding of Tic (and perhaps of CUF-2 or CUF-3). This antagonism may constitute a molecular switch for the control of *cab2* transcription by the circadian clock. Alternatively, CUF-2 or CUF-3 are also candidate targets for regulation by the circadian oscillator. All of these DNA-binding activities are present throughout the circadian cycle (Anderson et al., 1994; I. Carré and S. Kay, unpubl.). It is possible, however, that Tac (CUF-2 or CUF-3) interacts with a partner that prevents it from binding DNA, or from entering the nucleus, at specific circadian times.

III. Interaction of Photoreceptors and Clocks

The interaction of photoreceptor and circadian signal transduction pathways is likely to be complex. Light controls the phase of the circadian clock, and synchronizes it to day–night environmental cycles. Single light pulses given under constant (or free-running) conditions reset the phase of the circadian clock in a phase-dependent manner (Pittendrigh, 1965), indicating that the action of photoreceptor pathways on the clock is modulated in a circadian fashion. The largest advances are usually observed in response to light pulses given near the middle of the subjective night, while signals given in the middle of the subjective day have no effect. This "gating" of light action on the clock may involve a feed-back from the central oscillator on its input pathway. Alternatively, the loss of sensitivity to light at specific phases of the circadian cycle may be a property inherent to the oscillator itself. In this latter model, oscillator components would oscillate in such a way that they become insensitive to the light signals during the subjective days.

A. The Mammalian Brain

Mammalian circadian rhythms are regulated by a pacemaker located in the suprachiasmatic nucleus (SCN) of the hypothalamus. Neurons of the SCN receive direct input from ganglion cells of the retina. Light signals were shown to induce the expression of the protooncogene *c-fos* in SCN cells. *C-fos* induction was restricted to times when light also induces a phase-shift of behavioral rhythms. *C-fos* is one of the immediate-early class of genes, that are induced rapidly in response to several extracellular signals. The Fos protein is a transcription factor that heterodimerizes with members of the Jun family of transcription factors to induce the transcription of late response genes. Thus, Fos may play a key role in light-induced phase-shifting of circadian rhythms. The cAMP-responsive element (CRE) present in the *c-fos* promoter mediates *c-fos* induction by cAMP and Ca^{2+}. These second messengers trigger the phosphorylation of a CRE-binding protein (CREB) on Ser 133 and this phosphorylation is necessary for transcriptional activation. Light stimuli that induce *c-fos* expression and phase-shift circadian rhythms also induce the phosphorylation of Ser 133 in SCN cells (Ginty et al., 1993). Light failed to induce CREB phosphorylation if the stimulus was given during the subjective day. These results indicate that phosphorylation of CREB plays a role in the regulation of *c-fos* expression by light. They also suggest that in mammals the circadian clock gates light responsiveness upstream of CREB on the signal transduction pathway (Ginty et al., 1993).

B. Higher Plants

Higher plants have several classes of photoreceptors. UV, blue light (cryptochrome) and red light (phytochrome) photoreceptors all can mediate resetting of the circadian clock by light (Edmunds, 1988). The phytochrome signal transduction pathway is the best characterized at this point, both at the genetic and at the molecular levels (Millar et al., 1994).

In nyctinastic legumes, Ca^{2+} has been proposed to be the second messenger for resetting the leaf movement rhythm by red light pulses, as well as for opening or closing the leaves in response to light signals (Satter and Galston, 1981). Since this hypothesis was formulated, evidence for a role of calcium in mediating phytochrome responses has accumulated. Brief (5–30 s) light signals were shown to activate phosphatidyl inositol turnover in *Samanea saman* leaf pulvini (Morse et al., 1989b; Crain, 1990), presumably inducing calcium release from cytoplasmic stores. In etiolated wheat protoplasts, cytoplasmic Ca^{2+} was shown to rise transiently in response to red light. The swelling induced by phytochrome activation was mimicked by the photolytic release of chemically caged calcium and inositol-1,4,5-trisphosphate (IP_3) (Shacklock et al., 1992). Furthermore, the release of caged Ca^{2+} or IP_3 induced the phosphorylation of a 70 kDa polypeptide, that was also phosphorylated in response to phytochrome (Fallon et al., 1993). Similarly, the microinjection of G-protein activators, Ca^{2+} or Ca^{2+}-calmodulin into phytochrome A deficient tomato cells was shown to induce the expression of a phytochrome-regulated *cab::gus* reporter gene. In addition, calmodulin antagonists inhibited the induction of *cab::gus* expression by the microinjection of phytochrome A, indicating that Ca^{2+}-calmodulin is necessary and sufficient for the induction of *cab* expression by phytochrome (Neuhaus et al., 1993). Calcium does not mediate all phytochrome responses, however, since neither Ca^{2+} nor Ca^{2+}-calmodulin injection induced anthocyanin biosynthesis and only induced partial chloroplast development. In contrast, microinjected cGMP triggered the production of anthocyanins, and together with calcium, the assembly of fully mature chloroplasts (Bowler et al., 1994a, b).

Thus, phytochrome phototransduction involves the activation of one or more G proteins that are coupled to at least two different pathways. Phase-resetting of the clock following phytochrome activation may involve calcium, but it may also involve cGMP, or other, not yet identified, signal transduction pathway(s) (Bowler and Chua, 1994). Calcium and agents that perturb the cellular Ca^{2+} or Ca^{2+}-calmodulin levels have been shown to phase-shift circadian rhythms of cell division in *Euglena* (Tamponnet and Edmunds, 1990), conidiation in *Neurospora* (Nakashima, 1984), phototaxis in *Chlamydomonas* (Goodenough and Bruce, 1980), and leaf movement in *Trifolium repens* (Bollig et al., 1978) suggesting that calcium may play a role on the input pathway from phytochrome or from another photoreceptor.

Several new systems have been described, that will be helpful to identify components of the input pathway for resetting the clock by light. Protoplasts prepared from the flexor and extensor regions of *Samanea saman* pulvini have been shown to exhibit circadian rhythms in the opening and closing of K^+ channels, that persisted for at least 21 h in constant darkness (Kim et al., 1993). Rhythms were 180° out of phases in flexor and extensor protoplasts. They were shifted in a similar manner by a 6 h delay of the onset of constant darkness, a property expected of true circadian rhythms. The protoplasts also maintained phytochrome and blue-light responsiveness, in a manner that was consistent with the behavior of whole plants. The protoplasts make possible a pharmacological approach for the study of signal transduction pathways for the control of leaf movement by light and by the circadian clock. In addition, patch clamp techniques should make possible the identification of ion channels involved in the rhythmic swelling and shrinking of motor cells. This system is limited, however, by the short-term viability of the protoplasts. A suspension culture from a higher plant would be a good alternative for these kind of studies. Robust, photosynthetic cell lines are now available, in which circadian rhythms of gene expression could be easily studied, as well as their response to signal transduction intermediates. For example, *cab* expression in the SBP cell line oscillates for up to 7 days in dim LL (S. Welsh and S. Kay, unpubl.). Transgenic lines carrying *cab2::luc* gene fusions are being generated, and will be a powerful tool to study *cab* gene regulation by light and the circadian clock.

Genetic approaches to light signal transduction pathways are also under way in *Arabidopsis* (Millar et al., 1995b). Transgenic plants carrying a *cab2::luc* gene fusion displayed a shorter period length when transferred to constant light (24.7 h) than when moved to constant darkness (30–36 h). This result indicated that light intensity modulates period length in *Arabidopsis*, as previously demonstrated for *Gonyaulax* (Roenneberg and Hastings, 1988), *Coleus* (Halaban, 1969), and *Drosophila* (Konopka et al., 1995). Many *Arabidopsis* mutants are now available, that are deficient in photomorphogenetic responses (Pepper et al., 1993; Millar et al., 1994). The first class of mutants (*hy*) are deficient in all or part of the de-etiolation response to red or blue light, and correspond to elements that mediate the response to light. Another class of mutants (*det, cop*) has characteristics of light-grown plants when grown in darkness. These mutants correspond to elements that normally repress the photomorphogenetic program in the absence of light. The *cab2::luc* transgene was introduced by crossing into these mutant lines, and the cycling of luciferase expression was followed under different light conditions (Millar et al., 1995b). The period length of wild-type plants was unchanged in either constant blue light or constant red light, indicating that both blue and red light signal transduction pathways shorten the period length to a similar extent. The phytochrome-deficient mutant *hy1-100* showed a period similar to that of the wild type under white light, indicating that blue light photoreceptors, together with very low amounts of phytochrome, are sufficient to maintain a short period length. The period length was significantly longer (26.5 h) under red light, providing the first genetic evidence that one or more phytochromes mediate the modulation of period length by red light. The period was unchanged under blue light, indicating that the *hy1-100* mutation selectively affected the red light (phytochrome) pathway. The constitutively photomorphogenetic mutation *det1-1* shortened the period length to a duration (18 h) in DD. This result suggested that the *det1-1* mutation mimics the effect of very high light intensities, and acts in the input pathway from both red and blue light photoreceptors to the clock.

A leaf movement rhythm has also been demonstrated in *Arabidopsis thaliana*, that persisted for up to 12 days in constant dim light, and could be phase-shifted by red light pulses (I. Carré and S. Kay, unpubl. results). This

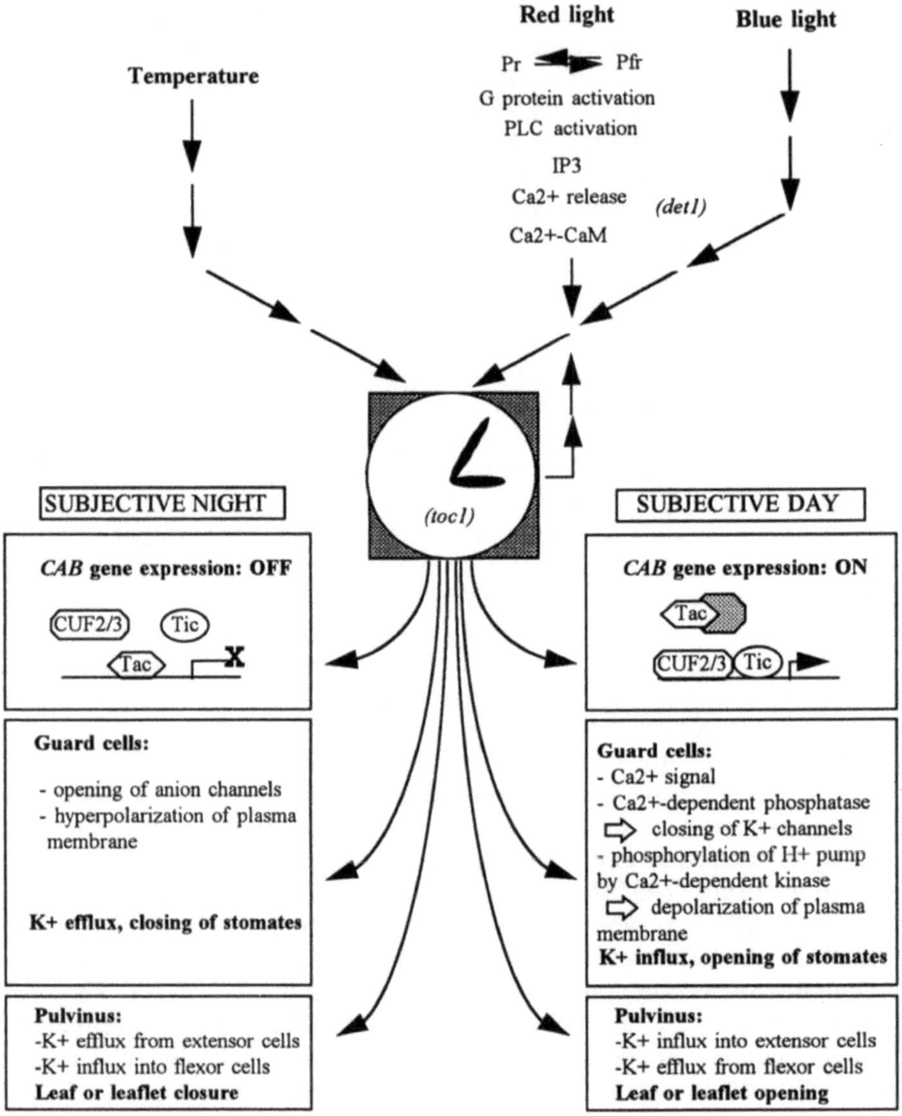

Fig. 2. Circadian input and output pathways in higher plants. The *Arabidopsis toc1* gene is shown tentatively as an element of the circadian oscillator, and the *det1* gene as a common element between the blue and the red light transduction pathways. Regulation of *cab* gene expression may be mediated by the periodic sequestering of Tac, a DNA-binding activity that antagonizes Tic and CUF-2 (or CUF-3) binding. Mechanisms for the opening and closing of guard cells are assumed to be identical to those characterized in response to blue light and ABA, respectively

new rhythmic phenotype does not require the introduction of a transgenic marker, and can be monitored using basic and relatively inexpensive imaging technology (Engelmann et al., 1994; I. Carré et al., unpubl.). This fully automated assay will allow the immediate testing of the effects of mutations that affect better characterized signal transduction pathways on the circadian clock. Such experiments may allow the identification of common components between signaling pathways, and further localize the clock within the cellular network of regulation.

IV. Conclusion

The circadian clock lies at the center of a complex network of signaling pathways, that mediate the entrainment of the oscillator to daily changes in environmental conditions, and, downstream of the clock, the control of many rhythmic outputs. Figure 2 summarizes our current understanding of input and output pathways for the circadian oscillator in higher plants. Recent genetic evidence indicates that one or more classes of phytochromes act on the input pathway for the modulation of period length by light intensity. Several of the early steps of the signal transduction pathway from phytochrome have been elucidated, although the genes that mediate these primary events have not been identified. The blue light phototransduction pathway is less well characterized, but the *det1* gene may be an element common to both blue and red light signaling cascades. The best characterized output pathways are those for the control of gene expression (and especially of the *Arabidopsis cab2* gene), leaf movements and stomatal opening. Once again, only the terminal steps (those furthest remote from the oscillator) have been elucidated, mostly using biochemical and pharmacological tools. *Arabidopsis* has recently proven to be a useful genetic system for the isolation of mutants with aberrant circadian rhythms of *cab* gene expression and leaf movements. We believe that a combination of biochemical, molecular, and genetic approaches will provide all the tools needed to identify the components of the input and output pathway, as well as elements of the biological clock(s) that regulate many vital processes in plants.

V. References

Anderson SL, Kay SA (1995) Functional dissection of circadian clock- and phytochrome-regulated transcription of the *Arabidopsis CAB2* gene. Proc Natl Acad Sci USA 92: 1500–1504

Anderson SL, Teakle GR, Martino-Catt SJ, Kay SA (1994) Circadian clock- and phytochrome-regulated transcription is conferred by a 78 bp cis-acting domain of the *Arabidopsis CAB2* promoter. Plant 6: 457–470

Aronson BD, Johnson KA, Loros JJ, Dunlap JC (1994) Negative feedback defining a circadian clock: autoregulation of the clock gene frequency. Science 263: 1578–1584

Assmann SM (1993) Signal transduction in guard cells. Annu Rev Cell Biol 9: 345–375

Bollig IC, Mayer K, Mayer W-E, Engelmann W (1978) Effects of cAMP, theophyllin, imidazole, and 4-(3,4-dimethoxybenzyl)-2-imidazole on the leaf movement rhythm of *Trifolium repens* test of the cAMP hypothesis. Planta 141: 225–230

Borello U, Ceccarelli E, Giuliano G (1993) Constitutive, light-responsive and circadian clock-responsive factors compete for the different I box elements in plant light-regulated promoters. Plant J 4: 611–619

Bowler C, Chua N-H (1994) Emerging themes of plant signal transduction. Plant Cell 6: 1529–1541

Bowler C, Neuhaus G, Yamagata H, Chua N-H (1994a) Cyclic GMP and calcium mediate phytochrome phototransduction. Cell 77: 73–81

Bowler C, Yamagata H, Neuhaus G, Chua N-H (1994b) Phytochrome signal transduction pathways are regulated by reciprocal control mechanisms. Genes Dev 8: 2188–2202

Chalmers DE, Kyriacou CP (1993) Glowing reports on biorhythm research. Bioessays 15: 755–756

Crain RC (1990) Phototransduction in the *Samanea* pulvinus: role of accelerated phosphatidyl inositol turnover. In: Satter RL (eds) The pulvinus: motor organ for leaf movement. American Society of Plant Physiologists, Rockville, MD, pp 175–188

Dunlap JC (1990) Closely watched clocks: molecular analysis of circadian rhythms in *Neurospora* and *Drosophila*. Trends Genet 6: 159–165

Dunlap JC (1993) Genetic analysis of circadian clocks. Annu Rev Physiol 55: 683–728

Edmunds LN (1988) Cellular and molecular bases of biological clocks. Springer, Berlin Heidelberg New York Tokyo

Engelmann W, Simon K, Phen CJ (1994) Leaf movement in *Arabidopsis thaliana*. Z Naturforsch 47: 925–928

Fallon KM, Shacklock PS, Trewavas AJ (1993) Detection in vivo of very rapid red light-induced calcium-sensitive protein phosphorylation in etiolated wheat (*Triticum aestivum*) leaf protoplasts. Plant Physiol 101: 1039–1045

Fejes E, Pay A, Kanevsky I, Szell M, Adam E, Kay S, Nagy F (1990) A 268 bp upstream sequence mediates the circadian clock-regulated transcription of the wheat *Cab-1* gene in transgenic plants. Plant Mol Biol 15: 921–932

Ginty DD, Kornhauser JM, Thompson MA, Bading H, Mayo KE, Takahashi JS, Greenberg ME (1993) Regulation of CREB phosphorylation in the suprachiasmatic nucleus by light and a circadian clock. Science 260: 238–241

Goodenough JE, Bruce VG (1980) The effects of caffeine and theophylline on the phototactic rhythm of *Chlamydomonas rheinhardii*. Biol Bull 159: 649–655

Gorton HL (1990) Stomates and pulvini: a comparison of two rhythmic, turgor-mediated movement systems. In: Satter RL (eds) The pulvinus: motor organ for leaf movement. The American Society of Plant Physiologists, Rockville, MD, pp 223–237

Gorton HL, Williams WE, Binns ME, Gemmell CN, Leheny EA, Shepherd AC (1989) Circadian stomatal rhythms in epidermal peels from *Vicia faba*. Plant Physiol 90: 1329–1334

Gorton HL, Williams WE, Assmann SM (1993) Circadian rhythms in stomatal responsiveness to red and blue light. Plant Physiol 103: 399–406

Green PJ, Kay SA, Chua N-H (1987) Sequence-specific interaction of a pea nuclear factor with light-responsive elements upstream of the rbcS-3A gene. EMBO J 6: 2543–2549

Halaban R (1969) Effect of light quality on the circadian rhythm of leaf movement of a short-day plant. Plant Physiol 44: 973–977

Hall JC (1990) Genetics of circadian rhythms. Annu Rev Genet 24: 659–697

Hall JC, Rosbash M (1993) Oscillating molecules and how they move circadian clocks across evolutionary boundaries. Proc Natl Acad Sci USA 90: 5382–5383

Hardin PE, Hall JC, Rosbash M (1990) Feedback of the *Drosophila period* gene product on circadian cycling of its messenger RNA levels. Nature 343: 536–540

Hennessey TL, Field CB (1992) Evidence of multiple circadian oscillators in bean plants. J Biol Rhythms 7: 105–113

Hiratsuka K, Wu X, Chua N-H ((1994) Molecular dissection of GT-1 from *Arabidopsis*. Plant Cell 6: 1805–1813

Huang ZJ, Edery I, Rosbash M (1993) PAS is a dimerization domain common to *Drosophila period* and several transcription factors. Nature 364: 259–262

Jacklet JW, Lotshaw DP (1986) Involvement of protein synthesis in the circadian clock of the Aplysia eye. Am J Physiol 250: 5–17

Kay SA (1993) Shedding light on clock controlled *cab* gene transcription in higher plants. Sem Cell Biol 4: 81–86

Kay SA, Millar AJ, Smith KW, Anderson SL, Brandes C, Hall JC (1994) Video imaging of regulated firefly luciferase activity in transgenic plants and *Drosophila*. Promega Notes 49: 22–27

Kim HY, Cote GC, Crain RC (1993) Potassium channels in *Samanea saman* protoplasts controlled by phytochrome and the biological clock. Science 260: 960–962

Kondo T, Tsinoremas NF, Golden SS, Johnson CH, Kutsuna S, Ishiura M (1994) Circadian clock mutants of cyanobacteria. Science 266: 1233–1236

Konopka RJ, Pittendrigh CS, Orr D (1995) Reciprocal behavior associated with altered homeostasis and photosensitivity of *Drosophila* clock mutants. J Neurogenet 6: 1–10

Kyriacou CP (1994) Clock research perring along: it's about time! Trends Genet 10: 69–71

Lam E, Chua N-H (1990) GT-1 binding site confers light responsive expression in transgenic tobacco. Science 248: 471–474

Lohse G, Hedrich R (1992) Characterization of the plasma membrane H^+-ATPase from *Vicia faba* guard cells. Planta 188: 206–214

Luan S, Li W, Rusnak F, Assmann SM, Schreiber SL (1993) Immunosuppressants implicate protein phosphatase regulation of K^+ channels in guard cells. Proc Natl Acad Sci USA 90: 2202–2206

McClung CR, Kay SA (1994) Circadian rhythms in *Arabidopsis thaliana*. In: Somerville CS, Meyerowitz E (eds) *Arabidopsis thaliana*. Cold Spring Harbor Press, Cold Spring Harbor, pp 615–637

McMahon DG, Block GD (1987) The Bulla circadian pacemaker I. Pacemaker neuron membrane potential controls phase through a calcium-dependent mechanism. J Comp Physiol A 161: 335–346

Millar AJ, Short SR, Chua N-H, Kay SA (1992a) A novel circadian phenotype based on firefly luciferase expression in transgenic plants. Plant Cell 4: 1075–1087

Millar AJ, Short SR, Hirasuka K, Chua N-H, Kay SA (1992b) Firefly luciferase as a reporter of regulated gene expression in higher plants. Plants Mol Biol Rep 10: 324–337

Millar AJ, McGrath RB, Chua N-H (1994) Phytochrome phototransduction pathways. Annu Rev Gen 28: 325–349

Millar AJ, Carré IA, Strayer CA, Chua N-H, Kay SA (1995a) Circadian clock mutants in *Arabidopsis* identified by luciferase imaging. Science 267: 1161–1163

Millar AJ, Straume M, Chory J, Chua N-H, Kay SA (1995b) The regulation of circadian period by phototransduction pathways in *Arabidopsis*. Science 267: 1163–1166

Mittag M, Lee D-H, Hastings W (1994) Circadian expression of the luciferin-binding protein correlates with the binding of a protein to the 3 untranslated region of its mRNA. Proc Natl Acad Sci USA 91: 5257–5261

Molina CA, Foulkes NS, Lalli E, Sassone-Corsi P (1993) Inducibility and negative autoregulation of CREM: an alternative promoter directs the expression of ICER, an early response repressor. Cell 75: 875–886

Moran N (1990) The role of ion channels in osmotic volume changes in *Samanea* motor cells analysed by patch-clamp methods. In: Satter RL (eds) The pulvinus: motor organ for leaf movement. American Society of Plant Physiologists, Rockville, MD, pp 142–159

Morse D, Milos PM, Roux E, Hastings JW (1989a) Circadian regulation of the synthesis of substrate binding protein in the *Gonyaulax* bioluminescent system involves translational control. Proc Natl Acad Sci USA 86: 172–176

Morse MJ, Crain RC, Cote GG, Satter RL (1989b) Light-stimulated inositol phospholipid turnover in *Samanea saman* pulvini: increased levels of diacyl glycerol. Plant Physiol 89: 724–727

Nagy F, Kay SA, Chua N-H (1988) A circadian clock regulates transcription of the wheat *Cab-1* gene. Genes Dev 2: 376–382

Nakashima H (1984) Calcium inhibits phase-shifting of the circadian conidiation rhythm of *Neurospora crassa* by the calcium ionophore A23187. Plant Physiol 74: 268–271

Neuhaus G, Bowler C, Kern R, Chua N-H (1993) Calcium/calmodulin-dependent and -independent phytochrome signal transduction pathways. Cell 73: 937–952

Page TL (1994) Time is the essence: molecular analysis of the biological clock. Science 263: 1570–1572

Pepper A, Delaney T, Chory J (1993) Genetic interaction in plant photomorphogenesis. Sem Dev Biol 4:

Pittendrigh CS (1965) On the mechanism for the entrainment of a circadian rhythm by light cycles. In: Ashoff J (ed) Circadian clocks. North-Holland, Amsterdam, pp 227–297

Ralph MR, Foster RG, Davis FC, Menaker M (1990) Transplanted suprachiasmatic nucleus determines circadian period. Science 247: 975–978

Roenneberg T, Hastings JW (1988) Two photoreceptors control the circadian clock of a unicellar alga. Naturwissenschaften 75: 206–207

Roenneberg T, Morse D (1993) Two circadian oscillators in one cell. Nature 362: 362–364

Satter RL (1990) Leaf movement, an overview of the field. In: Satter RL (ed) The pulvinus: motor organ for leaf movement. American Society of Plant Physiologists, Rockville, MD, pp 1–9

Satter RL, Galston AW (1981) Mechanisms of control of leaf movements. Annu Rev Plant Phys 32: 83–110

Satter RL, Morse MJ (1990) Light-modulated, circadian rhythmic leaf movements in nyctinastic legumes. In: Satter RL (ed) The pulvinus: motor organ for leaf movements. American Society of Plant Physiologists, Rockville, MD, pp 10–24

Satter RL, Geballe GT, Applewhite PB, Galston AW (1974) Potassium flux and leaf movement in *Samanea saman*. I. Rhythmic movement. J Gen Physiol 64: 413–430

Schaller GE, Sussman MR (1988) Phosphorylation of the plasma membrane H^+-ATPase of oat roots by a calcium stimulated protein kinase. Planta 173: 509–518

Schroeder JI, Fang HH (1991) Inward-rectifying K^+ channels in guard cells provide a mechanism for low affinity K^+ uptake. Proc Natl Acad Sci USA 88: 11583–11587

Schroeder JI, Hagiwara S (1989) Cytosolic calcium regulates ion channels in the plasma membrane of *Vicia faba* guard cells. Nature 338: 427–430

Sehgal A, Price JL, Man B, Young MW (1994) Loss of circadian behavioral rhythms and *per* RNA oscillations in the *Drosophila* mutant *timeless*. Science 263: 1603–1605

Shacklock PS, Read ND, Trewavas AJ (1992) Cytosolic free calcium mediates red light induced photomorphogenesis. Nature 358: 753–755

Shimazaki KI, Kinoshita T, Nishimura M (1992) Involvment of calmodulin and calmodulin-dependent myosin light chain kinase in blue light-dependent H^+ pumping by guard cells protoplasts from *Vicia faba* L. Plant Physiol 99: 1416–1421

Stehle JH, Foulkes NS, Molina CA, Simonneaux V, Pevet P, Sassone-Corsi P (1993)

Adrenergic signals direct rhythmic expression of transcriptional repressor CREM in the pineal gland. Nature 365: 314–320

Sweeney BM (1987) Rhythmic phenomena in plants. Academic Press, San Diego

Takahashi JS (1991) Circadian rhythms: from gene expression to behavior. Curr Opin Neurobiol 1: 556–561

Takahashi JS (1993) Circadian clocks á la CREM. Nature 365: 299–300

Takahashi JS, Pinto LH, Vitaterna MH (1994) Forward and reverse genetic approaches to behavior in the mouse. Science 264: 1724–1733

Tamponnet C, Edmunds LNJ (1990) Entrainment and phase-shifting of the circadian rhythm of cell division by calcium in synchronous cultures of the wild type Z strain and of the ZC achlorophyllous mutant of *Euglena gracilis*. Plant Physiol 93: 425-431

Thiel G, MacRobbie EAC, Blatt MR (1992) Membrane transport in stomatal guard cells: the importance of voltage control. J Membr Physiol 126: 1–18

Vitaterna MH, King DP, Chang A-M, Kornhauser JM, Lowrey PL, McDonald JD, Dove WF, Pinto LH, Turek FW, Tahakashi JS (1994) Mutagenesis and mapping of a mouse gene, *Clock*, essential for circadian behavior. Science 264: 719–725

Vosshall LB, Price JL, Sehgal A, Saez L, Young MW (1994) Block in nuclear localization of *period* protein by a second clock mutation, *timeless*. Science 263: 1606–1609

Wuarin J, Schibler U (1990) Expression of the liver-enriched transcriptional activator protein DBP follows a stringent circadian rhythm. Cell 63: 1257–1266

Wuarin J, Falvey E, Lavery D, Talbot D, Schmidt E, Ossipow V, Fonjallaz P, Schibler U (1992) The role of the transcriptional activator protein DBP in circadian liver gene expression. J Cell Sci 103 [Suppl 16]: 123–127

Zeng H, Hardin PE, Rosbash M (1994) Constitutive overexpression of the *Drosophila period* protein inhibits *period* mRNA cycling. EMBO J 13: 3590–3598

Role of Plasmodesmata and Virus Movement Proteins in Spread of Plant Viruses

Csilla A. Fenczik, Bernard L. Epel, and Roger N. Beachy

Contents

I. Introduction

In higher plants the direct cytoplasm to cytoplasm intercellular movement of water, nutrients, small signaling molecules, and, in certain cases, of macromolecules is via plasmodesmata (singular plasmodesma), highly specialized gatable cytoplasmic trans-wall channels that interconnect contiguous cells. In addition to their normal function, plasmodesmata are apparently exploited by plant viruses in order to spread from cell to cell. Since the cell wall which encases the plant protoplast constitutes an impermeable barrier, plant viruses have developed strategies which allow them to exploit the natural channels connecting contiguous cells, the plasmodesmata, as conduits for cell-to-cell spread.

Plasmodesmata were thought for many years to be nonselective pores, passively allowing the bidirectional movement of molecules between adjacent cells. However, during the last few years the concept of static plasmodesmata has been changing and it is now clear that plasmodesmata are dynamic, selective entities with the capacity to "gate". In this review, we briefly consider new studies on plasmodesmatal structure, composition, on the regulation of gating, and review in depth recent studies on mechanisms of viral movement. For recent reviews on plasmodesmatal biology, see Epel (1994), Lucas et al. (1993), and Oparka (1993).

II. Plasmodesmata

A. Plasmodesmatal Structure

Early EM studies of plasmodesmata led to a simple consensus model describing the plasmodesma as a wall-embedded plasmalemma-lined unbranched cylinder about 40–80 nm in diameter that contains a central axial component, generally termed the desmotubule. The desmotubule was considered to be derived from and continuous with the endoplasmic reticula of adjoining cells. Recent studies suggest that the endoplasmic reticulum (ER) component of a plasmodesma is a derivative of and continuous with cortical ER (Hepler, 1982; Hepler et al., 1990; Oparka, 1993). In some cells, the outer regions of the plasmodesma are constricted and form the neck region, a structure which is not a general feature of all plasmodesmata (Robinson-Beers and Evert, 1991). The cylindrical space between the desmotubule and the plasmalemma is referred to as the cytoplasmic annulus or sleeve. Most models consider the cytoplasmic annulus to be the pathway through which transport occurs. Electron micrographs show that in the orifice, the ER is apparently constricted and lacks a lumen and, therefore, could not function as a transport channel. However, we consider this conclusion to be premature, as the ER within the plasmodesmatal channel may be a dynamic structure, constricting and dilating as signaling and function dictate.

The substructural detail of plasmodesmata (PD) in higher plants has been examined by various workers and a number of models have been proposed (for review, see Robards and Lucas, 1990; Botha et al., 1993; Ding et al., 1992b; Tilney et al., 1991). Recently, two models of plasmodesmatal structure were proposed based on computer-enhanced digital imaging analysis of electron micrographs of plasmodesmata from a dicotyledonous plant (Ding et al., 1992b) and from a C_4 grass (Botha et al., 1993). Ding et al. (1992b) examined the substructure of the plasmodesmata of a C_3 dicot, *Nicotiana tabacum*, following cryofixation and freeze substitution. According to their model (Fig. 1), the plasmodesma is depicted as a complex pore containing proteinaceous par-

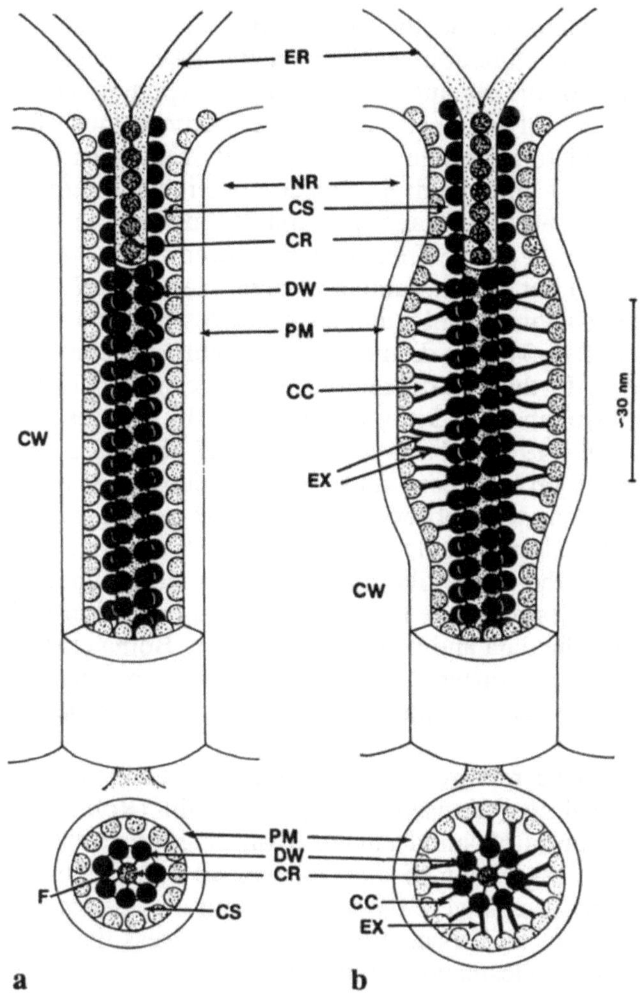

Fig. 1. Model of structure of plasmodesmata without (a) and with (b) a central cavity, based on computer enhanced digital imaging analysis of electron micrographs of leaf mesophyll in *N. tabacum* (from Ding et al., 1992b). ER, endoplasmic reticulum; NR, neck region; CS, cytoplasmic sleeve; CR, central rod; DW, desmotubulewall; PM, plasma membrane; CC, central cavity; EX, spoke-like extensions; CW, cell wall

ticles embedded in the inner leaflet of the plasma membrane and in the outer leaf of the desmotubule. Spoke-like filamentous strands apparently connect the globular proteins of the outer leaflet of the desmotubule to the proteins embedded in the encasing PD plasma membrane. The central region of the desmotubule, the "rod", is electron dense and is depicted as a series of particles, probably protein, that are embedded in the lipid of the fused inner leaflet of the ER membrane. In their model, the desmotubule is constricted and lacks a lumen. It was suggested that the electron dense particles embedded in the inner leaflet of the plasmalemma and the outer leaflet of the desmotubule form a convoluted channel that functions as a molecular sieve, determining the size exclusion properties of the channel. It was further hypothesized that the radial spokes may function to dynamically alter the effective radius of the cytoplasmic sleeve, thus regulating the size exclusion limit of the plasmodesma.

Botha et al. (1993) employed similar computer enhancement techniques with plasmodesmata at the Kranz mesophyll bundle sheath interface of *Thermeda triandra*, a C_4 grass. Their model showed many similarities to that of Ding et al. (Fig. 1) but with a number of significant differences. Associated with the inner plasmalemma leaflet and the outer desmotubule wall were particles approximately 2.5–3 nm in diameter, presumably proteins that abutted the cytoplasmic sleeve. In the central region of the desmotubule were electron-dense particles, also about 2.5 nm in diameter. In contrast to Ding et al. (1992b), Botha et al. (1993) observed a lumen of about 2.5 nm in diameter between the central protein core and the desmotubule wall. Furthermore, filamentous connections were observed between the central rod particles of the desmotubule and the desmotubule wall. In contrast, the model of Ding et al. (1992b) predicts that the spoke-like filamentous strands connected proteins of the outer leaf of the desmotubule to the proteins embedded in the encasing PD plasma membrane. The data and model presented by Botha et al. (1993) brings into question the general consensus that desmotubules are constricted and cannot function in cell-to-cell transport. Additional studies are needed to address this question.

Other workers have reported the presence of structures, termed "sphincters", either external to (Olensen, 1979) or within (Robinson-Beers et al., 1991) the neck region of plasmodesmata. It was suggested that these "sphincters" function in the regulation of plasmodesmatal conductance. It should be noted that these conclusions were based on structural evidence alone and since these structures are not universal, it is difficult to assess their involvement in the regulation of plasmodesmatal conductance.

Two types of plasmodesmata are described, the simple or primary plasmodesmata that are formed during the process of cell division and secondary plasmodesmata that may be formed later in development either by modification of existing plasmodesmata or de novo across existing cell walls (Lucas et al., 1993). During tissue development many simple plasmodesmata apparently undergo modification and change from a simple cylinder to highly branched structures, often with large open central cavities (Lucas et al., 1993).

B. Transport Through Plasmodesmata

Plasmodesmata were once thought to be passive pores that nonselectively allow passage of molecules from one cell to the next. More recent studies have altered this view, revealing plasmodesmata as dynamic channels which can be regulated by Ca^{2+} levels (Baron-Epel et al., 1988; Erwee and Goodwin, 1983), and components of the phosphoinositol cycle (Tucker, 1988). Transport through plasmodesmata has generally been studied by measuring the cell-to-cell movement of plasma membrane-impermeable fluorescent dyes of different sizes and properties. Such studies (Erwee et al., 1983; Goodwin, 1983; Terry and Robards, 1987; Tucker, 1982) led to a consensus that only molecules less then about 800–1000 Dalton pass freely through plasmodesmata. Other studies, however, are leading to a re-evaluation of the size exclusion limit (SEL) and selectivity of plasmodesmata. These new studies indicate not only that the plasmodesma can be gated, but that the size exclusion limit can be modulated by environmental and developmental signals.

Tucker and Tucker (1993) employed fluorescein (F) labeled mono-, di-, tri-, and quatra-amino acids to show that the plasmodesmata of the stamenal hairs of *Setcreasea purpurea* select for small hydrophilic molecules of less then 800 Da with −2 to −4 charge. Small peptides with attached aromatic amino acid such as phenylalanine and tryptophan exhibited low mobility through plasmodesmata with kinetics that generally did not fit a model of simple diffusion. From experiments in which mobile and non-mobile aromatic amino acid probes were co-injected, it was concluded that the non-mobile probes had neither blocked nor closed the permeation pores but that the plasmodesmata exhibited specific selectivity against these molecules irrespective of their size. However, following azide treatment the conductance properties of *Setcreasea purpurea* plasmodesmata were markedly altered (Tucker, 1993). In untreated cells, probes such as F-Phe, F-Try, F-$(meth)_2$, F-$(his)_2$ and FITC labeled Angiotensin II, an octa-peptide, did not pass through plasmodesmata or passed with non-diffusion kinetics. However, following azide treatment all probes passed with ease with diffusion coefficients that increased by about 400%. The data indicated that azide treatment may have caused the plasmodesmatal pore to dilate.

Further support for the notion that the size exclusion limit of plasmodesmata is much more extended then previously believed and that size selectivity can be modulated comes from studies by Cleland et al. (1994). These workers found that under aerobic conditions or in the absence of inhibitors of respiration, the SEL of wheat root PD as probed with Lucifer Yellow (LYCH) or F-dextrans was about 1 kDa. However, after either azide (1 or 10 mM) treatment or flushing with nitrogen, the SEL increased within 30–60 min to 3–4 kDa and sometimes as high as 7–10 kDa. Thus, natural mechanisms exist for modulating the size of the plasmodesmatal pore; the extent that the pore can be dilated is unclear. Nor has it been determined whether or not all plasmodesmata dilate to the same extent. It has been suggested that phosphorylation may be involved in regulating plasmodesmatal SEL. Epel et al. (1994) suggested that phosphorylation of plasmodesmatal components results in the down regulation of conductance and dephosphorylation results in a dilation of the conducting channel and an up-regulation of conductance.

C. Regulation of Closure by Environmental Signals

Differential pressure. The gating of plasmodesmata has been shown to be affected by various environmental and development signals (Cote et al., 1987;

Epel and Erlanger, 1991; Oparka, 1993; Oparka and Prior, 1992; Reid and Overall, 1992; Shepherd and Goodwin, 1992a, b). In giant algae such as *Chara* and *Nitella*, pressure gradients between cells increased electrical resistance across the node and inhibited intercellular transport of radioactively labeled assimilates (Cote et al., 1987; Reid et al., 1992). In higher plants, a pressure gradient between adjacent cells of leaf tricomes of *Nicotiana clevelandii* in excess of 0.2 MPa completely closed plasmodesmata. The degree of plasmodesmatal closure was dependent on the magnitude of the pressure differential between the two cells.

Light. Plasmodesmatal gating can also be regulated by light. Epel and Erlanger (1991) reported that lateral symplastic transport from the stele into the mesocotyl cortex of etiolated maize seedlings was inhibited by prior white-light irradiation. The inhibitory effect of white-light was completely photomodulatable by terminal far-red and far-red/red irradiations, suggesting the involvement of phytochrome. It was suggested that the modulation of plasmodesmatal conductance by light, and possibly by other environmental and/or development signals, might modulate growth and development, in part by establishing or altering symplastic domains and by channeling cell-to-cell transport of nutrients and growth regulators.

D. Cell and Tissue Specific Differences in Size Exclusion Limit

There is accumulating evidence to indicate that the SEL of plasmodesmata may differ in different cells and tissues. It was found that when fresh cut sections of the crease tissue of developing wheat grains were incubated in fluorochrome solutions of normally impermeable apoplastic probes such as LYCH, dye was rapidly absorbed via unsealed plasmodesmata (Wang and Fisher, 1994). Once absorbed, the dye moved symplastically. Using a size-graded series of probes, the SEL for the post-phloem pathway was estimated. In most, perhaps all cells of the crease tissue with the exception of the pericarp, the effective molecular diameter of the conducting channel was estimated to be about 6.2 nm vs. about 2.5–3 nm from other studies with other plants and tissues.

It has been suggested that plasmodesmata between intermediary and mesophyll cells in the sieve-element–companion-cell complex (SE-CCC) in the minor veins of plants which are symplastic loaders may have different diffusion characteristics than plasmodesmata between other cell types (Grama and Mayor, 1990; M. Grusak et al., unpubl.). In plants which are putative symplastic loaders, companion cells in minor veins exhibit distinct ultrastructural features and have been termed "intermediary cells". One obvious distinguishing feature of the intermediary cells are the presence of numerous plasmodesmata in tissue where loading presumably occurs, i.e., between intermediary cells and bundle sheath cells. These plasmodesmata are quite different than those between mesophyll cells. Minor vein structure apparently correlates with the type of sugar transported. Plants with SE-CCC with intermediary type companion cells, translocate in the phloem oligosaccharides such as raffinose (565 Da) and stachyose (666 Da) plus sucrose (342 Da). Turgeon (1990, 1991) proposed a model to describe the diffusion of sucrose from the bundle sheath into intermediary cells through plasmodesmata that connect the two cell types. Raffinose (a trisaccharide) and stachyose (a tetrasaccharide) are synthesized from sucrose in the intermediary cells, but are not able to diffuse back into the mesophyll cells because of their size. The trapping of oligosaccharides in the SE-CCC would require that the molecular SEL of the plasmodesmata between bundle sheath and intermediary cells be somewhat smaller than is common in plasmodesmata at other interfaces. Biochemical data are now available which support this model (M. Grusak et al., unpubl.).

A number of recent studies suggest that plasmodesmata between sieve elements and their companion cells may traffic in proteins (Bostwick et al., 1992; Fisher et al., 1992). In higher plants, mature sieve elements are generally enucleated and lack ribosomes. Since these cells can function for prolonged periods, either the proteins in these cells are very long lived or proteins are somehow transported into these cells, most likely from adjacent companion cells via plasmodesmata. Fisher et al. (1992) investigated protein turnover in sieve tubes of wheat, and showed by radiolabeling studies that about 200 different soluble proteins were present in the sieve tube, many of which were rapidly labeled. Furthermore, the data indicated that most proteins were loaded at the source (the leaf) and unloaded at the sink (the grain). Certain proteins showed very rapid turnover suggesting some selectivity of movement from companion cells into sieve tubes.

Further evidence that plasmodesmata between companion cells and sieve elements can selectively transport proteins was provided by experiments with the phloem exudate from pumpkin. The phloem exudate from pumpkin contains two abundant basic proteins, termed PP1 and PP2, both of which are expressed exclusively in developing and mature sieve elements and their companion cells (Smith et al., 1987). PP1 and PP2 mRNAs were localized by in situ hybridization (PP1: Bostwick et al., 1992; PP2: G. A. Thompson, pers. comm.) only in companion cells within the phloem of hypocotyl tissues. However, the exact site of P-protein synthesis in differentiating phloem is still unclear. Thompson (pers. comm.) suggests three possible scenarios: (1) protein synthesis occurs exclusively in the companion cells and is transported into sieve elements; (2) immature nucleated sieve elements synthesize stable proteins which are also synthesized in companion cells; and (3) a combination of 1 and 2 where protein synthesis occurs in both cell types and protein transport occurs from the companion cells to the enucleated mature sieve elements. Temporal developmental studies using in situ localization and/or using transgenic plants containing reporter constructs under the transcriptional regulation of the PP promoter should aid in determining the correct model. If P-proteins are synthesized in companion cells and transported unidirectionally via plasmodesmata to contiguous sieve elements, this could indicate that the plasmodesmata between the companion cells and sieve element have different properties and possibly different composition then other plasmodesmata. Alternatively, it could be hypothesized that plasmodesmata connecting the companion cells with sieve element have a unique regulatory mechanism for passage of large molecules. If movement through these plasmodesmata is specific for particular proteins, then it must be concluded that these plasmodesmata do not function as simple molecular sieves.

It is unclear whether the potential to transport proteins between cells is a unique property of sieve tubes and their companion cells or whether protein transport can normally occur between all cells. Molecular studies must be designed to test whether the potential for protein transport requires special targeting sequences. Micro-injection experiments with fluorescently labeled phloem proteins could also be employed to test whether plasmodesmata in other tissues are competent to transport phloem resident proteins.

Since membrane lipids and proteins within the sieve tubes probably also undergo turnover, there must be some mechanism for the transport of membranes or membrane components, both lipids and proteins, into sieve tubes. This could be accomplished by migration within the plasma membrane of the plasmodesmata or via budding off of desmotubule membranes and incorporation into the plasmamembrane of the sieve tube. A recent study using the technique of fluorescence redistribution after photobleaching (FRAP) has provided evidence that the ER membranes of plasmodesmata can serve as a dynamic diffusion pathway for the movement of lipids and lipid signaling molecules between contiguous cells (Graski et al., 1993). However, no detectable intercellular movement was observed for fluorescent probes residing exclusively in the plasma membrane despite the fact that probes located on the plasma membrane showed considerable lateral mobility within the plasma membrane of a single cell.

E. Protein Components of Plasmodesmata

The first biochemical and molecular information on plasmodesmata came from studies that employed antibodies directed against gap-junction proteins and suggested the presence of proteins homologous to animal connexin proteins in plants (Hunte et al., 1992, 1993; Kotlizky et al., 1992; Meiners et al., 1988, 1991; Meiners and Schindler, 1987, 1989; Schulz et al., 1992; Xu et al., 1990; Yahalom, 1991). Epel and co-workers provided more definitive evidence for the presence of connexin homologous proteins in plants in immunological studies in maize (Yahalom, 1991). These workers showed by indirect immunogold labeling that these plasmodesmata contain two different proteins that crossreact with antibodies against connexins 32 and 43 (Yahalom, 1991). However, the biological relevance of these apparent similarities has not yet been determined.

Further characterization of plasmodesmata-associated proteins (PAPs) was made possible by the development of techniques for plasmodesmata isolation. Initial characterization of putative PAPs was performed with preparations of cell walls that contain embedded plasmodesmata (Kotlizky et al., 1992; Monzer and Kloth, 1991; Turner et al., 1994; White et al., 1994; Yahalom, 1991). Plasmodesmata isolated from maize walls (Epel, 1994; Epel et al.,

1995) following treatment with cellulase contained peptides with apparent mol. wts. of 64, 51, 41, 32, 26, 21, 17, and 15 kDa. Epel and coworkers generated polyclonal antibodies against the 17, 26, 32, 41, 51, and 64 kDa wall associated polypeptides and showed by Western blot analysis that the 17, 26, 32, and 51 kDa proteins were detectable only in the wall fraction of cell extracts. Ongoing immunolocalization studies have so far shown that the 41 kDa protein is associated with the plasmodesmata as well as Golgi membranes. Similar studies with antibodies against other putative plasmodesmata-associated proteins are in progress.

White et al. (1994) recently presented immunocytochemical data to suggest the presence of actin in plasmodesmata. Li and Zhang (1994) employing embeddment-free sections for electron microscopy have observed clear images of 5–7 nm filaments, presumably actin, which traverse enlarged modified plasmodesmata in wheat nucellus and integument cells. In these specialized cells the actin filaments are apparently involved in the intercellular movement of organelles through these modified plasmodesmata. In light of the reports by White et al. (1994), we speculate that actin filaments may be involved in the regulation of plasmodesmatal conductivity and the selective movement of macromolecules through plasmodesmata.

III. Virus Movement in Infected Plants

Most, if not all plant viruses spread from cell to cell via plasmodesmata (Atabekov and Taliansky, 1990; Hull, 1989; Maule, 1991). Several studies over the past few years that focused on the process of viral cell-to-cell spread have demonstrated the involvement of virus-encoded movement proteins (MP). Studies with different classes of viruses indicate at least two distinct mechanisms of movement. One mechanism, exemplified by tobacco mosaic virus (Doem et al., 1987), involves one or more virus-encoded movement proteins that facilitate the passage of a non-virion form of the virus. The second mechanism, exemplified by the comovirus cowpea mosaic virus (CPMV), involves the movement of the complete virus particles through tubular structures to adjacent cells (van Lent et al., 1990, 1991; Wellink et al., 1993). It is still an open question whether existing plasmodesmata are modified or whether new connections between cells are induced by the virus.

The first evidence to suggest that viruses encode specific transport functions came from temperature sensitive mutants of tomato mosaic tobamovirus (Nishiguchi et al., 1978). The Ls1 strain of TOMV, a spontaneous mutant of TOMV strain L is defective in cell-to-cell movement at nonpermissive temperatures (32 °C) in inoculated leaves. Leonard and Zaitlin (1982) used two dimensional peptide analysis to show that there was a minor difference between the mobilities of the 30 kDa protein of the Ls1 strain and the L strains. Sequence analysis of the 30 kDa open reading frame showed a single base change which resulted in substitution of a serine in Ls1 for proline 154 in the L strain (Ohno et al., 1983).

More direct evidence that the 30 kDa protein is involved in movement was obtained by two distinct yet complementary approaches. Meshi et al. (1987) introduced the Pro to Ser change in the 30 kDa protein into a full length infectious cDNA clone of the L strain. Infectious transcripts of the mutated viral cDNA were replicated in inoculated leaves but did not move from cell-to-cell at the non-permissive temperature. Constructs with frameshift mutations in the gene that encodes the 30 kDa protein were replication and assembly competent but failed to move from cell to cell. Deom et al. (1987) used transgenic tobacco plants that expressed a gene encoding the 30 kDa protein to complement the movement deficient Ls1 virus. Transgenic plants that accumulated the 30 kDa protein of tobacco mosaic tobamovirus (TMV) supported both local and systemic spread of Ls1 at 32 °C. Based on these result the 30 kDa protein of TMV was designated the movement protein (MP).

MPs from several different viruses are currently being studied by a variety of approaches. The most direct approach is to use an infectious clone of the virus in question, i.e., a cloned cDNA of the virus that is itself infectious or from which infectious transcripts can be derived in vitro. Mutations in the cDNA enable the researcher to determine the function of each viral gene, including those that are required for cell-to-cell and long-distance movement. TMV (Meshi et al., 1987; Saito et al., 1990), tobacco rattle tobravirus (TRV) (Harrison and Robinson, 1986; Ziegler-Graff et al., 1991), and red clover necrotic mosaic dianthavirus (RCNMV) (Xiong et al., 1993) each encode a protein that is essential for movement between cells; these viruses do not require the coat protein for cell-to-cell movement, indicating that virions do not move from cell to cell. Some viruses require more than one protein for cell-to-cell movement. The proteins encoded by three genes encoded in overlapping ORFs (referred to as triple block proteins) in potexviruses (Beck et al., 1991), furoviruses (Gilmer et al., 1992), and hordeoviruses (Petty et al., 1990) are each required for cell-to-cell movement of these viruses. Turnip crinkle carmovirus encodes two small proteins that are essential for cell-to-cell movement (Hacker et al., 1992).

The viral encoded proteins that are involved in the formation of tubules through which viruses apparently pass have also been identified. In the case of cauliflower mosaic caulimovirus (CaMV; Perbal et al., 1993), cowpea mosaic comovirus (CPMV; van Lent et al., 1990) and tomato ringspot nepovirus (TomRSV; Wieczorek and Sanfacon, 1993), it was demonstrated that MPs were associated with the tubules. In the case of CPMV, two MPs and the coat protein are needed for cell-to-cell movement (Wellink and Van Kammen, 1989). In addition to these viruses, MPs have been identified in tombusviruses (Dalmay et al., 1993), cucumoviruses (Suzuki et al., 1991), bromoviruses (Allison et al., 1990), and ilarviruses (Dore et al., 1991). Table 1 lists the MPs that have been identified to date and summarizes information concerning their functions. From these and other studies it is apparent that plant viruses have evolved different mechanisms that modify the structure and/or function of plasmodesmata so that local and long-distance spread of infection is achieved.

Table 1. A list of viruses for which an MP gene has been identified. Summary of information about the MPs from representative members of virus families in which the MP function has been determined; names of genes or size of MPs are indicated

Virus name	Virus family	Nuclei acid binding	MP nomenclature	Requirement of CP for cell-to-cell movement	Long-distance movement	Tubules in infected tissue
Alfalfa mosaic	ilar-	yes	3a	yes	yes	yes[a]
Barley stripe mosaic	hordeo-	n.d.	triple block genes	n.d.	n.d.	n.d.
Beet necrotic yellow vein	furo-	n.d.	triple block genes	no	n.d.	n.d
Brome mosaic	bromo-	n.d.	3a	somewhat	yes	no
Cauliflower mosaic	caulimo-	yes	P1	n.d.	n.d.	yes
Cowpea mosaic	como-	n.d.	58 kDa and 48 kDa	yes	yes	yes
Cucumber mosaic	cucomo-	yes	3a	yes	yes	yes[a]
Cymbidium ringspot	tombus-	n.d.	22 kDa	no	yes	no
Red clover necrotic	diantho-	yes	35 kDa	no	yes	no
Squash leaf curl	gemini-	yes	BR1 and BL1	no	no	n.d.[b]
Tobacco mosaic	tobamo-	yes	30 kDa	no	yes	no
Tobacco rattle	tobra-	n.d.	29 kDa	no	yes	no
Tomato ringspot	nepo-	n.d.		n.d.	n.d.	yes
Turnip crinkle	carmo-	n.d.	p8 and p9	no	yes	n.d.
White clover mosaic	potex-	n.d.	triple block genes	yes	yes	no

[a]Single report
[b]For Euphorbia mosaic geminivirus tubules have been observed in plasmodesmata in infected leaf mesophyll cells
n.d., not determined

A. *Biochemistry and Genetics of Movement Proteins*

1. Genetic Studies

The MPs of tobamoviruses have been studied for a number of years and to date, the sequence of a least 14 different tobamovirus MPs has been determined. A comparison of the amino acid sequences of tobamovirus MPs derived from nucleotide sequences shows three distinct regions of sequence similarity within the proteins. Regions I and II comprise the sequences with the greatest degree of similarity (Saito et al., 1988). Region III, which lies near the carboxy-terminus of the MP shows wide sequence divergence, while retaining certain structural similarities. The region can be divided into three sub-regions, A, B, and C. A and C contain high proportions of acidic amino acids while region B is highly basic in charge. Biochemical and genetic analyses of tobamovirus MPs have begun to reveal the functional significance of these structural domains.

Several naturally occurring TMV mutants with altered MP function have been identified. Two mutations in region II cause temperature-sensitive defects in the MP; Ls1 and Ni2519 abolish cell-to-cell movement of TMV at restrictive temperatures (Ohno et al., 1983; Zimmern and Hunter, 1983). Meshi et al. (1989) reported that mutations within region I allow TOMV, a tobamovirus closely related to TMV, to overcome the resistance conferred by the *Tm-2* gene in tomatoes. While TOMV replicates in initially infected cells in resistant plants but cannot move from cell to cell (Motoyoshi and Oshima, 1977), mutations that cause certain amino acid substitutions in region I (Cys-68 to Phe and Glu-133 to Lys) enable TOMV to overcome *Tm-2* resistance. This result may suggest that this region of the MP interacts in some way with one or more host components to overcome the effects of the *Tm-2* gene. The resistance gene *Tm-2*^{2}, an allele of *Tm-2*, can be overcome by two concurrent mutations within the COOH terminal end of the TOMV MP (Ser-238 to Arg and Lys-244 to Glu; Weber et al., 1993).

Naturally occurring mutations within MPs have also been found in other viruses. A single missense mutation found in the BR1 movement protein of squash leaf curl geminivirus (SqLCV) alters the host range of this virus (Ingham and Lazarowitz, 1993). The mutant virus is no longer able to infect one of its natural host plants while retaining the ability to infect other host species. While this mutation may reveal a portion of the MP that is important in host range, most naturally occurring mutations simply render the MP inactive. Until specific biochemical functions of the MP are determined, the cause for these genetic lesions will remain unknown.

Directed mutagenesis has been performed on the MP genes of several viruses as a method to identify functional domains within the protein. The effects of mutations on the function of the TMV MP were analyzed by two different methods. The mutant MP genes were either expressed in transgenic plants that were subsequently assayed for the ability of the transgene to complement TMV mutants or they were used to replace the wild-type gene in an infectious CDNA clone of TMV. When up to 55 amino acids were deleted from the carboxy-terminal end of the MP, virus infection spread from cell to cell, although some mutants resulted in poor cell-to-cell movement as indicated by the size of necrotic local lesions on Xanthi NN tobacco plants (Berna et al., 1991). Deleting 73 amino acids from the C-terminal end of the MP resulted in a dysfunctional protein. Transgenic plants that express a gene carrying this mutation did not complement the movement of a MP(–) mutant of TMV and the SELs of plasmodesmata in these plants were not altered (described further in Sect. IV.A). This mutant movement protein was not localized to the cell wall fraction, and presumably not to the plasmodesmata. Based upon these results, Berna et al. (1991) proposed that amino acids 195–213 contain sequences that are important for localization to plasmodesmata. Deletions in the NH_2-end of the MP resulted in nonfunctional proteins (Gafny et al., 1992). Fine structure deletion mutagenesis was performed on the TMV MP gene by introducing in-frame deletions which resulted in the loss of three internal amino acids at 10 amino acid intervals (M. Lapidot and R. N. Beachy, unpubl.). When each of the mutations was tested for biological activity in TMV, only those in which the mutations were near the carboxy-terminal end of the MP allowed cell-to-cell movement of the virus. However, the effect of these deletions on other function(s) of the MP have yet to be determined.

Directed mutagenesis of genes from viruses other than TMV has been done primarily to identify those open reading frames that encode MPs. Mutagenesis of the MP of alfalfa mosaic ilarvirus (AlMV) identified a domain that is necessary for association of the protein with cell walls of transgenic plants (Erny et al., 1992). Directed mutagenesis throughout the MP of red clover necrotic mosaic dianthavirus (RCNMV) led to the identification of three distinct regions of the protein; one domain is necessary for in vitro RNA binding, another for cooperative binding of RNA in vitro, and a third domain that is necessary for cell-to-cell movement (Giesman-Cookmeyer and Lommel, 1993). These mutants are discussed further in the following sections.

2. Post-translational Modifications of Movement Proteins

There have been limited numbers of published studies that examined the role of post-translational modifications of MPS on their function; all involved the MP of TMV. There is evidence that the MP of TMV (the MP contains 268 amino acids) is phosphorylated when expressed in insect cells (Atkins et al., 1991b), in TMV infected protoplasts (Watanabe et al., 1992), and in in vitro reactions (Citovsky et al., 1993). The sites of in vitro phosphorylation were localized to the ten amino acids near the carboxy-terminal end of the MP by Citovsky et al. (1993); Ser-258, Ser-265, and Thr-261 were phosphorylated by a protein kinase activity associated with isolated cell walls. They also showed that when the last 43 amino acids were deleted from the MP, it was not phosphorylated in their in vitro assay. Watanabe et al. (1992) demonstrated that the MP of ToMV was phosphorylated between amino acids 234–261. If this region is deleted, the MP does not become phosphorylated.

The role of phosphorylation in the function of the TMV MP is not clear. Berna et al. (1991) reported that a mutant MP which is truncated to amino acid 213 retains its function. This result implies that phosphorylation of amino acids downstream of residue 213 is not essential for biological activity of the MP, at least under the conditions used in these studies.

The TMV MP appears not to be modified by N-linked glycosylation when it is expressed in insect cells (Atkins et al., 1991b) or in transgenic tobacco plants (C. M. Deom, pers. comm.).

3. Subcellular Localization of Movement Proteins

The primary approaches used to determine the distribution of MPs have been immuno-electron microscopy, and biochemical subcellular fractionation using virus infected tissues and transgenic plants that express genes encoding MPs. The TMV MP was associated with a fraction enriched for plasma membrane, but a proportionately greater fraction was extracted from isolated cell walls (Doem et al., 1990; Moore et al., 1992; Moser et al., 1988). Furthermore, the TMV MP that associated with the plasma membrane behaved as a highly hydrophobic integral membrane protein (Moore et al., 1992). A small amount of soluble MP in TMV infected cells and in MP(+) transgenic plants was in a high molecular weight complex with an approximate molecular weight above 550 kDa. The soluble high molecular weight complex was apparently comprised solely of proteins and contained no nucleic acids (Fenczik, 1994).

Immunocytochemical studies using electron microscopy of TMV infected plants (Tomenius et al., 1987) and transgenic plants that express the MP gene (Atkins et al., 1991; Moore et al., 1992) localized the TMV MP to branched secondary plasmodesmata. In TMV-infected protoplasts, the MP was associated with electron-transparent structures present in the cytosol (Meshi et al., 1992); however, such structures have not been observed in cells of infected tissues, and their relevance to the infection process is unknown. The MP may also be associated with simple plasmodesmata, but at concentrations below detection limits. It is not clear whether the MP detected in compound plasmodesmata is of physiological significance.

The MPs of RCNMV, AlMV, CaMV, and SqLCV were each shown to accumulate in cell wall fractions of infected cells (Albrecht et al., 1988; Godefroy-Colburn et al., 1986; Osman and Buck, 1991; Pascal et al., 1993). Two of the three "triple block" proteins of beet necrotic yellow vein furovirus (BNYVV), p42 and p13, each of which are required for cell-to-cell movement, accumulated in a membrane-rich fraction (Neisbach-Klosgen et al., 1990). In contrast, p25, one of the proteins encoded by the triple block genes in potato virus X (PVX), was found in a soluble fraction in association with cytoplasmic inclusion bodies (Davies et al., 1993).

The BL1 protein of SqLCV was in a soluble fraction, the cell wall fraction, and the plasma membrane enriched fraction extracted from transgenic tobacco plants and infected plants (Pascal et al., 1993). Interestingly, transgenic plants that express the BL1 gene displayed disease symptoms similar to those caused by virus infection. The transgenic plants that express the BL1 gene, however, did not functionally complement a mutant SqLCV that contains a dysfunctional BL1 gene. These results suggest either that BL1 does not play a role in virus spread, or that transgenic expression of BL1 does not produce a protein that is capable of functioning independent of other events in virus replication. Another explanation could be that the levels of expression of the BL1 protein in the transgenic plant are not sufficiently high in appropriate cell types for functional complementation of the mutant virus.

The methods employed in the subcellular fractionation studies described above are sufficiently different to preclude direct comparisons among them, and thus general conclusions from such research are limited until it is shown in which fraction(s) an MP functions to effect the spread of virus infection. However, the variety of locations in which different MPs accumulate may indicate that there are different mechanisms for local virus spread. Thus, although some MPs do not localize to plasmodesmata or to the cell wall fraction, they may nevertheless interact with plasmodesmata in a manner that cannot be determined by the methods of the experiment.

4. Binding of Nucleic Acids by Movement Proteins

The MPs from AlMV (Schoumacher et al., 1992), CaMV (Citovsky et al., 1992), TMV (Citovsky et al., 1990), and RCNMV (Osman et al., 1992) each bind to nucleic acids in a cooperative manner in in vitro binding assays. In each case, the binding of MP to nucleic acid showed no sequence specificity, and only in the case of AlMV and RCNMV were viral nucleic acids used in the assays. The TMV MP was shown to unfold single stranded RNA and DNA, forming thin extended structures 1.5–2 nm in diameter (Citovsky et al., 1992). These authors suggested that it is this form of viral RNA that moves from cell to cell. The RCNMV MP, on the other hand, does not unfold viral RNA in vitro (Fujiwara et al., 1993). This may be reflective of the different ways that these MPs function.

In a recent study, alanine substitution mutations were produced in the RCNMV MP cistron to study the different functions of the MP (Giesman-Cookmeyer et al., 1993). This study showed that the ability to cooperatively bind RNA in vitro did not correlate with the ability of the mutant MPs to potentiate viral movement. Furthermore, there was no strong correlation between virus movement, the ability to alter the SEL of plasmodesmata, or the ability to traffic RNA from one cell to the next when microinjected into plant cells (Fujiwara et al., 1993). From these results one can conclude that the result of the in vitro protein-RNA binding assays may not reflect the mechanisms of RCNMV MP function in vitro.

Perbal et al. (1993) suggested that the ability of the CaMV MP to bind to single stranded nucleic acids in vitro is not reflective of its role in vivo. CaMV, a double stranded DNA virus, induced the formation of tubular structures in infected tissues in which virions were visualized. This result supports the hypothesis that CaMV is transported from cell to cell as a virion in CaMV infection rather than as a nucleic acid–MP complex as previously predicted by Citovsky et al. (1992) based upon in vitro studies.

B. Mechanisms of Short-distance Movement

1. Functional Modification of Plasmodesmata

The movement proteins of several plant viruses have been shown to modify plasmodesmata both structurally and functionally. Transgenic plants that express the TMV MP gene were used to show that the MP has an effect on the SEL of plasmodesmata (Wolf et al., 1989). Microinjection of fluorescently labeled dextrans into individual cells of MP(+) transgenic plants showed that 10 kDa molecules could move through plasmodesmata of these plants, while the largest able to move out of the injected cell in control plants was a 0.75 kDa dextran. In a later study, it was reported that when purified TMV MP was co-injected with FITC-labeled dextrans into cells of a nontransgenic plant, there was an increase of plasmodesmatal permeability that allowed the movement of a 20 kDa dextran from initially injected cells (Waigmann et al., 1994). Furthermore, the labeled dextrans moved 10–20 cells distal to the injected cell, suggesting that the MP can either potentiate SEL changes in nearby cells or that the MP itself moves from one cell to the next. These studies also identified a domain between amino acids 126 and 224 of the MP (MP contains 268 amino acids) that was necessary for altering SEL.

The plasmodesmata of MP(+) transgenic plants are also structurally altered. The TMV MP was immunologically detected in secondary but not primary plasmodesmata of transgenic MP(+) tobacco plants (Ding et al., 1992a; Moore et al., 1992). As defined by Ding et al. (1992a), primary plasmodesmata are simple non-branched structures that are present in early stages of leaf development and secondary plasmodesmata are branched with an open center cavity and are characteristic of later stages of leaf development. In MP(+) plants, the open cavity is filled with fibrous material of unknown composition (Ding et al., 1992a; Moore et al., 1992) and is apparently labeled with anti-MP antibodies. Fibrous material in the central cavity of plasmodesmata of TMV infected plants was previously observed (Weintraub et al., 1976) and was presumed to be viral rods. This is probably an incorrect interpretation, since such fibers are seen in the absence of virus infection in transgenic plants that express the MP gene.

Although the results of these studies are intriguing and show that the movement protein is capable of altering plasmodesmata function, they have not established a direct correlation between viral movement, increased SELs, presence of fibrous material in the plasmodesmata, and antibody labeling. Both Moore et al. (1992) and Ding et al. (1992a) performed an analysis of transgenic tobacco leaves of different developmental ages in an attempt to correlate these characteristics with movement of a MP(–) mutant of TMV. In both studies there were inconsistencies in the data. For example, movement of the MP(–) virus occurred in cells that did not stain with anti-MP antibody or contain fibrous material in the central cavity of plasmodesmata. This result can be explained if the accumulation of the fibrous material resulted from an overabundance of MP in plasmodesmata, rather than MP that is involved in virus movement. The lack of antibody staining of the plasmodesmata of young leaves may be due to differences in the amounts of MP needed for detection compared with the amounts needed for virus movement.

A second inconsistency was presented by Ding et al. (1992a) who reported there was no movement of the 9.4 kDa dextran between cells at the tip of young leaves of MP(+) transgenic plants, although there was accumulation of fibrous material and labeling with anti-MP antibodies. Furthermore, MP(–) TMV can spread cell to cell in those tissues. It is possible that small amounts of MP can potentiate movement by altering the SEL, but the change is temporary and is not readily detected by dye coupling studies. This may also be the reason that Ding et al. (1992a) did not observe dye movement between bundle sheath cells into phloem parenchyma cells, despite the presence of TMV MP in the plasmodesmata of these cells. The movement of molecules through plasmodesmata between bundle sheath and phloem parenchyma cells is proposed to be more stringently regulated than those between mesophyll cells (Leisner and Turgeon, 1993). Perhaps the presence of MP in these cell types does not allow a permanent change in the SEL, and is therefore not detected by dye coupling studies.

In recent studies, purified movement proteins have been injected into individual cells to investigate their role in host–virus interactions. Fujiwara et al. (1993) showed that co-injecting purified RCNMV MP with 9.4 kDa FITC-labeled dextran into cells of cowpea (*Vigna unguiculata*) leaves caused an increase of the SEL of plasmodesmata and movement of the dye. It was also shown that fluorescently tagged MP moves out of the initially injected cell. These investigators also co-injected fluorescently labeled RCNMV RNA2 with the MP and showed that RNA moved into neighboring cells, whereas labeled DNA did not move. By injecting mutants of the RCNMV MP, it was possible to correlate the capacity to alter the SEL, movement of co-injected RNA, and the ability of the MP to potentiate movement of virus in inoculated plants. One of the MP mutants altered the plasmodesmata SEL, but was unable to move RNA from cell to cell, suggesting that these are separate functions of the MP.

In a study reported by Noueiry et al. (1994), the BL1 and BR1 proteins encoded by bean dwarf mosaic geminivirus (BDMV), a single-stranded DNA virus, were injected into plant cells and assayed for the ability of each protein to move into adjacent cells and for their capacity to alter plasmodesmatal SEL. This study showed that BL1 moved from cell to cell and altered the SEL so that a 10 kDa dextran moved to adjacent cells. The BL1 protein also potentiated the movement of single-stranded DNA but not RNA from cell to cell. The BR1 protein was shown to potentiate the movement of viral DNA out of the nucleus, the site of virus replication, into the cytoplasm. It was previously shown that virus particles and virus-induced abnormalities occurred only in phloem parenchyma cells of BDMV-infected bean leaves, indicating that this virus is phloem-limited (Morales et al., 1990). The observation that injected BL1 and BR1 can function in mesophyll cells apparently indicates that the phloem limitations of the virus may not be due to inability of the proteins to function in mesophyll cells.

Changes in plasmodesmata function that result during virus infection have been studied by injecting TRV into leaf trichome cells of *Nicotiana clevelandii* (Derrick, 1992). Injected trichomes were removed at various times after injection after which the remaining leaf tissues were observed for virus infection. In this way, these investigators determined the time required for the virus to move from injected cells into the leaf mesophyll cells. In this study, a minimum of four hours was required for the virus to move from the injected cell to adjacent mesophyll cells. FITC-labeled dextrans were injected at various times post inoculation to determine if the plasmodesmatal SELs were altered. In cells that were not injected with virus, dextrans of 0.54 kDa, but not 4.4 kDa moved from cell to cell. In cells that were infected, the 4.4 kDa dextran moved from cell to cell within 3 h post infection, consistent with the time required for spread of the infection. This report showed that alterations in SEL of plasmodesma occurred during virus infection and were not simply an artifact related to the expression of MP in transgenic plant lines, or to the techniques associated with the injection procedures.

2. Structural Modification of Plasmodesmata

In certain plant virus infections, viral transport involves the formation of tubules that alter the structure of plasmodesmata. In plants infected by CPMV, it was possible to observe virus particles in the tubules that passed through plasmodesmata (van Lent et al., 1991). However, the virus-like particles were not recognized by anti-coat protein (CP) antibodies (van Lent et al., 1991), perhaps because the antigenic sites of the CP were not accessible to the antibody. Similar results were obtained with TomRSV (Wieczorek et al., 1993).

Electron microscopy/immunogold labeling studies of tissues infected with TomRSV (Wieczorek et al., 1993), CPMV (van Lent et al., 1991), red clover mottle comovirus (RCMV; Shanks et al., 1989), and CaMV (Linstead et al., 1988) have shown that the respective movement proteins are associated with tubules in infected tissues. In TomRSV infected tissues, the tubular structures were found only in a zone of 1 to 4 cells in width immediately surrounding necrotic tissue (Wieczorek et al., 1993). Tubular structures were also observed in tissues infected with AlMV (Godefroy et al., 1990) and cucumber mosaic virus (CMV; Francki et al., 1985). However, unlike the tubules observed with comovirus infections, those observed with AlMV and CMV were transient and seen only at the leading edge of the infection.

Euphorbia mosaic geminivirus (EMV) is phloem limited in some hosts, such as *Euphorbia* spp. but not in others, such as *Datura stramonium* (Kim and Lee, 1992). When EMV infects *D. stramonium*, tubules form transiently at the leading edge of the infection and virus-like particles are found within the tubules. Tubules have not otherwise been visualized in infections of geminiviruses that are phloem-limited. There are several reasons why tubules may not have been visualized in infected tissues. First, an MP may not be essential for virus movement through the phloem, so tubules would not be produced. Second, since the tubules in EMV infected tissues are observed only at the leading edge of an infection, the chance of sectioning through a phloem cell with tubules may be slight. However, in light of the selective pressures on viral genomes, it seems unlikely that a nonessential gene would be maintained, suggesting that an EMV encoded protein may indeed have a role in movement of virus through phloem tissues, or that the virus is capable of infecting non-phloem tissues at some time during infection of the plant.

Recently it was demonstrated that tubules can form in the absence of a cell wall or plasmodesmata. Protoplasts that were infected with either CPMV or CaMV, each of which produced tubules in infected plants, produced elongated tubules that extended from the surface of the protoplast (Perbal et al., 1993; van Lent et al., 1991) and bound anti-MP antibodies. The tubules were longer than those seen in infected cells, indicating that their physical nature may be regulated by the presence of the cell wall and/or plasmodesmata. Since the tubules are formed in the absence of plasmodesmata, it was suggested that the tubules are not altered forms of the desmotubule. In the case of CaMV, tubules were isolated from the protoplasts and shown by Western blot analysis to contain the P1 protein (Perbal et al., 1993). Additional studies to determine the biochemical composition and ultrastructural nature of the tubules are necessary if the role of such structures in virus spread is to be determined.

3. Requirement of Virus Coat Protein for Cell-to-Cell Movement

While the MP is considered essential for cell-to-cell spread of virus infections, some viruses also require the CP for local movement. In many cases, the requirement of CP is taken as an indication that virions move from cell to cell, as is the case with CPMV (Wellink et al., 1989), although this remains to be confirmed for selected viruses. In several cases virions or virus-like particles were observed in tubules examined by electron microscopy (van Lent et al., 1990). The CP of cucumber mosaic virus (CVM) is required for cell-to-cell movement (Suzuki et al., 1991), and a mutant carrying a 26 amino acid deletion near the amino terminus of the CMV CP is capable of cell-to-cell spread, but not long-distance movement. Alfalfa mosaic ilarvirus requires the CP for spread in the inoculated leaf due, at least in part, to the role of the CP in replication of the virus (Bol et al., 1971). Finally, viruses may employ different mechanisms depending upon the host. For example, turnip crinkle carmovirus (TCV) requires the CP for cell-to-cell movement in *Chenopodium amaranticolor* and *Brassica campestris*, but not in *N. benthamiana* (Hacker et al., 1992; Heaton et al., 1989).

C. Long-distance Movement of Viral Infections

Role of coat protein in long-distance movement. It is generally the case that the CP is required for long distance movement for plant viruses in most hosts, the exception being the bipartite geminiviruses (Gardiner et al., 1988;

Hayes et al, 1988; Ward et al., 1988). Many viruses that require CP for long-distance movement in one host may, in some circumstances, not require it in another host. For instance, mutations of either the CP or the origin of assembly in TMV render the virus unable to move throughout the plant (Saito et al., 1990). However, TMV can systemically infect *Nicotiana benthamiana* without a functional CP (Dawson and Hilf, 1992). *N. benthamiana*, however, may be an unusual plant host; in addition to its susceptibility to viruses without a CP, it is also susceptible to a great range of plant viruses (Dawson et al., 1992). It would be interesting to know if the vascular tissue or the plasmodesmata between vascular and nonvascular tissues of these plants are different from those of other species of *Nicotiana*. Electron microscopy studies of the plasmodesmata connecting vascular and nonvascular cells and physiological studies of this plant may shed light on its unusual susceptibility to virus infections.

Role of virus movement proteins in long-distance spread of virus infections. The role of virus movement proteins in long-distance movement is only now beginning to be elucidated. Geisman-Cookmeyer and Lommel (1993) showed that two mutants in the MP of RCNMV were able to move from cell to cell in inoculated leaves but were unable to move systemically. This is the first case of a mutation in the MP gene that uncoupled cell-to-cell and long-distance movement.

The phloem-limited viruses may be especially useful for studying the role of MP in long-distance movement. It has been suggested that phloem-limited viruses such as the luteoviruses do not require an MP to infect plants (Atabekov and Taliansky, 1990). As discussed earlier, this is not the case for EMV which is able to infect leaf mesophyll cells in some hosts but is restricted to phloem cells in other hosts. Further study of the geminivirus MPs may reveal a distinct role of MP in phloem-mediated transport.

D. Interactions Among Transport Functions from Different Virus Systems

1. Effects of Mixed Virus Infections

Studies of mixed virus infections have shown that both local and long-distance transport of some viruses on non-host plants can be triggered by co-inoculation with a second virus. For example, Dodds and Hamilton (1972) showed that TMV which normally remains in the inoculated leaves of barley plants, could move systemically when co-inoculated with barley stripe mosaic hordeiovirus.

Mixed infection studies have also been performed with phloem limited and non-limited viruses (Atabekov and Dorokhov, 1984; Barker, 1987; Harrison et al., 1990). TMV promoted the penetration of bean golden mosaic geminivirus (BGMV) into tobacco mesophyll cells (Carr and Kim, 1983), and potato leafroll luteovirus (PLRV) spread from phloem specific tissues into mesophyll cells when co-inoculated with tobamo-, poty-, potex-, tobra-, and umbroviruses

(Harrison et al., 1990). However, PLRV did not spread to mesophyll cells by co-inoculation with nepo-, cucumo-, enamoviruses, or parsnip yellow fleck virus. These studies showed that movement functions are sufficiently similar among certain viruses to effect cooperation among distantly related viruses. There seems to be no correlation between the apparent mechanism of virus movement and the ability of one virus to affect the movement of another, since viruses that form tubules in infected cells can promote the movement of non-tubule-forming viruses, and vice versa (Atabekov and Taliansky, 1990; Malyshenko et al., 1988b). Perhaps when more is known about the different functions of various movement proteins these relationships will be more clearly understood.

2. Interactions Between Different Movement Proteins in Transgenic Plants

To study the functions or phenotypes mediated by movement proteins apart from the effects of other viral proteins in the results of mixed infections, several studies have been performed with transgenic plants that express genes encoding MPs. Tobacco plants that express a TMV MP transgene complemented movement of a mutant of TMV that lacks a functional MP (Holt and Beachy, 1991). These MP(+) plants also facilitated the spread of PLRV into mesophyll cells (Holt and Beachy, 1989); this is similar to the studies by Atabekov and Taliansky (1990) with double infections of PLRV and TMV. The transgene expressed in these plants also complemented a mutation of TRV that lacked a functional MP (Ziegler-Graff et al., 1991): the TRV mutant was able to spread in the TMV MP transgenic plants, but disease symptoms were not observed, suggesting a role for the TRV MP in symptom induction either in addition to systemic infection or as a result of systemic infection.

Transgenic plants that express the TMV MP gene were also used to study the role of the TMV MP in the movement of RCMV RNA B in the absence of RNA M, which carries the MP for RCMV (Taliansky et al., 1992). Co-inoculation of RNA B with TMV causes local infection by RCMV RNA B, while co-inoculation with a mutant TMV that lacks a functional MP does not. However, transgenic MP(+) tobacco plants were unable to potentiate the movement of RNA B unless co-infected with MP(–) TMV. The authors suggested that the modification of plasmodesmata by the TMV MP is necessary but not sufficient for the movement of RCMV RNA B, and that an additional TMV component produced during virus replication is needed for movement.

In summary, in the case of TRV and PLRV, neither of which apparently induces the formation of tubules, the TMV MP alone is capable of potentiating the spread of either virus. In contrast, in the case of RCMV, a virus that induces tubule formation, the TMV MP is necessary but not sufficient for spread of RNA B. These types of studies may reveal differences and similarities in the interactions of MPs of viruses that induce tubule formation and viruses that do not.

Transgenic plants have also been developed to accumulate MPs from viruses that are not pathogenic on the transformed host, or with dysfunctional

MPs (dMP). Malyshenko et al. (1993) produced transgenic plants that express a gene encoding the brome mosaic bromovirus (BMV) MP and showed that these plants were somewhat resistant to infection by TMV, indicating that the BMV MP interfered in some way with the movement of TMV. Likewise, transgenic plants that accumulated a ts (temperature sensitive) form of the TMV MP prevented the spread of TMV if the plants were held at the non-permissive temperature for a substantial period of time (~14 days) prior to infection (Malyshenko et al., 1993).

Lapidot et al. (1993) developed transgenic plants that express an MP from which amino acids 3–5 were deleted. This MP was unable to potentiate movement of TMV when expressed in an infectious cDNA clone of TMV (Gafny et al., 1992) or in transgenic plants (Lapidot et al., 1993) and was therefore described as a dMP (defective MP). Analysis of transgenic plants that express a gene encoding the dMP revealed that the dMP did not accumulate to the same levels in the cell wall fraction as wild type MP. Conversely, a greater proportion of the dMP was observed in the organelle/membrane fraction. Plasmodesmata of the transgenic plants that accumulate the dMP did not label with antibodies against the TMV MP and did not accumulate the fibrous material that is observed in transgenic plants that express the wild-type MP gene. The SEL of the plasmodesmata in these plants was increased to allow passage of a 3 kDa dextran but not passage of a 10 kDa dextran. It is interesting that these plants were resistant to a number of tobamoviruses (Lapidot et al., 1993) as well as viruses from other families such as TRV, peanut chlorotic streak caulimovirus (PClSV), and AlMV (Cooper et al., 1995). It is likely that the dMP protects transgenic plants from local and/or systemic infections by non-tobamoviruses by blocking at least one of the functions of the MP of the challenge virus. Analysis of transgenic plants that express mutant MPs may make it possible to further dissect the multiple functions of the MP.

3. The Use of Hybrid Viruses to Study Virus Spread

The specificity of interactions of MPs with host and other viral components has been studied by creating hybrid viruses. This approach has been applied to sun hemp mosaic tobamovirus (SHMV) and cowpea chlorotic mottle bromovirus (CCMV) by cloning the MP of SHMV into an infectious clone of RNA3 of CCMV, replacing the 3a gene of CCMV (Mise et al., 1993). The hybrid virus was capable of moving both cell to cell and systemically in cowpea plants, the natural host for both viruses. The fact that the tobamovirus MP can substitute for the 3a gene of CCMV led these authors to conclude that the function of the MP is not entirely dependent on RNA sequence, the nature of the capsid (SHMV is a rod shaped virus, CCMV is an icosahedral virus), or other viral components. However, the time course of infection of the hybrid virus was different from that of either parent virus, implying that in order to achieve optional rates of systemic infection there may be a requirement for specific interactions between the MP and other viral component(s).

4. Role of Movement Protein in Determining Virus Host Range

There is substantial evidence to conclude that the MP interacts with specific host components to cause the cell-to-cell and long-distance spread of infection (Meshi et al., 1989; Weber et al., 1993). Many plants are resistant to certain viruses because the viruses are unable to move from inoculated to adjacent cells. In such examples, the virus replicates in initially infected cells, but does not move to surrounding cells (Cheo, 1970; Cheo and Gerard, 1971; Sulzinski and Zaitlin, 1982). Studies of mixed infections were used to show that in some cases a second co-inoculated virus can potentiate the local spread of the first virus. Subsequently, studies were done with hybrid viruses in which the MP from one virus was exchanged with the MP of a related virus that has a different host range. Nejidat et al. (1991) replaced the TMV MP with that of tobacco mild green mosaic tobamovirus (TMGMV) and showed that the MP influences the size of local lesions on *N. tabacum* cv. Xanthi NN. Mise et al. (1993) showed that when the MPs of the bromoviruses BMV and CCMV were exchanged, the ability of the recombinant viruses to infect their natural host plants was also changed. In another study, Hilf and Dawson (1993) showed that the MP and CP of odontoglossum ringspot tobamovirus (ORSV) contains host range determinants by transferring the relevant gene sequences into an infectious cDNA clone of TMV.

The sequences near the carboxyl-terminal end of tobamovirus MPs are highly variable (Saito et al., 1988) and it was proposed that this region contains the host range determinants of the tobamovirus MP (Saito et al., 1988). This proposal is supported by a recent study of a tobamovirus that infects shallots (Kwon et al., 1994). The MP of this tobamovirus differs from the TMV MP by only eleven amino acids, eight of which are contained within the COOH terminal end of the MP. The role of the C-terminal region of the tobamovirus MP sequences in determining the host range of the virus was further confirmed by recent studies with ORSV and TMV (Fenczik et al., 1995; Holt et al., 1995). Whereas TMV causes only a local infection on *Vanilla planifolia*, TMV that carries MP from ORSV in place of the TMV MP caused a systemic infection on *V. planifolia*, but not in *N. tabacum* (Holt et al., 1995). A spontaneous mutant of the ORSV MP that resulted in a C-terminal truncation of 48 amino acids produced a virus capable of infecting *N. tabacum*, but unable to infect *V. planifolia*. Fenczik et al. (1995) constructed mutations in MP gene sequence that truncated the ORSV MP in the chimeric virus. Following inoculation of the mutant chimeric viral RNAs to plants, it was determined that deleting 11 amino acids of the ORSV MP resulted in a virus that could infect *N. tabacum* at a more rapid rate than the full length ORSV MP, but could no longer infect *V. planifolia*. These studies conclusively demonstrated that MP sequences can determine whether or not a virus causes a local and/or systemic infection on a host.

As discussed above, the MP plays a role both in cell-to-cell and long-distance movement of a virus in a potential host. Experiments that reveal how a virus is limited in a certain host may identify distinct MP functions that allow the protein to participate in both cell-to-cell and long-distance spread of infection.

IV. Conclusions

A full understanding of how plant viruses spread from their site of infection to adjacent cells and tissues distal to the infection will require knowledge both of the virus encoded movement proteins that are required for spread, and the components and functions of the structures through which spread occurs, the plasmodesmata. Although significant progress has been made to elucidate the biological and structural features of plasmodesmata, the isolation and characterization of components of the plasmodesmata have only just commenced. During recent years significant research effort has been directed toward characterizing the viral proteins that are essential for local and systemic spread of virus infections. While these studies have described interesting biological and biochemical features, the functional mechanisms of a single movement protein has yet to be discovered. Because of the intricate nature of the interactions between movement proteins and plasmodesmata in viral pathogenesis, and the importance of plasmodesmata in intercellular communication during plant growth and development, there is ample justification for continuing the studies that elucidate the structure and functions of these partners in pathogenesis. Furthermore, it is likely that a more complete understanding of these proteins and structures will lead to additional strategies to develop genetically modified plants with increased resistance to virus diseases and plants with altered intercellular communication and carbon allocation properties.

Acknowledgments

Work from the authors' laboratories was supported by grants from the National Science Foundation (MCB 9209532 and MCB 9317368) and the Israeli Science Foundation, by the Joint German–Israeli Research Program, MOST, NCRD (project no. DISNAT 1036), by the Dutch–Israeli Agricultural Research Program and the Karse-Epel Fund for Botanical Research at Tel Aviv University.

V. References

Albrecht J, Geldreich A, de Murcia JM, Kirchherr D, Mesnard J-M, Lebeurier G (1988) Cauliflower mosaic virus gene I product detected in a cell-wall-enriched fraction. Virology 163: 503–508

Allison R, Thompson C, Ahlquist P (1990) Regeneration of a functional RNA virus genome by recombination between deletion mutants and requirement for cowpea chlorotic mottle virus 3a and coat genes for systemic infection. Proc Natl Acad Sci USA 87: 1820–1824

Atabekov JG, Dorokhov YL (1984) Plant virus-specific transport function and resistance of plants to viruses. Adv Virus Res 29: 313–364

Atabekov JG, Taliansky ME (1990) Expression of a plant virus-coded transport function by different viral genomes. Adv Virus Res 38: 201–248

Atkins D, Hull R, Wells B, Riberts K, Moore P, Beachy RN (1991a) The tobacco mosaic virus 30K movement protein in transgenic tobacco plants is localized to plasmodesmata. J Gen Virol 72: 209–211

Atkins D, Roberts K, Hull R, Prehaud C, Bishop DHL (1991b) Expression of the tobacco mosaic virus movement protein using a baculovirus expression vector. J Gen Virol 72: 2831–2835

Barker H (1987) Invasion of non-phloem tissue in *Nicotiana clevelandii* by potato leafroll luteovirus is enhanced in plants also infected with potato Y potyvirus. J Gen Virol 68: 1223–1227

Baron-Epel O, Hernandez D, Jiang LW, Meiners S, Schindler M (1988) Dynamic continuity of cytoplasmic and membrane compartments between plant cells. J Cell Biol 106: 715–721

Beck DL, Guilford PJ, Voot DM, Andersen MT, Forster RLS (1991) Triple gene block proteins of white clover mosaic potexvirus are required for transport. Virology 183: 695–702

Berna A, Gafny R, Wolf S, Lucas WJ, Holt CA, Beachy RN (1991) The TMV movement protein: role of the C-terminal 73 amino acids in subcellular localization and function. Virology 182: 682–689

Bol JF, van Vloten-Doting L, Jaspars EMJ (1971) A functional equivalence of top component a RNA and coat protein in the initiation of infection by alfalfa mosaic virus. Virology 46: 73–85

Bostwick DE, Dannenhoffer JM, Skaggs MI, Lister RM, Larkin BA, Thompson GA (1992) Pumpkin phloem lectin genes are specifically expressed in companion cells. Plant Cell 4: 1539–1548

Botha CEJ, Hartley BJ, Cross RHM (1993) The ultra structure and computer-enhanced digital image analysis of plasmodesmata at the Kranz mesophyll-bundle sheath interface of *Thermeda triandra* var. imberbis (Retz) A. Camus in conventionally-fixed leaf blades. Ann Bot 72: 255–261

Carr RJ, Kim KS (1983) Evidence that bean golden mosaic virus invades non-phloem tissue in double infections with tobacco mosaic virus. J Gen Virol 64: 2489–2492

Cheo PC (1970) Subliminal infection of cotton by tobacco mosaic virus. Phytopathology 60: 41–46

Cheo PC, Gerard JS (1971) Differences in virus-replicating capacity among plant species inoculated with tobacco-mosaic virus. Phytopathology 61: 1010–1012

Citovsky V, Knorr D, Schuster G, Zambryski P (1990) The P30 movement protein of tobacco mosaic virus is a single-strand nucleic acid binding protein. Cell 60: 637–647

Citovsky V, Wong ML, Shaw AL, Venkataram PBV, Zambryski P (1992) Visualization and characterization of tobacco mosaic virus movement protein binding to single-stranded nucleic acids. Plant Cell 4: 397–411.

Citovsky V, McLean BJ, Zupan JR, Zambryski P (1993) Phosphorylation of tobacco mosaic virus cell-to-cell movement protein by a developmentally regulated plant cell wall-associated protein kinase. Genes Dev 7: 904–910

Cleland RE, Fujiwara T, Lucas WJ (1994) Plasmodesmatal-mediated cell-to-cell transport in wheat roots is modulated by anerobic stress. Protoplasma 178: 81–85

Cooper B, Lapidot M, Heick JA, Dodds JA, Beachy RN (1995) Multivirus resistance in transgenic tobacco plants expressing a dysfunctional movement protein of tobacco mosaic virus. Virology 206: 307–313

Cote R, Thain JF, Fensom DS (1987) Increase in electrical resistance of plasmodesmata of *Chara* induced by an applied pressure gradient across nodes. Can J Bot 40: 509–511

Dalmay T, Rubino L, Burgyan J, Kollar A, Russo M (1993) Functional analysis of cymbidium ringspot virus genome. Virology 194: 697–704

Davies C, Hills G, Baulcombe DC (1993) Sub-cellular localization of the 25-kDa protein encoded in the triple gene block of potato virus X. Virology 197: 166–175

Dawson WO, Hilf ME (1992) Host range determinants of plant viruses. Annu Rev Plant Physiol Plant Mol Biol 43: 527–555

Deom CM, Oliver MJ, Beachy RN (1987) The 30-kilodalton gene product of tobacco mosaic virus potentiates virus movement. Science 237: 389–394

Deom CM, Schubert KR, Wolf S, Holt CA, Lucas WJ, Beachy RN (1990) Molecular characterization and biological function of the movement protein of tobacco mosaic virus in transgenic plants. Proc Natl Acad Sci USA 87: 3284–3288

Derrick PM, Barker H, Oparka KJ (1992) Increase in plasmodesmatal permeability during cell-to-cell spread of tobacco rattle virus from individually inoculated cells. Plant Cell 4: 1450–1412

Ding B, Haudenshield JS, Hull RJ, Wolf S, Beachy RN, Lucas WJ (1992a) Secondary plasmodesmata are specific sites of localization of the tobacco mosaic virus movement protein in transgenic tobacco plants. Plant Cell 4: 915–928

Ding B, Turgeon R, Parthasarathy MV (1992b) Substructure of freeze-substituted plasmodesmata. Protoplasma 169: 28–41

Dodds A, Hamilton R (1972) The influence of barley stripe mosaic virus on the replication of tobacco mosaic virus in *Hordeum vulgare* L. Virology 50: 404–411

Dore JM, van Dun CMP, Pinck L, Bol JF (1991) Alfalfa mosaic virus RNA3 mutants do not replicate in transgenic plants expressing RNA3 specific genes. J Gen Virol 72: 253–258

Epel BL (1994) Plasmodesmata—composition, structure and trafficking. Plant Mol Biol 26: 1343–1356

Epel BL, Erlanger M (1991) Light regulates symplasmic communication in etiolated corn seedlings. Physiol Plant 83: 149–153

Epel B, Kuchuck B, Kotlizky G, Shurtz S, Erlanger M, Yahalom A (1995) Isolation and characterization of plasmodesmata. In: Galbraith DW, Bourque EP, Bohnert HJ (eds) Methods in cell biology, methods in plant cell biology, part B. Academic Press, San Diego, pp 237–253

Erny C, Schoumacher F, Godefroy-Colburn T, Stussi-Garaud C (1992) Nucleic acid binding properties of the 92-kDa replicase subunit of alfalfa mosaic virus produced in yeast. Eur J Biochem 203: 167–172

Erwee MG, Goodwin PB (1983) Characterization of *Egeria densa* Planch leaf symplast. Inhibition of intercellular movement of fluorescent probes by group II ions. Planta 158: 320–328

Fenczik C (1994) Characterization of two tobamovirus movement proteins and their role in host range determination. PhD thesis, Washington University, St. Louis, MO

Fenczik C, Padgett H, Holt C, Casper S, Beachy R (1995) Mutational analysis of the movement protein of odontoglossum ringspot virus to identify a host-range determinant. Mol Plant Microbe Interact 8: 666–673

Fisher DB, Wu Y, Ku MSB (1992) Turnover of soluble proteins in the wheat sieve tube. Plant Physiol 100: 1433–1441

Francki RIB, Milne RG, Hatta T (1985) Atlas of plant viruses. CRC Press, Boca Raton

Fujiwara T, Giesmann-Cookmeyer D, Ding B, Lommell SA, Lucas WJ (1993) Cell-to-cell trafficking of macromolecules through plasmodesmata potentiated by the red clover necrotic mosaic virus movement protein. Plant Cell 5: 1783–1794

Gafny R, Lapidot M, Berna A, Holt CA, Deom CM, Beachy RN (1992) Effects of terminal deletion mutations on function of the movement protein of tobacco mosaic virus. Virology 187: 499–507

Gardiner W, Sunter G, Brand L, Elmer JS, Rogers SJ, Bisaro DM (1988) Genetic analysis of tomato golden mosaic virus: the coat protein is not required for systemic spread or symptom development. EMBO J 7: 899–904

Giesman-Cookmeyer D, Lommel SA (1993) Alanine scanning mutagenesis of a plant virus movement protein identifies three functional domains. Plant Cell 5: 973–982

Gilmer D, Bouzoubaa S, Hehn A, Guilley H, Richards K, Jonard G (1992) Efficient cell-to-cell movement of beet necrotic yellow vein virus requires 3′ proximal genes located on RNA2. Virology 189: 40–47

Godefroy-Colburn T, Schoumacher F, Erny C, Berna A, Moser O, Gagey MJ, Stussi-Garaud C (1986) A non-structural protein of alfalfa mosaic virus in the walls of infected tobacco cells. J Gen Virol 67: 2233–2239

Godefroy-Colburn T, Schoumacher F, Erny C, Berna A, Moser O, Gagey MJ, Stussi-Garaud C (1990) The movement protein of some plant viruses. In: Fraser RSS (ed) Recognition and response in plant–virus interactions. Springer, Berlin Heidelberg New York Tokyo, pp 207–231 (NATO-ASI series, series H, vol 41)

Goodwin PB (1983) Molecular size limit for movement in the symplast of *Elodea* leaf. Planta 157: 124–130

Grama DP, Mayor IY (1990) Construction of a model of potexvirus coat proteins. Mikrobiol Zh 52: 59–66

Graski S, de Feijter AW, Schindler M (1993) Endoplasmic reticulum forms a dynamic continuum for lipid diffusion between contiguous soybean root cells. Plant Cell 5: 25–38

Hacker DL, Petty ITD, Wei N, Morris TJ (1992) Turnip crinkle virus genes required for RNA replication and virus movement. Virology 186: 1–8

Harrison BD, Robinson DJ (1986) Tobraviruses. In: Van Regenmortel MHV, Fraenkel-Conrat H (eds) The plant viruses, vol 2. Plenum, New York, pp 339–369

Harrison BD, Barker H, Derrick PM (1990) Intercellular spread of PLRV: effects of co-infection and plant resistance. In: Fraser RSS (ed) Recognition and response in plant–virus interactions. Springer, Berlin Heidelberg New York Tokyo, pp 405–414 (NATO-ASI series, series H, vol 41)

Hayes PJ, Petty ITD, Coutts RHA, Buck KW (1988) Gene amplification and expression in plants by a replicating geminivirus vector. Nature 334: 179–182

Heaton LA, Carrington JC, Morris TJ (1989) Turnip crinkle virus infection from RNA synthesized in vitro. Virology 170: 214–218

Hepler PK (1982) Endoplasmic reticulum in the formation of the cell plate and plasmodesmata. Protoplasma 111: 121–133

Hepler PK, Palevitz BA, Lancelle SA, McCauley MM, Lichtscheidl (1990) Cortical endoplasmic reticulum in plants. J Cell Biol 96: 355–373

Hilf ME, Dawson WO (1993) The tobamovirus capsid protein functions as a host-specific determinant of long-distance movement. Virology 193: 106–114

Holt CA, Beachy RN (1989) Accumulation of potato leafroll luteovirus is enhanced in transgenic tobacco plants expressing the 30 kDa movement protein of tobacco mosaic virus. Presented at the Symposium on Viral Genes and Plant Pathogenesis, Lexington, KY, October 16–17, 1989

Holt CA, Beachy RN (1991) In vivo complementation of infectious transcripts from mutant tobacco mosaic virus cDNAs in transgenic plants. Virology 181: 109–117

Holt C, Fenczik C, Casper S, Beachy R (1995) The carboxy-terminal region of the tobamovirus movement protein encodes a host-range determinant. Virology (in press)

Hull R (1989) The movement of viruses in plants. Annu Rev Phytopathol 27: 231–240

Hunte C, Schnabl H, Traub O, Willecke K, Schulz M (1992) Immunological evidence of connexinlike proteins in the plasma membrane of *Vicia faba* L. Bot Acta 105: 104–110

Hunte C, Janssen M, Schulz M, Traub O, Willecke K, Schnabl H (1993) Age-dependent modification and further localization of the cx 26-like protein from *Vicia faba* L. Bot Acta 106: 207–212

Ingham D, Lazarowitz S (1993) A single missense mutation in the BRI movement protein alters the host range of squash leaf curl virus. Virology 196: 694–702

Kim KS, Lee KW (1992) Geminivirus-induced macrotubules and their suggested role in cell-to-cell movement. Phytophathology 82: 664–669

Kotlizky G, Shurtz S, Yahalom A, Malik Z, Traub O, Epel BL (1992) Isolation and characterization of plasmodesmata embedded in clean maize cell walls. Plant J 2: 623–630

Kwon S, Sako N, Oshima K (1994) Nucleotide sequence of the coat protein and the 30K genes of TMV-Rakkyo strain locally infecting *Nicotiana tabacum* L. cv. Bright Yellow. Ann Phytopathol Soc Japan 60: 186–195

Lapidot M, Gafny R, Ding B, Wolf S, Lucas WJ, Beachy RN (1993) A dysfunctional movement protein of tobacco mosaic virus that partially modifies the plasmodesmata and limits virus spread in transgenic plants. Plant 2: 959–970

Leisner SM, Turgeon R (1993) Movement of virus and photoassimilate in the phloem: a comparative analysis. Bioessays 15: 741–748

Leonard D, Zaitlin M (1982) A temperature-sensitive strain of tobacco mosaic virus defective in cell-to-cell movement generates an altered viral-coded protein. Virology 117: 416–424

Li RQ, Zhang WC (1994) Three dimensional electron microscopy visualization of the cytoskeleton of integument cells and nuclear epidermis of *Triticum aestivum*. Chin J Bot 6: 118–122

Linstead PJ, Hills GJ, Plaskitt KA, Wilson IG, Harker CL, Maule AJ (1988) The subcellular location of the gene 1 product of cauliflower mosaic virus is consistent with a function associated with virus spread. J Gen Virol 69: 1809–1818

Lucas WJ, Ding B, Van der Schoot C (1993) Plasmodesmata and the supracellular nature of plants. New Phytol 125: 435–476

Malyshenko SI, Kondakova OA, Nazarova JV, Kaplan IB, Taliansky ME, Atabekov JG (1993) Reduction of tobacco mosaic virus accumulation in transgenic plants producing non-functional viral transport proteins. J Gen Virol 74: 1149–1156

Malyshenko SI, Lapchic LG, Kondakova OA, Kiznetzova LL, Taliansky ME, Atabekov JG (1988b) Red clover mottle comovirus B-RNA spreads between cells in tobamovirus-infected tissues. J Gen Virol 69: 407–412

Maule AJ (1991) Virus movement in infected plants. Crit Rev Plant Sci 9: 457–473

Meiners S, Schindler M (1987) Immunological evidence for gap junction polypeptide in plant cells. J Biol Chem 262: 951–953

Meiners S, Schindler M (1989) Characterization of a connexin homologue in cultured soybean cells and diverse plant organs. Planta 179: 148–155

Meiners S, Baron-Epel O, Schindler M (1988) Intercellular communication—filling in the gaps. Plant Physiol 88: 791–793

Meiners S, Xu A, Schindler M (1991) Gap junction protein homologue from *Arabidopsis thaliana*: evidence for connexins in plants. Proc Natl Acad Sci USA 88: 4119–4122

Meshi T, Watanabe Y, Saito T, Sugimoto A, Maeda T, Okada Y (1987) Function of the 30 kd protein of tobacco mosaic virus: involvement in cell-to-cell movement and dispensability for replication. EMBO J 6: 2557–2563

Meshi T, Motoyoshi F, Maeda F, Yoshiwoka S, Watanabe H, Okada Y (1989) Mutations in the tobacco mosaic virus 30-kD protein gene overcome Tm-2 resistance in tomato. Plant Cell 1: 515–522

Meshi T, Hosokawa D, Kawagishi M, Watanabe Y, Okada Y (1992) Reinvestigation of intracellular localization of the 30K protein in tobacco protoplasts infected with tobacco mosaic virus RNA. Virology 187: 809–813

Mise K, Allison RF, Janda M, Ahlquist P (1993) Bromovirus movement protein genes play a crucial role in host specificity. J Virol 67: 2815–2823

Monzer J, Kloth S (1991) The preparation of plasmodesmata from plant tissue homogenates:

access to the biochemical characterization of plasmodesmata-related polypeptides. Bot Acta 104: 82–84

Moore PJ, Fenczik CA, Beachy RN (1992) Developmental changes in plasmodesmata in transgenic plants expressing the movement protein of tobacco mosaic virus. Protoplasma 170: 115–127

Morales F, Niessen A, Ramirez B, Castano M (1990) Isolation and partial characterization of a geminivirus causing bean dwarf mosaic. Phytopathology 80: 96–101

Moser O, Gagey MJ, Godefroy-Colburn T, Stussi-Garaud C, Ellwart-Tschurtz M, Nitschko H, Mundry K-W (1988) The fate of the transport protein of tobacco mosaic virus in systemic and hypersensitive hosts. J Gen Virol 69: 1367–1373

Motoyoshi F, Oshima N (1977) Expression of genetically controlled resistance to tobacco mosaic virus infection in isolated tomato leaf protoplasts. J Gen Virol 34: 499–506

Neisbach-Klosgen U, Guilley H, Jonard G, Richards K (1990) Immuno-detection in vivo of beet necrotic yellow vein virus encoded proteins. Virology 178: 52–61

Nejidat A, Cellier F, Holt CA, Gafny R, Eggenberger A, Beachy RN (1991) Transfer of the movement protein gene between two tobamoviruses: influence on local lesion development. Virology 180: 318–26

Nishiguchi M, Motoyoshi F, Oshima N (1978) Behavior of a temperature sensitive strain of tobacco mosaic virus in tomato leaves and protoplasts. J Gen Virol 39: 53–61

Noueiry AO, Lucas WJ, Gilbertson RL (1994) Two proteins of a plant DNA virus coordinate nuclear and plasmodesmal transport. Cell 76: 925–932

Ohno T, Takamatsu N, Meshi T, Okada Y, Nishiguchi M, Kiho Y (1983) Single amino acid substitution in 30K protein of TMV defective in virus transport function. Virology 131: 255–258

Olensen P (1979) The neck constriction in plasmodesmata: evidence for a peripheral sphincter-like structure revealed by fixation and tannic acid. Planta 144: 349–358

Oparka KJ (1993) Signalling via plasmodesmata—the neglected pathway. Sem Cell Biol 4: 131–138

Oparka KJ, Prior DAM (1992) Direct evidence for pressure-generated closure of plasmodesmata. Plant J 2: 741–750

Osman TA, Buck KW (1991) Detection of the movement protein of red clover necrotic mosaic virus in a cell wall fraction from infected *Nicotiana clevelandii* plants. J Gen Virol 72: 2853–2856

Osman TAM, Hayes RJ, Buck KW (1992) Cooperative binding of the red clover necrotic mosaic virus movement protein to single stranded nucleic acids. J Gen Virol 73: 223–227

Pascal E, Goodlove PE, Wu LC, Lazarowitz SG (1993) Transgenic tobacco plants expressing the geminivirus BL1 protein exhibit symptoms of viral disease. Plant Cell 5: 795–807

Perbal M-C, Thomas CL, Maule AJ (1993) Cauliflower mosaic virus gene I product (P1) forms tubular structures which extend from the surface of infected protoplasts. Virology 195: 281–285

Petty ITD, French R, Jones RW, Jackson AO (1990) Identification of barley stripe mosaic virus genes involved in viral RNA replications and systemic movement. EMBO J 9: 3453–3457

Reid RJ, Overall RL (1992) Intercellular communication in Chara: factors affecting transnodal electrical and solution fluxes. Plant Cell Environ 15: 507–517

Robards AW, Lucas WJ (1990) Plasmodesmata. Annu Rev Plant Physiol 41: 369–419

Robinson-Beers K, Evert RF (1991) Fine structure of plasmodesmata in mature leaves of sugarcane. Planta 184: 307–318

Saito T, Imai Y, Meshi T, Okada Y (1988) Interviral homologies of the 30K proteins of tobamovirues. Virology 167: 653–656

Saito T, Yamanaka K, Okada Y (1990) Long-distance movement and viral assembly of tobacco mosaic virus mutants. Virology 176: 329–336

Schoumacher F, Erny C, Berna A, Godefroy-Colburn T, Stussi-Garaud C (1992) Nucleic acid-binding properties of the alfalfa mosaic virus movement protein produced in yeast. Virology 188: 896–899

Schulz M, Traub O, Knop M, Willecke K, Schnabl H (1992) Immunofluorescent localization of connexin 26-like protein at the surface of mesophyll protoplasts from *Vicia faba* L. and *Helianthus annuus* L. Bot Acta 105: 111–115

Shanks M, Tomenius K, Clapham D, Huskisson NS, Barker PJ, Wilson IG, Maule AJ, Lomonossoff GP (1989) Identification and subcellular localization of a putative cell-to-cell transport protein from red clover mottle virus. Virology 173: 400–407

Shepherd VA, Goodwin PB (1992a) Seasonal patterns of cell-to-cell communication in *Chara corallina* Klein ex Welld. I. Cell-to-cell communication in vegetative lateral branches during winter and spring. Plant Cell Environ 15: 137–150

Shepherd VA, Goodwin PB (1992b) Seasonal patterns of cell-to-cell communication in *Chara corallina* Klein ex Welld. II. Cell-to-cell communication during the development of antheridia. Plant Cell Environ 15: 151–162

Smith LM, Sabnis DD, Johnson RPC (1987) Immunochemical localization of phloem lectin from *Cucurbita maxima* using peroxidase and colloidal-gold labels. Planta 170: 461–470

Sulzinski MA, Zaitlin M (1982) Tobacco mosaic virus replication in resistant and susceptible plants: in some resistant species virus is confined to a small number of initially infected cells. Virology 121: 12–19

Suzuki M, Kuwata S, Kataoka J, Masuta C, Nitta N, Takanami Y (1991) Functional analysis of deletion mutants of cucumber mosaic virus RNA3 using an in vitro transcription system. Virology 183: 106–113

Taliansky ME, Malyshenko SI, Kaplan IB, O.A. K, Atabekov JG (1992) Production of the tobacco mosaic virus (TMV) transport protein in transgenic plants is essential but insufficient for complementing foreign virus transport: a need for the full-length TMV genome or some other TMV-encoded product. J Gen Virol 73: 471–474

Terry BR, Robards AW (1987) Hydrodynamic radius alone governs the mobility of molecules through plasmodesmata. Planta 171: 145–157

Tilney L, Cooke T, Connely P, Tilney M (1991) The structure of plasmodesmata as revealed by plasmolysis, detergent extraction and protease digestion. J Cell Biol 112: 739–747

Tomenius K, Clapham D, Meshi T (1987) Localization by immunogold cytochemistry of the virus-coded 30K protein in plasmodesmata of leaves infected with tobacco mosaic virus. Virology 160 : 363–371

Tucker EB (1982) Translocation in the staminal hairs of *Setcreasea purpurea* I. A study of cell ultrastructure and cell to cell passage of molecular probes. Protoplasma 113: 193–202

Tucker EB (1988) Inositol biphosphate and inositol triphosphate inhibit cell-to-cell passage of carboxyfluorescein in staminal hairs of *Setcreasea purpurea*. Planta 174: 358–363

Tucker EB (1993) Azide treatment enhances cell-to-cell diffusion in staminal hairs of *Setcreasea purpurea*. Protoplasma 174: 45–49

Tucker EB, Tucker JE (1993) Cell-to-cell selectivity in staminal hairs of *Setcreasea purpurea*. Protoplasma 174: 36–44

Turgeon R (1991) Symplastic phloem loading and sink–source transport in leaves: a model. In: Bonnemain JL, Delrot S, Lucas WJ, Dainy J (eds) Advances in phloem transport and assimilation compartmentation. Quest Editions, Nantes, pp 18–22

Turgeon R, Gowen E (1990) Phloem loading in coleur glumei in absence of carrier-mediated uptake of export sugar from the apoplast. Plant Physiol 94: 1244–1249

Turner A, Wells B, Roberts K (1994) Plasmodesmata of maize root tips: structure and composition. J Cell Sci 107: 3351–3361

van Lent J, Wellink J, Goldbach R (1990) Evidence for the involvement of the 58K and 48K proteins in the intercellular movement of cowpea mosaic virus. J Gen Virol 71: 219–223 van Lent J, Storms M, van der Meer F, Wellink J, Goldbach R (1991) Tubular structures involved in movement of cowpea mosaic virus are also formed in infected cowpea protoplast. J Gen Virol 72: 2615–2623

Waigmann E, Lucas WJ, Citovsky V, Zambryski P (1994) Direct functional assay for tobacco mosaic virus cell-to-cell movement protein and identification of a domain involved in increasing plasmodesmal permeability. Proc Natl Acad Sci USA 91: 1433–1437

Wang N, Fisher DB (1994) The use of fluorescent tracers to characterize the post-phloem transport pathway in maternal tissues of developing wheat grains. Plant Physiol 104: 17–27

Ward A, Etessami P, Stanley J (1988) Expression of a bacterial gene in plants mediated by infectious geminivirus DNA. EMBO J 7: 1583–1587

Watanabe Y, Ogawa T, Okada Y (1992) In vivo phosphorylation of the 30-kDa protein of tobacco mosaic virus. FEBS Lett 313: 181–184

Weber H, Schultze S, Pfitzner AJP (1993) Two amino acid substitutions in the tomato mosaic virus 30-kilodalton movement protein confer the ability to overcome the Tm-2-2 resistance gene in the tomato. J Virol 67: 6432–6438

Weintraub M, Ragetli HWJ, Leung E (1976) Elongated virus particles in plasmodesmata. J Ultrastruct Res 56: 351–364

Wellink J, Van Kammen A (1989) Cell-to-cell transport of cowpea mosaic virus requires both the 58K/48K proteins and the capsid proteins. J Gen Virol 70: 2279–2286

Wellink J, van Lent JWM, Verver J, Sujen T, Goldbach RW, van Kammen AB (1993) The cowpea mosaic virus M RNA-encoded 48-kilodalton protein is responsible for induction of tubular structures in protoplasts. J Virol 67: 3660–3664

White RG, Badett K, Overall RL, Vesk M (1994) Actin associated with plasmodesmata. Protoplasma 180: 169–184

Wieczorek A, Sanfacon H (1993) Characterization and subcellular localization of tomato ringspot nepovirus putative movement protein. Virology 194: 734–742

Wolf S, Deom CM, Beachy RN, Lucas WJ (1989) Movement protein of tobacco mosaic virus modifies plasmodesmatal size exclusion limit. Science 246: 377–379

Xiong A, Kim KH, Giesman-Cookmeyer D, Lommell SA (1993) The roles of the red clover necrotic mosaic virus capsid and cell to cell movement proteins in systemic infection. Virology 192: 27–32

Xu A, Meiners S, Schindler M (1990) Immunological investigations of relatedness between plant and animal connexins. In: Robards AW (ed) Parallels in cell to cell junctions in plants and animals. Springer, Berlin Heidelberg New York Tokyo, pp 171–183

Yahalom A, Warmbrodt RD, Laird DW, Traub O, Revel JP, Willecke K, Epel BL (1991) Maize mesocotyl plasmodesmata proteins cross-react with connexin gap junction protein antibodies. Plant Cell 3: 407–147

Ziegler-Graff V, Guilford PJ, Baulcombe DC (1991) Tobacco rattle virus RNA-1 29K gene product potentiates viral movement and also affects symptom induction in tobacco. Virology 182: 145–155

Zimmern D, Hunter T (1983) Point mutation in the 30K open reading frame of TMV implicated in temperature-sensitive assembly and local lesion spreading of mutant Ni2519. EMBO J 2: 1893–1900

Ex-Planta and In-Planta Signals in Legume–*Rhizobium* Interaction

Zonglie Hong and Desh Pal S. Verma

Contents

I. Introduction

Nodule organogenesis is a highly-programmed developmental process triggered by rhizobial signal molecules and controlled by both *Rhizobium* and plant (Verma, 1992; Fisher and Long, 1992; Hirsch, 1992). Formation of *Rhizobium*–legume symbiosis provides a unique experimental system for genetic and molecular studies for communication between bacteria and host plants. Plant flavonoids and rhizobial Nod factors (modified oligosaccharides) are two classes of unique signal molecules that represent direct communication between these two organisms. The host roots exude flavonoids that act as chemotactic agents to attract the rhizobia to move towards the root surface and, more importantly, as strong inducers to initiate rhizobial *nod* gene expression. The *nod* genes encode enzymes responsible for biosynthesis and secretion of Nod factors (Carlson et al., 1994). Induction of the *nod* gene expression results in the production of Nod factors that elicit morphological changes in the host roots. These changes include root hair curling, formation of nodule primodia in the cortical cell and nodule organogenesis.

It is apparent that more than these two classes of signals are required, from both host plant and rhizobial partners, for a successful establishment of the symbiosis (see Verma, 1992). A body of evidence from genetic and biochemical studies on rhizobia has suggested that bacterial factors (BFs) such as diglycosyl diacylglycerol, and surface components of rhizobial cell may play important roles in the interaction between the two symbiotic partners (see below). The exact function of these rhizobial factors in the formation of symbiosis, however, is not clear. Phytohormones, one of the key regulators in plant development, are likely to act as mediators in regulating nodule organogenesis, the balance of which could be altered by both partners.

II. Plant Flavonoid Signals and the Response of Rhizobia

Many phenolic compounds are produced by plants under normal growth conditions as well as in response to pathogen invasion and wounding. Flavonoids, a distinct group of plant phenolics, share a common structure of two aromatic rings joined by a three carbon unit (C_6-C_3-C_6; see Fig. 1). Most flavonoids are water soluble colored compounds and are generally found in the vacuoles and chloroplasts in leaves and flowers. A group of flavonoids are also synthesized in legume roots, exuded into the rhizosphere and act as plant signals in inducing expression of *Rhizobium nod* genes.

Biochemical studies on flavonoids biosynthesis have revealed that the first ring (ring A) of a flavonoid is formed by condensation of three molecules of malonyl-CoA with one molecule of acetyl-CoA. The phenylpropane residue (ring B and the C_3 unit) is derived from coumaric acid that is formed via the shikimate pathway (Van Der Meer et al., 1993). Different flavonoids are produced by modifications on the core flavonoid molecule via oxidation, hydroxylation, methylation or glycosylation. Oxidation and substitution in the C ring result in the formation of chalcones, flavanones, flavonols, isoflavonoids, flavones and anthocyanins. Extensive studies in this area over the past years

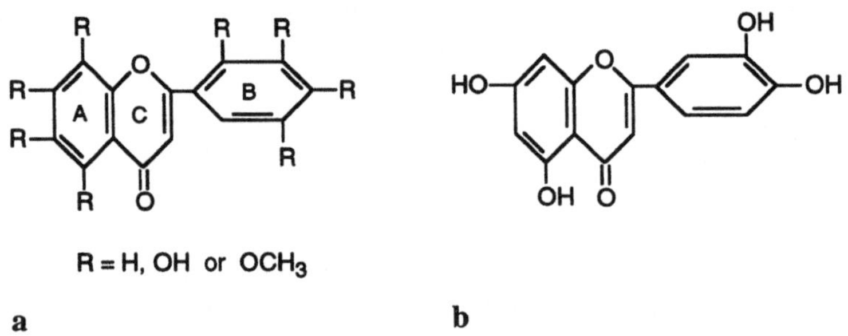

Fig. 1. Chemical structure of flavonoids (a). Luteolin (b) is the first flavonoid compound that was shown to be an inducer of rhizobial *nod* genes

have focused on flavonoid biosynthesis in flowering plants where they act as flower colors. Mutagenesis has identified at least 35 genetic loci in *Petunia* and 10 genes in maize, that are involved in flavonoid biosynthesis (Dooner et al., 1991). Some of these genes, encoding both biosynthetic enzymes and regulatory proteins, have been cloned and characterized (see Van Der Meer et al., 1993). Induction of these genes is regulated by a number of factors such as UV light, pathogens and *Rhizobium* infection (Estabrook and Sengupta-Gopalan, 1991; Lawson et al., 1994). Genetic studies have suggested that synthesis of specific flavonoids and concentration of these compounds in the rhizosphere are two important factors for induction of *Rhizobium nod* gene expression. Both traits are believed to be controlled by the host genes (Kapulnik et al., 1987).

The ability of a flavonoid compound to induce *Rhizobium nod* gene expression is usually assayed by monitoring the induction level of a fused gene that contains the bacterial *lacZ* coding region under the control of a *nod* gene promoter. Because expression of *Rhizobium nodABC* genes is induced during early interaction of rhizobia with legume roots, promoters of these genes should be able to sense possible plant inducers. By measuring the expression level (β-galactosidase activity) of *lacZ* fused to *nodABC* in *Rhizobium* cells in the presence of legume exudates, many plant flavonoids acting as signals have been identified. By using this assay, it was first demonstrated that extracts of several legume plants are able to induce *Rhizobium nod* genes (Rossen et al., 1985; Mulligan and Long, 1985; Innes et al., 1985). The first plant flavonoid compound identified to be active as inducer was luteolin (Fig. 1) from the alfalfa seed extracts (Peters et al., 1986). Many flavonoids are found to be conjugated with sugars. The luteolin may be released enzymatically from its glyco-conjugate, luteolin-7-glucoside (Hartwig and Phillips, 1991). It was later found that luteolin, though existing in alfalfa seed extracts, is not present in alfalfa root exudates. Instead, the active compouds that are actually exuded into the rhizosphere are chalcone derivatives (Maxwell et al., 1989). The chalcone compounds are more active than luteolin as the former induce *nod* genes at a concentration (about 10 nM), one order of magnitude lower than the latter (100 nM). Detailed studies have revealed that different legume species may release specific flavonoid molecules and an individual species can exude several different flavonoids. Soybean exudates contain at least two active isoflavones, daidzein and genistein (Kosslak et al., 1987; Sadowsky et al., 1988). Six flavanones and two chalcones have been identified to be active for induction of *Rhizobium nod* gene expression (Recourt et al., 1991). Common bean roots exude at least three flavonoids that induce transcription of *Rhizobium nod* genes at a nanomolar concentration (Hungria et al., 1991b), while seeds of the plants release up to eleven active flavonoids that act as inducers of *nod* genes (Hungria et al., 1991a).

Production and release of flavonoids seem to be enhanced during rhizobial infection (Recourt et al., 1991; Van Brussel et al., 1990). Chalcone synthase (CHS) catalyzes the stepwise condensation of malonyl-CoA with coumaroyl-CoA to produce a chalcone molecule and is regarded as the key enzyme of the

Table 1. Location and possible functions of *nod* genes

Gene	Group	Cellular location	Possible functions	Ref.
nodD		inner membrane	transcriptional activator	1, 2
nodA	common	cytoplasm	*N*-acyltransferase	3–7
nodB	common	cytoplasm	chitooligosaccharide deacetylase	3–8
nodC	common	cytoplasmic membr.	*N*-acetylglucosaminyl-transferase	9–11
nodI	common	cytoplasmic membr.	ATP-binding transport protein	12–14
nodJ	common	cytoplasmic membr.	transmembrane protein	12–14
nodE	host specific	cytoplasmic membr.	β-ketoacyl synthase	15, 16
nodF	host specific	cytoplasm	acyl carrier protein	17–19
nodG	host specific	cytoplasm	β-ketoacyl reductase	15, 17
nodH	host specific	cytoplasmic membr.	sulphotransferase	20–22
nodL	host specific	cytoplasmic membr.	acetyl transferase	23
nodM	host specific	cytoplasm	glucosamine synthase	24
nodN	?	?	possible GlcNAc-P uridyl transferase	25, 26
nodO	?	secreted	Ca^{2+}-regulated ion channel	27–29
nodP	host specific	cytoplasm	ATP sulfurylase	30
nodQ	host specific	cytoplasm	adenosine 5′-phosphosul-phate kinase	30
nodS	host specific	?	methyltransferase	31
nodT	host specific	outer membrane	membrane protein	32
nodV	?	?	sensor protein of the two-component system	33
nodW	?	?	activator of the two-component system	33
nodX	host specific	?	hydrophobic membrane protein	34, 35
nodZ	host specific	?	2-*O*-methylfucosyl transferase	36
SymM	?	?	transcriptional activator	37
nolR	?	?	repressor of *nodD*	38

References: 1, Mulligan and Long, 1985; 2, Honma and Ausubel, 1987; 3, Atkinson et al., 1994; 4, Eglehoff et al., 1985; 5, J. Schmidt et al., 1986; 6, J. Schmidt et al., 1988; 7, Rohrig et al., 1994; 8, John et al., 1993; 9, John et al., 1985; 10, Johnson et al., 1989; 11, Geremia et al., 1994; 12, Evans and Downie, 1986; 13, Schlaman et al., 1990; 14, Vazquez et al., 1993; 15, Horvath et al., 1986; 16, Bibb et al., 1989; 17, Debelle et al., 1986; 18, Shearman et al., 1986; 19, Spaink et al., 1991; 20, Faucher et al., 1989; 21, Roche et al., 1991; 22, P. Schmidt et al., 1994; 23, Downie, 1989; 24, Baev et al., 1991; 25, Surin and Downie, 1988; 26, Carlson et al., 1994; 27, Economou et al., 1990; 28, Scheu et al., 1992; 29, Sutton et al., 1994; 30, Schwedock and Long, 1990; 31, Geelen et al., 1993; 32, Surin et al., 1990; 33, Göttfert et al., 1990; 34, Davis et al., 1988; 35, Firmin et al., 1993; 36, Stacey et al., 1994; 37, Demont et al., 1994; 38, E. Kondorosi et al., 1991.

flavonoid biosynthesis. Several copies of CHS genes have been identified and it has been demonstrated that expression of these genes is differentially induced during *Rhizobium* infection of soybean roots (Estabrook and Sengupta-Gopalan, 1991). Induction of CHS gene expression by rhizobial occurs before root hair deformation (Lawson et al., 1994). Activation of isoflavonoid biosynthesis pathway by *Rhizobium* Nod factors also occurs in alfalfa cell suspensions (Savoure et al., 1994). This enhancement may be a result of a general response to stress caused by rhizobial infection, a phenomenon often seen during pathogen invasion. It is also possible that during rhizobial infection, the bacteria secrete elicitor-like substances (along with Nod factor) that in turn enhance the production of active plant flavonoids (Recourt et al., 1992). Flavonoid accumulation in soybean root exudate can be induced by Nod factors from *Bradyrhizobium japonicum* (P. Schmidt et al., 1994).

III. Synthesis of Nod Factors by *Rhizobium*

Rhizobial genes that are essential for initiation of nodulation on host roots are known as *nod* genes. Since all the letters of the alphabet have been used up to designate this group of genes, the newly identified *nod* genes were assigned as *nol* genes. Mutation in these genes results in delay or abolishment of nodule formation, or change in host specificity. The *nod* genes are involved in the production of Nod factors. A list of possible biochemical functions of these genes is presented in Table 1. The *nod* genes in the fast-growing *Rhizobium* species like *R. meliloti* are clustered on a large plasmid, called pSym, while in the slow-growing species such as *B. japonicum*, they are found on the chromosome.

Early studies indicated that the sterile supernatants of rhizobial cultures that had been treated by root extracts were able to cause changes in root or root hair morphology (see Denarie et al., 1992). Culture supernatants extracted with organic solutions were subjected to chromatography followed by a test

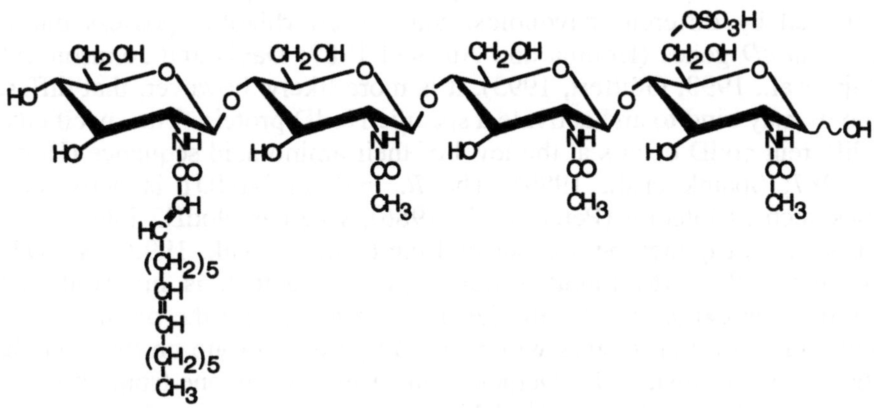

Fig. 2. Structure of the Nod factor NodRm-1 from *Rhizobium meliloti*

for their ability to induce root hair deformation (known as Had phenotype). Using this bioassay, Nod factors were purified from *R. meliloti* strains over expressing the *nod* genes (Lerouge et al., 1990; Roche et al., 1991). Structural studies on the purified Nod factors have demonstrated that these compounds represent a new family of lipo-oligosaccharides (Fig. 2). The basic structure is a tetra- or pentamer of β-1,4-linked *N*-acetylglucosamine residues. On the terminal non-reducing sugar residue, the N-acetyl group is substituted with an N-fatty acyl (C16-18) chain. Nod factors from *R. meliloti* are also *O*-sulfated on the reducing end. Length of the oligossacharide backbone, and length and saturation degree of the fatty acid chain may vary depending on the rhizobial strains. An individual strain may produce a family of these factors with slightly different structures (Relic et al., 1994). Active Nod factors have now been purified from different rhizobia including *Rhizobium*, *Bradyrhizobium* and *Azorhizobium* spp. They all share similar chemical structure.

A. NodD—the Sensor of Plant Flavonoids

While most of the *nod* genes are expressed only in the presence of a host-specific flavonoid, the *nodD* gene is expressed constitutively in free-living cells. The *nodD* gene encodes a transcriptional activator that, with the help of plant inducers, binds to the nod box, a highly conserved DNA sequence often present in the promoter regions of the *nod* operons (Fisher et al., 1988; Horvath et al., 1987). This binding greatly promotes the expression of *nod* genes leading to the synthesis of the Nod factors. Although there has not been direct evidence for an interaction of plant flavonoids with the NodD proteins, it is believed that they interact directly in the *Rhizobium* inner membrane where both NodD protein and plant flavonoids were found to accumulate (Schlaman et al., 1989; Recourt et al., 1989). This has been supported by the identification of the C-terminal regions of the NodD proteins that are responsible for flavonoid recognition and activation (Spaink et al., 1989; Horvath et al., 1987).

How do NodD proteins sense various active flavonoids produced by an individual legume species? It was thought that a NodD protein can bind to and be activated by different flavonoids. Since most rhizobia possess multiple copies of *nodD* genes (Honma and Ausubel 1987; Davis and Johnson 1990; Vanrhijn et al., 1993; Göttfert, 1993), it is more likely, however, that different flavonoids may bind to and activate a specific NodD protein. The specificity of each different nodD occurs at the level of their amino acid sequence (Horvath et al., 1987; Spaink et al., 1989). The *R. meliloti* NodD1 is activated by flavones such as luteolin (Peters et al., 1986), whereas NodD2 interacts with betains such as trigonelline and stachydrine (Phillips et al., 1992). NodD3, in conjunction with SyrM (another transcriptional factor), is involved in the control of gene expression by the level of ammonia and the production of a particular class of Nod factors with a (ω-1)-hydroxyl group on the acyl chain (Dusha and Kondorosi, 1993; Demont et al., 1994). Chalcone from alfalfa root exudates activates NodD1 and NodD2 proteins whereas other flavonoids activate only the NodD1 protein (Hartwig et al., 1990; Honma et al., 1990). In contrast to most *nodD* genes that are expressed constitutively, *B. japonicum nodD1* gene is inducible only by soybean flavonoids (genistein and daidzein) and is essential for flavonoid induction of other *nod* genes (Banfalvi et al., 1988; Smit et al., 1992). It has also been shown that clover secretes two types of phenolic compounds: one stimulates and the other represses *nod* gene expression in *R. trifolii* (Djordjevic et al., 1987b; see also Peters and Verma, 1988). It is likely that different combinations of plant flavonoid inducers with *Rhizobium* NodD proteins might represent a mechanism regulating the early interaction between legume roots and rhizobia, and determining the first level of control of host specificity.

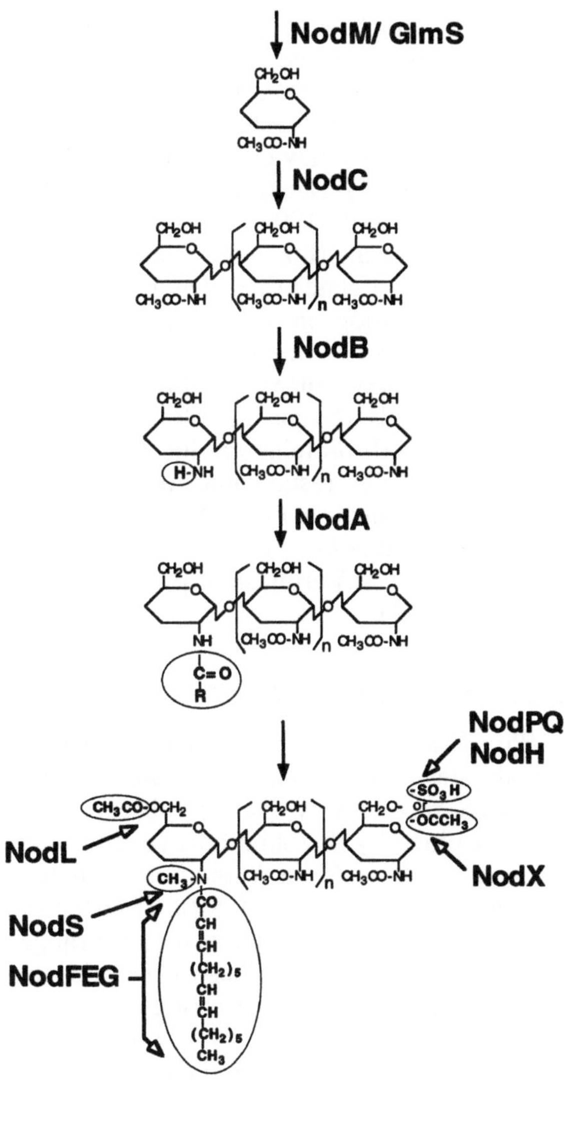

Fig. 3. Proposed pathway of Nod factor synthesis in *Rhizobium*. Glucosamine, the substrate for Nod factor backbone, is produced by NodM and GlmS glucosamine synthases. The chitooligosaccharide chain of Nod factors is assembled by NodC *N*-acetylglucosaminyl transferase. NodB is a chitooligosaccharide deacylase that removes the N-acetyl group of the nonreducing end of the chitooligosaccharide. NodA, possessing acyltransferase activity, is responsible for the addition of fatty acyl tail to the deacetylated nonreducing end of the chitooligosaccharide. Host specificity of Nod factors is often determined by proper modification of the chitooligosaccharide backbone (see text for detail)

B. Common nod Genes and Synthesis of the Nod Factor Backbone

The *nodABC* genes are found to be responsible for the synthesis of the lipo-oligosaccharide backbone (Spaink et al., 1991). Extensive genetic and biochemical studies have allowed deciphering of several key steps in the synthesis of the Nod factor backbone. These include synthesis of the oligosaccharide chain, deacetylation and transfer of an acyl moiety (Fig. 3). The oligosaccharide chain is produced by NodC. The NodC protein expressed in *E. coli* possesses the *N*-acetylglucosaminyl transferase activity and is responsible for the assembly of the chitooligosaccharide chain, the first step in biosynthesis of Nod factors (Geremia et al., 1994). NodB protein is a chitooligosaccharide deacetylase that removes the N-acetyl group at the nonreducing end of chitooligosaccharides (John et al., 1993). This deacetylation is necessary for attachment of the fatty acyl chain. The last step, N-acylation on the deacetylated nonreducing end of chitooligosaccharides, is carried out by NodA that is an *N*-acyltransferase (Rohrig et al., 1994; Atkinson et al., 1994). NodB and NodA proteins may possibly act as a complex in vivo, as has been suggested based on localization studies (Johnson et al., 1989). They could function separately because NodB is not required for NodA acyltransferase activity if the acetyl group of the substrate is absent (Atkinson et al., 1994).

The *nodIJ* genes have also been detected in all rhizobial species (Evans and Downie, 1986; Surin et al., 1990). These two genes of *R. leguminosarum* bv. *viciae* do not seem to be required for nodulation on pea, but are essential for nodulation on vetch (Evans and Downier, 1986). Mutations in *nodIJ* result in a delay in nodulation. The *nodIJ* genes encode an ATP-binding transport protein and a transmembrane protein, respectively (Evans and Downie, 1986; Schlaman et al., 1990; Vazquez et al., 1993). Both proteins are located on the inner membrane and may form a transport complex responsible for the secretion of Nod factors to the periplasm. It remains unknown how Nod factors are secreted from the periplasm to the surface of rhizobia. It has been proposed that an unknown protein with homology to *E. coli* periplasmic protein KpsD may be involved in this step (Carlson et al., 1994).

C. Host-specific nod Genes and Modification of Nod Factors

The symbiotic interaction between *Rhizobium* species and their legume host plants is highly specific. *R. meliloti* nodulates alfalfa, while *R. leguminosarum* bv. *viciae* forms nodules on pea and vetch, respectively. Genetic observations have indicated that the host range is determined by rhizobial host-specific *nod* genes, for example, *nodFEG*, *nodH, nodP*, and *nodQ* in *R. meliloti* (Faucher et al., 1989; Lerouge et al., 1990; Schwedock and Long, 1990). Transfer of these genes from one *Rhizobium* species to the other can change the host range (Cervantes et al., 1989; Roche et al., 1991). Mutations in these genes do not totally block nodulation, but cause a nodulation delay or a change, in some cases, in the host range. Some of these host-specific genes can be present in a particular species or biovar, but not in another one. Studies on molecular structure of Nod factors demonstrate that the host-specificity of *R. meliloti* is basically determined by the presence of a sulphate group on the reducing end and modifications of the acyl chain on the other end of the Nod factor molecule (Lerouge et al., 1990; Spaink et al., 1991). It remains largely unknown when these modifications take place during Nod factor biosynthesis. Roles of individual host-range *nod* genes have been biochemically correlated with the modifications of the Nod factors.

Sulfation reaction in *E. coli* and eukaryotes includes activation of sulfate, generating adenosine 5′-phosphosulphate (APS) and 3′-phosphoadenosine 5′-phosphosulphate (PAPS), and transfer of the activated sulfate group to a target. APS and PAPS are generated by ATP sulfurylase which is composed of two subunits encoded by *cysN* and *cysD* in *E. coli*. *NodP* and *nodQ* have been shown to be homologs of *cysN* and *cysD* and possess ATP sulfurylase activity both in vitro and in vivo, and are apparently involved in production of activated sulphate, PAPS (Schwedock and Long, 1990). Transfer of the activated sulfate to Nod factor precusors is apparently catalyzed by a sulphotransferase encoded by *nodH* (Roche et al., 1991). Mutations in *nodH* or *nodQ* resulted in production of an unsulphated Nod factor that can no longer induce nodule structure on the respective host (alfalfa) roots, but is still active on vetch (Roche et al., 1991). In *R. meliloti*, there are two additional homologs of *nodPQ*, of which $nodP_2Q_2$ genes are able to partially complement $nodP_1Q_1$ deletion, while the third set does not (Roche et al., 1991). *NodX* from *R. leguminosarum* bv. *viciae* TOM strain has homology to transacetylases and catalyses the addition of an acetyl group at O-6 of the acetylglucosamine residue (Firmin et al., 1993). This modification is required for nodulation on Afghanistan peas. *NodZ* from *B. japonicum* may encode a fucosyl transferase and is responsible for fucosylation at the O-6 of the reducing acetylglucosamine residue (Stacey et al., 1994). *NodS* has homology to *S*-adenosylmethionine-dependent methyl transferase and may be responsible for the addition of an N-methyl group at the nonreducing end (Geelen et al., 1993).

Biochemical functions of host-specific *nod* genes implicated in modification of the acyl chain of Nod factors are primarily based on sequence homology with the known proteins. *NodL* shares homology with *E. coli* acetyltransferase (Downie, 1989) and is able to add in vitro O-acetyl groups to lipo-oligosaccharides, chitin fragments and *N*-acetylglucosamine (Bloemberg et al., 1994). It is proposed that NodL is responsible for 6-O-acetylation of Nod factor precursors. Mutation in *R. meliloti nodL* results in the production of non-O-acetylated Nod factors and a significantly reduced capability to elicit infection thread formation in alfalfa, suggesting that O-acetylation of Nod factors is required for bacterial entry into the host roots (Ardourel et al., 1994). The deduced amino acid sequences of *nodFEG* genes are homologous to acyl carrier proteins, β-ketoacyl synthase and β-ketoacyl reductase, respectively. These proteins have been implicated in synthesis and modification of the specific fatty acyl chains that are then transferred by NodA to the nonreducing end of Nod factor precursors (Horvath et al., 1986; Bibb et al., 1989; Debelle et al., 1986; Shearman et al., 1986). Mutation in *R. meliloti nodF* results in a low efficiency of infection thread formation in alfalfa. Double *nodF/nodL* mutants are unable to penetrate into host roots (Ardourel et al., 1994). Although *nodFE* genes are conserved between *R. meliloti* and *R. leguminosarum* bv. *viciae*, they are not functionally equivalent. The *R. meliloti nodFE* genes determine the synthesis of unsaturated acyl chain (C16:2 or C16:3), whereas the allelic genes in *R. leguminosarum* bv. *viciae* direct the synthesis of C18:4 acyl moiety (Demont et al., 1993). Synthesis of a specific fatty acyl chain for Nod factor precursors thus represents another mechanism of host-range determination.

A class of host specific *nod* genes appear to be important for nodulation on a host, but are disposable for interaction with other host legumes. *NodV* and *nodW* are suggested to encode a two-component regulatory system in *B. japonicum*, which is essential for nodulation on bean and cowpea but contributes only marginally to the ability to nodulate soybean (Göttfert et al., 1990). This system seems to sense the presence of host flavonoid compounds and activates full expression of the common *nod* genes (Sanjuan et al., 1994). *NodT* and *nodX* from *R. leguminosarum* and *nolYZ* from *B. japonicum* have also been described to be required for efficient nodulation on a particular host (Surin et al., 1990; Davis et al., 1988; Dockendorff et al., 1994).

In *R. meliloti, nodO* gene encodes a secreted protein that can form a Ca^{2+}-regulated ion channel in an artificial membrane (Economou et al., 1990; Sutton et al., 1994). This host-specific gene has been implicated in the early recognition between rhizobia and the host. It is likely that this secreted rhizobial protein forms ion channels in the plant plasma membrane or the peribacteroid membrane and facilitates the response of the plant to *Rhizobium* signals by enhancing ion fluxes (Scheu et al., 1992).

The *nodM* is more closely related to the common *nod* genes than to host-specific *nod* genes, although it is currently classified to the latter group. This gene encodes an amidotransferase and is involved in the production of glucosamine, the precursor for the Nod factor backbone. The *R. meliloti nodM* gene can complement a glucosamine (*glmS*) auxotroph of *E. coli*, suggesting that NodM is involved in glucosamine biosynthesis (Baev et al., 1991). There are two genes (*nodM* and *glmS*) coding glucosamine synthases in *R. leguminosarum* (Marie et al., 1992). *NodM* and *glmS* are functionally homologous, but expressed differently. *GlmS* is expressed constitutively and is required for normal growth whereas *nodM* gene is induced during the interaction between the bacteria and legume roots. Induction of *nodM* gene expression may result in the increase of glucosamine concentration that is required for Nod factor synthesis.

IV. Other Rhizobial Signals Affecting Nodule Formation

Purified NodRm-1 at concentration of 10^{-11} M triggers root hair deformation and at higher than 10^{-7} M it induces root cortical cell divisions as well (Spaink et al., 1991). However, development of the nodule-like tissues induced by Nod factors is limited with the exception of *Glycine soja* in which Nod factors seem to induce complete nodule-like structures (Stokkermans and Peters, 1994). No infection threads are formed and no late-nodulin gene expression has been detected in these nodule-like tissues. This suggests that other rhizobial factors may be required for nodule development. Among these factors, bacteria-produced phytohormones, bacterial factors (BF) and polysaccharides have attracted some attention. In addition, specific proteins are known to be secreted by bacteria that may interact with the host. The latter may be more important once the bacteria come in direct contact with the host plasma membrane.

A. *Rhizobia-produced Phytohormones*

Rhizobium species are known to produce physiologically significant levels of plants hormones such as auxins, gibberellins and cytokinins during free-living growth and in association with legumes. Two biosynthetic routes involving indolepyruvate or indoleacetamine have been implicated in IAA synthesis in rhizobia. Enzymes of the indolepyruvate pathway were found to be expressed constitutively in *R. meliloti* and play an important role in IAA synthesis when high levels of exogenous tryptophan are present (Kittell et al., 1989). Biochemical evidence has also suggested that *B. japonicum* can synthesize IAA via an indoleacetamine intermediate, a pathway that is also used to produce IAA in *Agrobacterium tumefaciens*. This pathway uses indole-3-acetonitrile as a substrate and is catalyzed by nitrile hydratase and amidase (Kobayashi et al., 1995). A different pathway for IAA synthesis via nitrilase has been found in some phytopathogenic microorganisms but has not been shown to be present in *Rhizobium* species (Kobayashi et al., 1995). Genes controling auxin biosynthesis pathway may play an important role during infection and nodule development because IAA production by *R. meliloti* is stimulated by the addition of flavonoids (Prinsen et al., 1991). Treatment of *R. meliloti* with flavonoids induces production of new cytokinins, suggesting that these phytohormones may be involved in the development of nodule structure (Sturtevant and Taller, 1989; Taller and Sturtevant, 1991). Although these results are not conclusive, it is possible to demonstrate the role of rhizobia-produced hormones in nodule organogenesis by mutation of the corresponding genes and hence blocking the hormone biosynthesis. It has recently been shown that a *B. elkanii* mutant strain with reduced IAA production produces significantly less nodules on soybean roots and this deficiency can be suppressed by exogenous IAA application (Fukuhara et al., 1994). Another strategy has recently been taken by overexpression of an *Agrobacterium trans*-zeatin secretion (*tzs*) gene in a *Rhizobium* nod^{-} mutant (Cooper and Long, 1994). Overproduction of *trans*-zeatin was able to complement *nod* gene mutations and conferred the mutant strain to produce functioning nodules.

B. *Bacterial Factors*

In addition to the Nod factors and phytohormones, *Rhizobium* species also produce a number of bacterial factors (BF) that are active on legume roots,

BF-5

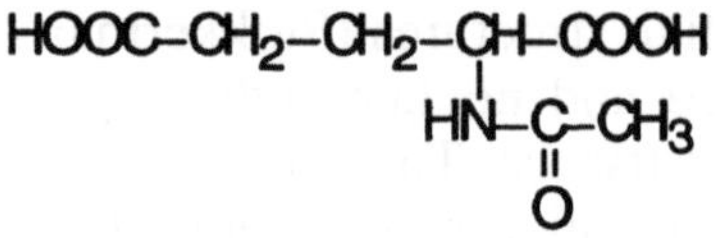

BF-7

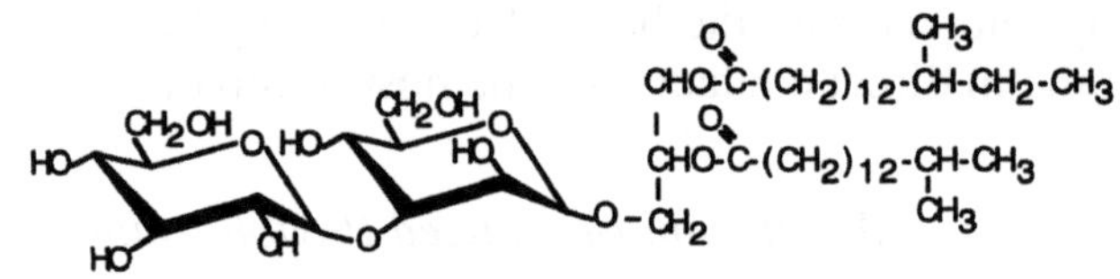

Fig. 4. Structures of the bacterial factors, *N*-acetylglutamic acid (BF-5) and 1,2-diacyl-3-*O*-(α-D-glucopyranosyl-(1→3)-*O*-α-D-mannopyranosyl)-glycerol (BF-7) from *Rhizobium leguminosarum* biovar *trifolii*

causing root hair proliferation, branching or deformation, cortical cell divisions or thick-short-root response. Of these factors, BF-5 has been identified as *N*-acetylglutamic acid (Philip-Hollingsworth et al., 1991) and BF-7 as diglycosyl diacylglycerol, a membrane glycolipid (Orgambide et al., 1994) (Fig. 4). The latter can induce root hair deformation and root cortical cell division at concentrations of 10^{-11} M, similar to the concentration at which the lipooligosaccharide Nod factors are effective. Moreover, production of these two factors is dependent on *nod* gene induction by flavonoids and their biological activity is host specific. It is, however, not known whether the genes responsible for BF production are essential for the symbiosis.

C. *Rhizobial Polysaccharides*

Following rhizobial infection, tubular structures, known as "infection threads", are initiated from the root hair and reach to the dividing cortical cells. The infection threads provide specialized "tunnels" for *Rhizobium* cells to enter into the cortical cells. As rhizobia enter the host cells, they are enclosed in plasma membrane envelopes, the peribacteroid membrane. Several rhizobial genes, *ndv*, *lps*, and *exo*, are involved in the development of the infection threads and affect the entry of rhizobia into the plant cells. Mutations in these genes lead to the retention of bacteria in the infection threads or the formation of empty nodules devoid of bacteria, a phenotype termed nodule development defective (Ndv⁻). These genes encode bacterial surface components and are usually required for infection thread formation and bacterial release, in addition to eliciting root hair deformation (Noel, 1992).

1. Exopolysaccharides

Exopolysaccharides (EPSs) are accumulated during late stages of rhizobial growth and are a major component of the extracellular material. The *exo* mutants fail to synthesize acidic EPSs and induce small, white nodules (Finan et al., 1985; Leigh et al., 1987). Different *exo* mutants were isolated by selecting for nonfluorescent colonies in the presence of Calcofluor that binds to polysaccharides. These mutants usually contain functional *nod* genes and are able to induce root hair deformation and initiate infection thread formation (Klein et al., 1988a, b). However, the infection threads abort early, leading to the formation of empty nodules. The Ndv⁻ phenotype caused by *R. leguminosarum exo* mutants can be suppressed by the addition of exogenous high-molecular-weight EPS (Djordjevic et al., 1987). For *R. meliloti*, only low-molecular-weight EPS is effective in restoring *exoA* mutation to wild type phenotype, suggesting that only some types of EPS can be effective signals for the formation of infection thread (Battisti et al., 1992). Two types of EPS are distinguished: EPS I that binds Calcofluor and EPS II that does not bind this dye (Leigh et al., 1985; Glazebrook and Walker, 1989). A number of genetic loci related to EPS synthesis and secretion have been identified and some of these loci have been sequenced from *R. meliloti* (Leigh and Walker, 1994; Petrovics et al., 1993). It is still not clear what role EPS plays during infection. EPSs are proposed to act as signals for eliciting necessary plant responses or as suppressors of deleterious plant reactions (Halverson and Stacey, 1986). The acidic EPSs may serve as a protective layer that blocks diffusion of bacterial "toxins" to the host plant cells or plant phytoalexins to the bacteria (O'Connell et al., 1990) and prevents elicitation of host defense (Verma and Nadler, 1984).

2. Lipopolysaccharides

Lipopolysaccharides (LPSs) are components of the bacterial outer membrane. They consist of a membrane-anchoring lipid moiety (lipid A) and a polysaccharide chain that is composed of a core oligosaccharide and an outer, strain-dependent *O*-antigen. LPS is required for normal nodule development in most of the legume plants except alfalfa on which *R. meliloti lps* mutants induce functional nodules (Clover et al., 1989). Defects in LPS synthesis in *R. leguminosarum* result in the induction of empty nodules in which the rhizobia are not released from the infection threads (Priefer, 1989) or the infection threads abort at early stage (Noel et al., 1986; Diebold and Noel, 1989). The *lps* mutants were originally isolated from *R. leguminosarum* as Ndv⁻ mutants, of which many were found to be defective in LPS (Noel et al., 1986). Five genetic loci comprising more than a dozen *lps* genes have been identified (Noel, 1992; Brink et al., 1990). Mutations in these genes have been found to affect the structures of the core-oligosaccharide of the *O*-antigen. A substantial variation in LPS structure appears to be tolerated for its role in nodule development. Moreover, the structure of LPS itself seems to change during nodule formation (Wood et al., 1980) or in the presence of host flavonoids (Reuhs et al., 1994), which has made the genetic dissection of LPS biosynthesis more difficult and complicated. Despite the high variability in structure, the *O*-antigen portion of the LPS does not seem to serve as host-range determinant because replacing one type of the *O*-antigen in a bacterial strain by another type from a different rhizobial strain does not change the host specificity of the strain (Brink et al., 1990). This result does not support the hypothesis that the *O*-antigen polysaccharide chain may be released from LPS and serve as diffusible signal molecule (Diebold and Noel, 1989). It has been proposed that LPS may play a protective role, as does EPS (see above), to mask bacterial elicitors or plant phytoalexins (Djordjevic et al., 1987a). The observation that LPS-defective mutants trigger host defense responses in pea nodules suggested that LPS plays a role in reducing host defense response during infection by rhizobia (Perotto et al., 1994). It is also possible that LPS provides an appropriate bacterial surface for interaction with the plant plasma membrane during endocytosis.

Lipopolysaccharides may stabilize the outer membrane of the rhizobia against physiological stresses (pH and osmotic conditions) during endocytosis (Kannenberg and Brewin, 1989). It is also possible that these bacterial surface components provide specific carbohydrate groups that interact with the plant membrane (Ho et al., 1990). A *Rhizobium Tn5*-induced mutant seems to fail to interact with the host at the stage of endocytosis, leading to the formation of nodules largely devoid of bacteria (Morrison and Verma, 1987). Several other mutants of this type have been isolated by G. Stacey's and H. Hennecke's groups.

3. β-Glucans

Another group of rhizobial surface components that are required for nodule development are β-glucans that are found mainly in the periplasm but also are secreted by bacteria. Most *Rhizobium* species synthesize β-1,2-glucan consisting of about 20 glucose residues, while *B. japonicum* also produces β-1,3-glucan and β-1,6-glucan. A class of rhizobial Ndv$^-$ mutants are found to be β-1,2-glucan deficient. Two genes, *ndvA* and *ndvB*, have been indentified in *R. meliloti* to be involved in β-1,2-glucan synthesis and export. The former encodes a hemolysin homolog and may be involved in the ATP-dependent transport of β-1,2-glucan to the outer membrane (Stanfield et al., 1988). The latter encodes a 319 kDa membrane protein that binds to UDP-glucose, an intermediate in β-1,2-glucan synthesis (Ielpi et al., 1990). Mutations in *ndvA* and *ndvB* genes result in the formation of incompletely developed nodules that are largely devoid of bacteria (Dylan et al., 1986). Mutants that suppress Ndv$^-$ mutations have been isolated. These pseudorevertants allow almost normal root hair attachment and infection thread formation, but are still deficient in adaptation to low osmotic conditions that is one of the characteristic phenotype of Ndv$^-$ mutants (Dylan et al., 1990b). This result does not support the hypothesis that β-1,2-glucans may function for physiological adjustment of the bacteria to osmotic stress that may occur in the infection thread (Dylan et al., 1990a). Although the production of β-1,2-glucans has been positively correlated with symbiotic efficiency, what exact role these molecules play is not clear.

V. Signals Produced by the Host in Reponse to Nod Factors

Nodule-like structures appear, though at a low frequency, spontaneously in the absence of *Rhizobium* (Truchet et al., 1989), indicating that the legume plants possess all the essential components of nodule formation machinery. This nodule organogenesis program is triggered by bacterial infection (Verma, 1992). Induction of nodule-like structures on alfalfa roots by auxin-transport inhibitors (Hirsch et al., 1989) further suggests that the nodule-formation machinery can be activated by changing the hormone balance of the root cells. Compared to those secreted by rhizobia, hormones produced by the host seem to play a more important role in nodule organogenesis. Adjustment of relative levels of endogenous phytohormones may represent the basic response of the host to rhizobial Nod factors. The role of auxins in legume nodulation was proposed half a century ago (Thimann, 1936) and since then all five classes of phytohormones have been implicated in nodule development (Grobbelaar et al., 1971; Philips, 1971). There have been a number of observations showing that treatment of legume roots with exogenous auxins and cytokinins induces cortical cell divisions (Libbenga et al., 1973). Although rhizobia produce GA_1 and GA_4, the gibberellins detected in the nodule extracts seem to be generated by the host cells (Atzorn et al., 1988).

The role of phytohormone balance in nodule formation has been studied by expressing *Agrobacterium rhizogenes nolABC* genes that are known to change hormone balance in transgenic tissues. Transgenic plants expressing *nolABC* formed nodules faster and produced more nodules than the control plants (A. Kondorosi et al., 1993).

Application of inhibitors of phytohormone biosynthesis and transport has generated useful data for evaluation of roles of phytohormones as mediators for nodule initiation. Treatment of legume roots with auxin transport inhibitors, 2,3,5-triisobenzoic acid (TIBA) or *N*-(1-naphthyl)phthalamic acid (NPA), induces the formation of nodule-like structures, suggesting that nodule initiation can be mimicked by changing relative levels of phytohormones (Hirsch et al., 1989; Scheres et al., 1992). In these bacteria-free pseudonodules, early nodulin (*enod2* and *enod12*) genes are expressed and their expression patterns are very close to those observed on normal nodules or Nod factor-induced pseudonodules.

During *Rhizobium* infection, legume roots produce a marked amout of ethylene and exogenous addition of this hormone strongly inhibits nodulation (Goodlass and Smith, 1979). Treatment with the ethylene inhibitor aminoethoxyvinylglycine (AVG) has been shown to stimulate nodule formation in alfalfa (Peters and Crist-Estes, 1989). AVG treatment can even overcome nitrate inhibition of nodulation, indicating that the inhibitory effect of nitrate is mediated through the phytohormone ethylene (Ligero et al., 1991). The restoration of pea *sym5* mutant phenotype to wild type by AVG also provides evidence for a role of ethylene in nodulation (Fearn and LaRue, 1991).

VI. Early Host Response to Bacterial Signals

Early host response to rhizobial signals has been studied by using microelectrodes. Application of Nod factors induces a rapid depolarization of membrane potential in legume root hairs (Ehrhardt et al., 1992). This depolarization response is specific because extracts from a Nod$^-$ *Rhizobium* strain failed to induce this response and roots of non-legume plants did not respond to purified Nod factors. Since signal transduction in mammalian cells often couples with a membrane potential depolarization, these results suggest that the host responds to Nod factors by transducing a signal across the plasma membrane (Ehrhardt et al., 1992). Studies on early nodulin gene expression have suggested that lithium, an inhibitor of phosphoinositide metabolism and an inducer of mitogenesis in yeast and animals, may also be involved in early host response to bacterial signals (T. Bisseling, pers. comm.).

Expression of early nodulin genes, *enod5* and *enod12,* has been shown to be induced in Nod factor-treated root hairs (Horvath et al., 1993). Two *enod12* genes are differentially induced during early infection, *enod12b* expression is related to the action of Nod factors whereas *enod12a* expression is associated with the invasion process in nodules (Bauer et al., 1994). However, it is not clear how Nod factor signals are transduced inside the plant cells and how these signals trigger the nodule morphogenesis cascade.

VII. Plant Phosphatidylinositol-3-kinase

Given a high degree of specificity in legume *Rhizobium* infection, it is reasonable to assume that these bacteria interact with some receptor(s) in the host plasma membrane. Such receptors are expected to be different from the potential Nod factor receptors (Verma, 1992). The basic mechanism of signal transduction in plants appears to be similar to that in yeast and animal cells. In root nodules, the bacteria send various signals to the host cell to initiate cell division, endocytosis and membrane proliferation.

One of the central pathways of signal transduction in eukaryotes employs phosphatidylinositol-3-kinase (PI-3-kinase; Cantley et al., 1991). PI-3-kinase was originally found to be required for viral transformation of cultured mammalian cells and for tumorigenesis in animals (Whitman et al., 1985, 1988). This enzyme has now been shown to be a component of a wide range of receptor tyrosine kinase complexes that transduce growth signals for cell division and membrane ruffling (Valius and Kazlauskas, 1993; Stephens et al., 1994). Two yeast genes have been identified to encode PI-3-kinases; *TOR2* is involved in signal transduction that controls cell cycle (Kunz et al., 1993) and *VPS34* is essential for vesicular targeting to the vacuole (Schu et al., 1993).

Recently, two soybean cDNAs encoding PI-3-kinases have been isolated (Hong and Verma, 1994). Primer extension results showed that *PI3K-1* gene is expressed in young nodules in parallel with membrane proliferation but its

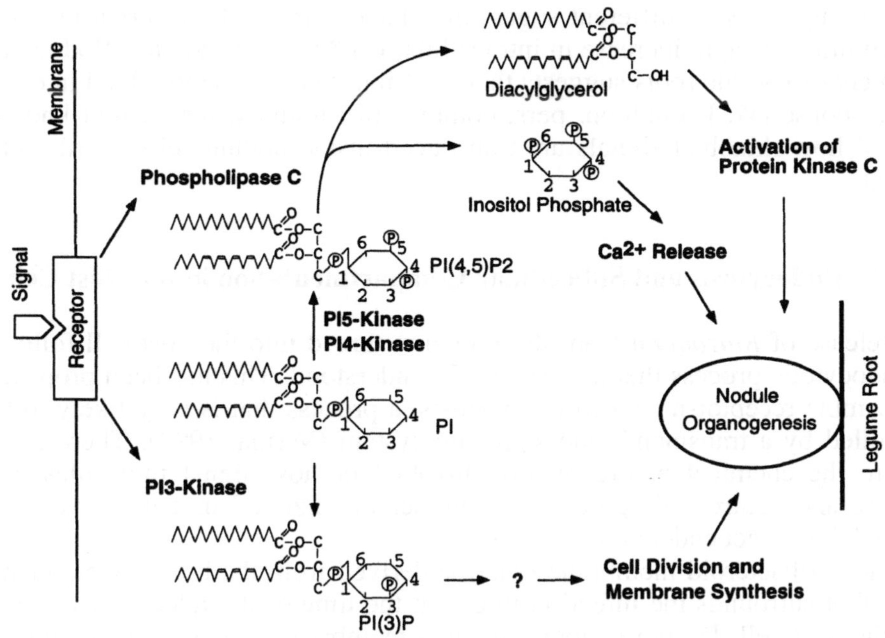

Fig. 5. Proposed role of phosphatidylinositol in signal transduction pathways leading to nodule organogenesis

expression is reduced in mature nodules and roots. The root form of the PI-3-kinase CDNA (*PI3K-5*) is expressed in roots and mature nodules but not in young nodules. High levels of PI-3-kinase activity were detected in membrane fractions of young nodule extracts (Hong and Verma, 1994). It is proposed that PI-3-kinase may play a pivotal role in signal transduction pathway that controls membrane proliferation during nodule organogenesis (Fig. 5).

PI-4-kinase and PI-specific phospholipase C have been characterized in plants (Xu et al., 1994; Irvine et al., 1980; Biffen and Hanke, 1990). Mammalian protein kinase C (PKC) is a Ca^{2+}-activated, phospholipid-dependent enzyme. The binding of mammalian PKC to Ca^{2+} can be greatly increased by diacylglycerol derived from $PI(4,5)P_2$ hydrolysis. This enzyme plays an important role in signal transduction in mammalian cells. Several reports have provided biochemical and molecular evidence for the presence of PKC homologs in plant cells (see Drobak, 1992). However, because the predominant protein kinases associated with the plant plasma membrane are not PKC and the levels of diacylglycerol are very high in plasma membranes of unstimulated plant cells compared to those in animal cells, the role of PKC in signal transduction in plants is not clear. Rhizobial diglycosyl diacylglycerol (BF-7) differs from wheat diglycosyl diacylglycerol (DGDG) in the sugar composition. The former has the ability to elicit morphological responses in legume roots, but the latter does not (Orgambide et al., 1994). This implies that diacylgycerols conjugated with different sugars may have completely different biological activities. A rapid increase in intracellular Ca^{2+} levels following Nod factor treatment of legume roots suggests that Ca^{2+} ion flux may be involved in early host response (W. Broughton, pers. comm.). In summary, the current knowledge of how rhizobial signals are transduced inside nodule cells is still very limited.

VIII. Endocytosis and Subcellular Compartmentation in the Host Cell

The release of *Rhizobium* from the infection thread into the host cell follows an endocytotic process that is very poorly understood. This has been proposed to resemble receptor-mediated endocytosis, a process that is very likely to be controlled by a transmembrane signaling system (Verma, 1992). There is no clue to the chemical nature of any rhizobial or host signal molecules that mediate and control this process. A number of rhizobial mutants have been isolated that affect endocytotic process.

The peribacteroid membrane (PBM) is derived from the host plasma membrane that surrounds the infection thread at the time of the release of bacteria into the host cell. During endocytosis, this membrane undergoes a significant change in composition. These changes along with the appropriate alterations in the peribacteroid fluid allow "internalization" of bacteria in the host cell cytoplasm (Verma, 1990). Electron microscopic studies showed a high rate of fusion of smooth and coated vesicles with the tip of the infection thread (Robertson and Lyttleton, 1982). Biochemical and molecular biological studies have clearly demonstrated that a number of nodulins are targeted to the newly-formed PBM (Fortin et al., 1985, 1987; Katinakis and Verma, 1985; Cheon et al., 1994; Miao et al., 1992). The ATPase found in the PBM resembles that of the plasma membrane (Verma et al., 1978; Day et al., 1990). A vacuolar marker enzyme (α-mannosidase) and a vacuolar protease inhibitor have been found to be present in the PBM compartment (Kinnback et al., 1987; Manen et al., 1991). This membrane, therefore, appears to be a mosaic membrane with properties common to both the plasma and vacuolar membranes. PBM also controls the transport of dicarboxylates (malate and succinate) that serve as carbon sources for the bacteroids, and allows diffusion of oxygen, nitrogen, carbon dioxide and ammonia across the membrane (Day et al., 1990). At the same time, it prevents entry of any substance from bacteria that may harm the plant cell, in any way.

Although only one PBM protein has so far been suggested to be of bacteroid origin (Fortin et al., 1987), rhizobial genes have been shown to play important roles in endocytosis and the formation of the PBM. Many *Rhizobium* mutants including *lps, exo* and *ndv* are unable to enter the host cell (see above). ER proliferation and vesicle formation in the cytoplasm of the infected host cells are a prerequisite step for the endocytosis of rhizobia. Mutations in the *Rhizobium* genome can also affect this step (Roth and Stacey, 1989), suggesting that signals produced by the bacteria are involved in the control of host secretory pathway that targets the host proteins to PBM (Cheon et al., 1994; Miao et al., 1992). Recently, Werner's group has isolated some protease mutants of rhizobia and demonstrated that proper processing and secretion of certain bacterial proteins is essential for the development of effective PBM.

IX. Conclusions

Nodule organogenesis requires well-regulated exchange of defined signal molecules from both symbiotic partners (Fig. 6). The discovery of the roles of plant flavonoids and rhizobial Nod factors in nodule organogenesis has advanced our knowledge of a two-way molecular conversation between plants and microbes. This also provided opportunity for further exploration of signal perception and transduction in plant cells.

Rhizobia produce a spectrum of active Nod factors with different modified groups on the same chemical backbone, the oligoglucosamine. It is not known what is the chemical nature of the plant receptors that directly interact with Nod factors and perceive the signals from *Rhizobium.* It is very likely that the receptors are localized on the plant plasma membrane. A difference in structures of

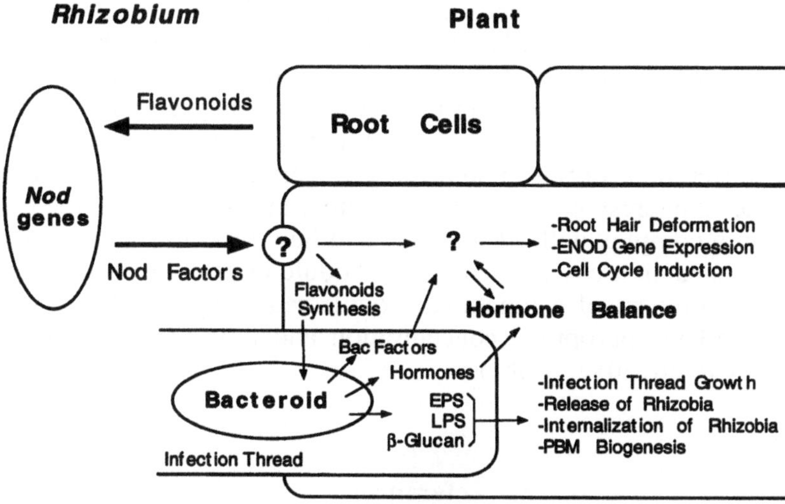

Fig. 6. Molecular communication between *Rhizobium* and legume roots during nodule organogenesis. The host roots secrete flavonoids that act as chemotactic agents to the rhizobia and as inducer for rhizobial *nod* gene expression. The *nod* genes encode enzymes catalyzing synthesis and secretion of Nod factors. The Nod factors are active in eliciting a cascade of responses in the host cells. Phytohormone balance of the host cell appears to be an important factor for nodule organogenesis. Rhizobial surface components including exopolysaccharides (EPS), lipopolysaccharides (LPS) and β-glucans may play essential roles in infection thread growth and endocytosis of the bacteria. Bacterial (Bac) factors such as diglycosyl diacylglycerol (BF-7) and *N*-acetylglutamic acid (BF-5) are also active at nanomolar concentrations in eliciting specific host responses

the receptors from different legumes might exist to allow host-specific recognition of Nod factors from different rhizobia. It is possible to isolate these putative Nod factor-binding proteins by affinity-labeling of purified Nod factors, or by using antibodies to Nod factors. Identification of the receptors would help understand the unique interaction between *Rhizobium* and legume plants.

Expression of *Rhizobium nodA* and *nodB* genes in tobacco plants causes abnormalities in growth and development of the plants, indicating that plants contain the necessary substrates for production of Nod factor-like compounds (J. Schmidt et al., 1993). Nod factors, when added to tomato cell culture, induce a transient alkalinization of the medium, an early reaction of plant cells to various elicitors (Staehelin et al., 1994a, b). Nod factors are also active in regulating other plant developmental processes like embryogenesis as addition of Nod factors rescues a carrot somatic embryo mutant (De Jong et al., 1993) and activates the cell cycle machinery in callus suspension (Savoure et al., 1994). It would not be surprising to find a new class of endogenous plant growth-regulating substances that resemble Nod factors in structure and biological activity. Oligosaccharides have been shown to induce flower development.

In addition to Nod factors, other rhizobial signals also seem to be required for the formation of infection threads, endocytosis of the bacteria and maintenance of the symbiosis. Little is known how the signals carried by rhizobial Nod factors and other molecules are transduced inside the plant cell. PI-3-kinases have recently been shown to be present in plant membrane fractions and are likely to be associated with the signal transduction pathway leading to membrane proliferation during nodule organogenesis (Hong and Verma, 1994). It will be of interest to determine the membrane components (the receptors?) that form a complex with PI-3-kinase and how this enzyme participates in the signal transduction cascade leading to nodule development. It is apparent that the development of a functional nodule requires several factors that must be produced in appropriate concentration and time to form this novel structure leading to effective symbiosis.

X. References

Ardourel M, Demont N, Debelle F, Maillet F, deBilly F, Prome J-C, Denarie J, Truchet G (1994) *Rhizobium meliloti* lipooligosaccharide nodulation factors: different structural requirements for bacterial entry into target root hair cells and induction of plant symbiotic developmental responses. Plant Cell 6: 1357–1374

Atkinson E, Palcic M, Hindsgaul L, Long S (1994) Biosynthesis of *Rhizobium meliloti* lipooligosaccharide Nod factors: NodA is required for an N-acyltransferase activity. Proc Natl Acad Sci USA 91: 8418–8422

Atzorn R, Crozier A, Wheeler C, Sandberg G (1988) Production of gibberellins and indole-3-acetic acid by *Rhizobium phaseoli* in relation to nodulation of *Phaseolus vulgaris* roots. Planta 175: 532–538

Baev N, Endre G, Petrovics G, Banfalvi Z, Kondorosi A (1991) Six nodulation genes of nod box locus-4 in *Rhizobium meliloti* are involved in nodulation signal production—*nodM* codes for D-glucosamine synthetase. Mol Gen Genet 228: 113–124

Banfalvi Z, Nieuwkoop AJ, Schell M, Besl L, Stacey G (1988) Regulation of *nod* gene expression in *Bradyrhizobium japonicum*. Mol Gen Genet 214: 420–424

Battisti L, Lara JC, Leigh JA (1992) Specific oligosaccharide form of the *Rhizobium meliloti* exopolysaccharide promotes nodule invasion in alfalfa. Proc Natl Acad Sci USA 89: 5625–5629

Bauer P, Crespi M, Szecsi J, Allison L, Schultze M, Ratet P, Kondorosi E, Kondorosi A (1994) Alfalfa *Enod12* genes are differentially regulated during nodule development by Nod factors and *Rhizobium* invasion. Plant Physiol 105: 585–592

Bibb MJ, Biro S, Motamedi H, Collins JF, Hutchinson CR (1989) Analysis of the nucleotide sequence of the *Streptomyces glaucescens tmcI* gene provides information about the enzymology of polyketide antibiotic biosynthesis. EMBO J 8: 2727–2736

Biffen M, Hanke D (1990) Polyphosphoinositidase activity in soybean membranes is Ca^{2+}-dependent and shows no requirement for guanine-nucleotides. Plant Sci 69: 147–155

Bloemberg GV, Thomas OJ, Lugtenberg B, Spaink HP (1994) Nodulation protein NodL of *Rhizobium leguminosarum O*-acetylates lipo-oligosaccharides, chitin fragments and N-acetylglucosamine in vitro. Mol Microbiol 11: 793–804

Brink BA, Miller J, Carlson RW, Noel KD (1990) Expression of *Rhizobium leguminosarum* CFN42 genes for lipopolysaccharides in strains derived from different *R. leguminosarum* soil isolates. J Bacteriol 172: 548–555

Cantley L, Auger K, Carpenter C, Duckworth B, Graziani A, Kapller R, Soltoff S (1991) Oncogenes and signal transduction. Cell 64: 281–302

Carlson R, Price N, Stacey G (1994) The biosynthesis of rhizobial lipo-oligosaccharide nodulation signal molecules. Mol Plant Microbe Interact 7: 684–695

Cervantes E, Sharma SB, Maillet F, Vasse J, Truchet G, Rosenberg C (1989) The *Rhizobium meliloti* host range *nodQ* gene encodes a protein which shares homology with translation elongation and initiation factors. Mol Microbiol 3: 745–755

Cheon C, Hong Z, Verma DPS (1994) Nodulin-24 follows a novel pathway for integration into the peribacteroid membrane in soybean root nodules. J Biol Chem 269: 6598–6602

Clover RH, Kieber J, Signer ER (1989) Lipopolysaccharide mutants of *Rhizobium meliloti* are not defective in symbiosis. J Bacteriol 171: 3961–3967

Cooper JB, Long SR (1994) Morphogenetic rescue of *Rhizobium meliloti* nodulation mutants by *trans*-zeatin secretion. Plant Cell 6: 215–225

Davis EO, Johnston AWB (1990) Regulatory functions of the 3 *nodD* genes of *Rhizobium leguminosarum* biovar *phaseoli*. Mol Microbiol 4: 933–941

Davis EO, Evans EJ, Johnston AWB (1988) Identifications of *nodX*, a gene that allows *Rhizobium leguminosarum* biovar *viciae* strain TOM to nodulate Afganistan peas. Mol Gen Genet 212: 531–535

Day DA,Yang L-JO, Udardi MK (1990) Nutrient exchange across the peribacteroid membrane of isolated symbiosomes. In: Gresshoff P, Roth J, Stacey G, Newton WE (eds) Nitrogen fixation: achievements and objectives. Chapman and Hall, New York pp 219–226

De Jong AJ, Heidstra R, Spaink HP, Hartog MV, Meijer EA, Hendriks T, Lo Shiaovo F, Terzi M, Bisseling T, Van Kammen A, De Vries SC (1993) *Rhizobium* lipo-oligosaccharides rescue a carrot somatic embryo mutant. Plant Cell 5: 615–620

Debelle F, Rosenberg C, Vasse J, Maillet F, Martinez E, Denarie J, Truchet G (1986) Assignment of symbiotic developmental phenotypes to common and specific nodulation (*nod*) genetic loci of *Rhizobium meliloti*. J Bacteriol 168: 1057–1086

Demont N, Debelle F, Aurelle H, Denarie J, Prome J (1993) Role of the *Rhizobium meliloti nodF* and *nodE* genes in the biosynthesis of lipo-oligosaccharidic nodulation factors. J Biol Chem 268: 20134–20142

Demont N, Ardourel M, Maillet F, Prome D, Ferro M, Prome J-C, Denarie J (1994) The *Rhizobium meliloti* regulatory *nodD3* and *syrM* genes control the synthesis of a particular class of nodulation factors *N*-acylated by (ω-1)-hydroxylated fatty acids. EMBO J 13: 2139–2149

Dénarié J, Debellé F, Rosenberg C (1992) Signaling and host range variation in nodulation. Annu Rev Microbiol 46: 497–531

Diebold R, Noel KD (1989) *Rhizobium leguminosarum* exopolysaccharide mutants: biochemical and genetic analysis and symbiotic behavior on three hosts. J Bacteriol 171: 4821–4830

Djordjevic M, Gabrielle D, Rolfe B (1987a) *Rhizobium*—the refined parasite of legumes. Annu Rev Phytopathol 25: 145–168

Djordjevic M, Redmond JW, Batley M, Rolfe BG (1987b) Clovers secrete specific phenolic compounds which either stimulate or repress *nod* gene expression in *Rhizobium trifolii*. EMBO J 6:1173–1179

Djordjevic M, Chen H, Batley M, Redmond JW, Rolfe BG (1987) Nitrogen fixation ability of exopolysaccharide synthesis mutants of *Rhizobium* sp. strain NG234 and *Rhizobium trifolii* is restored by addition of homologous exopolysaccharide. J Bacteriol 169: 53–60

Dockendorff T, Sharma A, Stacey G (1994) Identification and characterization of the *nolYZ* genes of *Bradyrhizobium japonicum*. Mol Plant Microbe Interact 7: 173–180

Dooner HK, Robbins TP, Jorgensen RA (1991) Genetic and developmental control of anthocyanin biosynthesis. Annu Rev Genet 25: 173–199

Downie JA (1989) The *nodL* gene from *Rhizobium leguminosarum* is homologous to the acetyl transferases encoded by *iacA* and *cysE*. Mol Microbiol 3: 1649–1651

Drobak B (1992) The plant phosphoinositide system. Biochem J 288: 697–712

Dusha I, Kondorosi A (1993) Genes at different regulatory levels are required for the ammonia control of nodulation in *Rhizobium meliloti*. Mol Gen Genet 240: 435–444

Dylan T, Ielpi L, Stanfield S, Kashyap L, Douglas C, Yanofsky M, Nester E, Helinski DR, Ditta G (1986) *Rhizobium meliloti* genes required for nodule development are related to chromosomal virulence genes in *Agrobacterium tumefaciens*. Proc Natl Acad Sci USA 83: 4404–4407

Dylan T, Helinski DR, Ditta GS (1990a) Hypoosmotic adaptation in *Rhizobium meliloti* required β-(1→2)-glucan. J Bacteriol 172: 1400–1408

Dylan T, Nagpal P, Helinski DR, Ditta GS (1990b) Symbiotic pseudorevertants of *Rhizobium meliloti ndv* mutants. J Bacteriol 172: 1409–1417

Economou A, Hamilton WDO, Johnston AWB, Downier JA (1990) The *Rhizobium* nodulation gene *nodO* encodes a Ca^{2+}-binding protein that is exported without *N*-terminal cleavage and is homologous to haemolysin and related proteins. EMBO J 9: 349–354

Egelhoff TT, Long SR (1985) *Rhizobium meliloti* nodulation genes: identification of *nodDABC* gene products, purification of NodA protein, and expression of *nodA* in *Rhizobium meliloti*. J Bacteriol 164: 591–599

Ehrhardt DW, Atkinson EM, Long SR (1992) Depolarization of alfalfa root hair membrane potential by *Rhizobium meliloti* Nod factors. Science 256: 998–1000

Estabrook EM, Sengupta-Gopalan C (1991) Differential expression of phenylalanine ammonia-lyase and chalcone synthase during soybean nodule development. Plant Cell 3: 299–308

Evans IJ, Downie JA (1986) The *nodI* gene product of *Rhizobium leguminosarum* is closely related to ATP-binding bacterial transport proteins: nucleotide sequence analysis of the *nodI* and *nodJ* genes. Gene 43: 95–101

Faucher C, Camut S, Dénarié J, Truchet G (1989) The *nodH* and *nodQ* host range genes of *Rhizobium meliloti* behave as avirulence genes in *R. leguminosarum* bv. *viciae* and determine changes in the production of plant-specific extracellular signals. Mol Plant Microbe Interact 2: 291–300

Fearn JC, Larue TA (1991) A temperature-sensitive nodulation mutant (Sym-5) of *Pisum sativum* L. Plant Cell Environ 14: 221–227

Finan TM, Hirsch AM, Leigh JA, Johnson E, Kuldau GA, Deegan S, Walker GC, Signer ER (1985) Symbiotic mutants of *Rhizobium meliloti* that uncouple plant from bacterial differentiation. Cell 40: 869–877

Firmin J, Wilson K, Carlson R, Davies A, Downier J (1993) Resistance to nodulation of cv. Afghanistan peas is overcome by *nodX*, which mediates an *O*-acetylation of the *Rhizobium leguminosarum* lipo-oligosaccharide nodulation factor. Mol Microbiol 10: 351–360

Fisher RF, Long SR (1992) *Rhizobium*–plant signal exchange. Nature 357: 655–660

Fisher RF, Egelhoff TT, Mulligan JT, Long SR (1988) Specific binding of proteins from *Rhizobium meliloti* cell-free extracts containing NodD to DNA sequences upstream of inducible nodulation genes. Genes Dev 2: 282–293

Fortin MG, Zelechowska M, Verma DPS (1985) Specific targeting of membrane nodulins to the bacteroid-enclosing compartment in soybean nodules. EMBO J 4: 3041–3046

Fortin MG, Morrison NA, Verma DPS (1987) Nodulin-26, a peribacteroid membrane nodulin is expressed independently of the development of the peribacteroid compartment. Nucleic Acids Res 15: 813–824

Fukuhara H, Minakawa Y, Akao S, Minamisawa K (1994) The involvement of indole-3-acetic acid produced by *Bradyrhizobium elkanii* in nodule formation. Plant Cell Physiol 35: 1261–1265

Geelen D, Mergaert P, Geremia RA, Goormachtig S, Van Montagu M, Holsters M (1993) Identification of *nodSUIJ* genes in *nod* locus 1 of *Azorhizobium caulinodans*: evidence that *nodS* encodes a methyltransferase involved in Nod factor modification. Mol Microbiol 9: 145–154

Geremia RA, Mergaert P, Geelen D, Van Montagu M, Holsters M (1994) The NodC protein of *Azorhizobium caulinodans* is an *N*-acetylglucosaminyltransferase. Proc Natl Acad Sci USA 91: 2669–2673

Glazebrook J, Walker GC (1989) A novel exopolysaccharide can function in place of the calcofluor-binding exopolysaccharide in nodulation of alfalfa by *Rhizobium meliloti*. Cell 56: 661–672

Goodlass G, Smith KA (1979) Effect of ethylene on root extension and nodulation of pea (*Pisum sativum* L.) and white clover (*Trifolium repens* L.). Plant Soil 51: 387–395

Göttfert M (1993) Regulation and function of rhizobial nodulation genes. FEMS Microbiol Rev 104: 39–64

Göttfert M, Grob P, Hennecke H (1990) Proposed regulatory pathway encoded by the *nodV* and *nodW* genes, determinants of host specificity in *Bradyrhizobium japonicum*. Proc Natl Acad Sci USA 87: 2680–2684

Grobbelaar N, Clarke B, Hough MC (1971) The nodulation and nitrogen fixation of isolated roots of *Phaseolus vulgaris* L. Plant Soil Special: 215–223

Hahlbrock K, Grisebach H (1979) Enzymatic control in the biosynthesis of lignin and flavonoids. Annu Rev Plant Physiol 30: 105–130

Halverson LJ, Stacey G (1986) Signal exchange in plant–microbe interactions. Microbiol Rev 50: 193–225

Hartwig UA, Phillips DA (1991) Release and modification of *nod* gene-inducing flavonoids from alfalfa seeds. Plant Physiol 95: 804–807

Hartwig UA, Maxwell CA, Joseph CM, Phillips DA (1990) Effects of alfalfa *nod* gene inducing flavonoids on *nodABC* transcription in *Rhizobium meliloti* strains containing different *nodD* genes. J Bacteriol 172: 2769–2773

Hirsch A (1992) Developmental biology of legume nodulation. New Phytol 122: 211–237

Hirsch A, Bhuvaneswari T, Torrey J, Bisseling T (1989) Early nodulin genes are induced in alfalfa root outgrowths elicited by auxin transport inhibitors. Proc Natl Acad Sci USA 86: 1244–1248

Ho SC, Schindler M, Wang JL (1990) Carbohydrate binding activities of *Bradyrhizobium japonicum*. 2. Isolation and characterization of a galactose-specific lectin. J Cell Biol 111: 1639–1643

Hong Z, Verma DPS (1994) A phosphatidylinositol 3-kinase is induced during soybean nodule organogenesis and is associated with membrane proliferation. Proc Natl Acad Sci USA 91: 9617–9621

Honma MA, Ausubel FM (1987) *Rhizobium meliloti* has three functional copies of the *nodD* symbiotic regulatory gene. Proc Natl Acad Sci USA 84: 8558–8562

Honma MA, Asomaning M, Ausubel FM (1990) *Rhizobium meliloti nodD* genes mediate host-specific activation of *nodABC*. J Bacteriol 172: 901–911

Horvath B, Kondorosi E, John M, Schmidt J, Torok I, Györgypal Z, Barabas I, Wieneke U, Schell J, Kondorosi A (1986) Organization, structure and symbiotic function of *Rhizobium meliloti* nodulation genes determining host specificity for alfalfa. Cell 46: 335–343

Horvath B, Bachem CW, Schell J, Kondorosi A (1987) Host-specific regulation of nodulation genes in *Rhizobium* is mediated by plant-signal, interacting with the *nodD* gene product. EMBO J 6: 841–848

Horvath B, Heidstra R, Lados M, Moerman M, Spaink HP, Promé JC, Van Kammen A, Bisseling T (1993) Lipo-oligosaccharides of *Rhizobium* induce infection-related early nodulin gene expression in pea root hairs. Plant J 4: 727–733

Hungria M, Joseph CM, Phillips DA (1991a) Anthocyanidins and flavonols, major *nod* gene inducers from seeds of a black seeded common bean (*Phaseolus vulgaris* L). Plant Physiol 97: 751–758

Hungria M, Joseph CM, Phillips DA (1991b) *Rhizobium nod* gene inducers exuded naturally from roots of common bean (*Phaseolus vulgaris* L). Plant Physiol 97: 759–764

Ielpi L, Dylan T, Ditta GS, Helinski DR, Stanfield SW (1990) The *ndvB* locus of *Rhizobium meliloti* encodes a 319-kDa protein involved in the production of β-(1→ 2)-glucan. J Biol Chem 265: 2843–2851

Innes R, Kuempel P, Plazinski J, Cremers H, Rolfe B, Djordjevic M (1985) Plant factors induce expression of nodulation and host-range genes in *Rhizobium trifolii*. Mol Gen Genet 201: 426–432

Irvinc R, Letcher A, Dawson R (1980) Phosphatidylinositol phosphodiesterase in higher plants. Biochem J 192: 279–283

John M, Schmidt J, Wieneke U, Kondorosi E, Schell J (1985) Expression of the nodulation gene *nodC* of *Rhizobium meliloti* in *Escherichia coli*: role of the *nodC* gene product in nodulation. EMBO J 4: 2425–2430

John M, Rohrig H, Schmidt J, Wieneke U, Schell J (1993) *Rhizobium* NodD protein involved in nodulation signal synthesis is a chitooligosaccharide deacetylase. Proc Natl Acad Sci USA 90: 625–629

Johnson D, Roth LE, Stacey G (1989) Immunogold localization of the NodC and NodA protein of *Rhizobium meliloti*. J Bacteriol 171: 2425–2430

Kapulnik Y, Joseph CM, Phillips DA (1987) Flavone limitations to root nodulation and symbiotic nitrogen fixation in alfalfa. Plant Physiol 84: 1193–1196

Katinakis P, Verma DPS (1985) Nodulin-24 gene of soybean codes for a peptide of the peribacteroid membrane and was generated by a tandem duplication of a sequence resembling an insertion element. Proc Natl Acad Sci USA 82: 4157–4161

Kinnback A, Mellor R, Werner D (1987) α-Mannosidase II isoenzyme in the peribacteroid space of *Glycine max* root nodules. J Exp Bot 38: 1373–1377

Kittell BL, Helinski DR, Ditta GS (1989) Aromatic aminotransferase activity and indoleacetic acid production in *Rhizobium meliloti.* J Bacteriol 171: 5458–5466

Klein S, Hirsch AM, Smith CS, Signer ER (1988a) Interaction of *nod* and *exo Rhizobium meliloti* in alfalfa nodulation. Mol Plant Microbe Interact 1: 94–100

Klein S, Walker GC, Signer ER (1988b) All *nod* genes of *Rhizobium meliloti* are involved in alfalfa nodulation by *exo* mutants. J Bacteriol 170: 1003–1006

Kobayashi M, Suzuki T, Fujita T, Masude M, Shimizu S (1995) Occurrence of enzymes involved in biosynthesis of indole-3-acetic acid from indole-3-acetonitrile in plant-associated bacteria, *Agrobacterium* and *Rhizobium.* Proc Natl Acad Sci USA 92: 714–718

Kondorosi A, Schultze M, Savoure A, Hoffmann B, Dudits D, Pierre M, L. A, Bauer P, Kiss G, Kondorosi A (1993) Control of nodule induction and plant cell growth by Nod factors. In: Nester EW, Verma DPS (eds) Advances in molecular genetics of plant–microbe interactions. Kluwer, Dordrecht, pp 143–150

Kondorosi E, Pierre M, Cren M, Haumann U, Buire M, Hoffmann B, Schell J, Kondorosi A (1991) Identification of *NolR,* a negative transacting factor controlling the nod regulon in *Rhizobium meliloti.* J Mol Biol 222: 885–896

Kosslak RM, Bookland R, Barkei J, Paaren HE, Appelbaum ER (1987) Induction of *Bradyrhizobium japonicum* common *nod* genes by isoflavones isolated from *Glycine max.* Proc Natl Acad Sci USA 84: 7428–7432

Kunz J, Henriquez R, Schneider U, Deuter-Reinhard M, Movva NR, Hall MN (1993) Target of rapamycin in yeast, TOR2, is an essential phosphatidylinositol kinase homolog required for G_1 progression. Cell 73: 585–596

Lawson C, Djordjevic M, Weinman J, Rolfe B (1994) *Rhizobium* inoculation and physical wounding result in the rapid induction of the same chalcone synthase copy in *Trifolium subterraneum.* Mol Plant Microbe Interact 7: 498–507

Leigh JA, Coplin DL (1992) Exopolysaccharides in plant–bacterial interactions. Annu Rev Microbiol 46: 307–346

Leigh JA, Walker GC (1994) Exopolysaccharides of *Rhizobium*: synthesis, regulation and symbiotic function. Trends Genet 10: 63–67

Leigh JA, Signer ER, Walker GC (1985) Exopolysaccharide-deficient mutants of *Rhizobium meliloti* that form ineffective nodules. Proc Natl Acad Sci USA 82: 6231–6235

Leigh JA, Reed JW, Hanks JF, Hirsch AM, Walker GC (1987) *Rhizobium meliloti* mutants that fail to succinylate their calcofluor-binding exopolysaccharide are defective in nodule invasion. Cell 51: 579–587

Lerouge P, Roche P, Faucher C, Maillet F, Truchet G, Promé JC, Dénarié J (1990) Symbiotic host-specificity of *Rhizobium meliloti* is determined by a sulphated and acylated glucosamine oligosaccharide signal. Nature 344: 781–784

Libbenga KR, Van Iren F, Bogers RJ, Schraag-Lamers MF (1973) The role of hormones and gradients in the initiation of cortex proliferation and nodule formation in *Pisum sativum* L. Planta 114: 29–39

Ligero F, Caba JM, Lluch C, Olivares J (1991) Nitrate inhibition of nodulation can be overcome by the ethylene inhibitor aminoethoxylglycine. Plant Physiol 97: 1221–1225

Long S (1992) Genetic analysis of *Rhizobium* nodulation. In: Stacey G, Burris R, Evans H (eds) Biological nitrogen fixation. Chapman and Hall, New York, pp 560–597

Manen JF, Simon P, Vanslooten JC, Osteras M, Frutiger S, Hughes GJ (1991) A nodulin specifically expressed in senescent nodules of winged bean is a protease inhibitor. Plant Cell 3: 259–270

Marie C, Barny MA, Downie JA (1992) *Rhizobium leguminosarum* has two glucosamine synthases, *glmS* and *nodM*, required for nodulation and development of nitrogen-fixing nodules. Mol Microbiol 6: 843–851

Maxwell CA, Hartwig UA, Joseph CM, Phillips DA (1989) A chalcone and 2 related flavonoids released from alfalfa roots induce *nod* genes of *Rhizobium meliloti*. Plant Physiol 91: 842–847

Miao GH, Hong Z, Verma DPS (1992) Topology and phosphorylation of soybean nodulin-26, an intrinsic protein of the peribacteroid membrane. J Cell Biol 118: 481–490

Morrison N, Verma DPS (1987) A block in the endocytosis of *Rhizobium* allows cellular differentiation in nodules but affects the expression of some peribacteroid membrane nodulins. Plant Mol Biol 9: 185–196

Mulligan J, Long S (1985) Induction of *Rhizobium meliloti nodC* expression by plant exudate requires *nodD*. Proc Natl Acad Sci USA 82: 1–5

Noel KD (1992) Rhizobial polysaccharides required in symbioses with legumes. In: Verma DPS (ed) Molecular signals in plant–microbe communications. CRC Press, Boca Raton, pp 341–357

Noel KD, Vandenosch KA, Kalpaca B (1986) Mutations in *Rhizobium phaseoli* that lead to arrested development of infection threads. J Bacteriol 168: 1392–1401

O'Connell KP, Araujo RS, Handelsman J (1990) Exopolysaccharide-deficient mutants of *Rhizobium* sp. strain CIAT899 induce chlorosis in common bean (*Phaseolus vulgaris*). Mol Plant Microbe Interact 3: 424–428

Orgambide GG, Philip-Hollingsworth S, Hollingsworth RI, Dazzo FB (1994) Flavone-enhanced accumulation and symbiosis-related biological activity of a diglycosyl diacylglycerol membrane glycolipid from *Rhizobium leguminosarum* biovar *trifolii*. J Bacteriol 176: 4338–4347

Perotto S, Brewin NJ, Kannenberg EL (1994) Cytological evidence for a host defense response that reduces cell and tissue invasion in pea nodules by lipopolysaccharide-defective mutants of *Rhizobium leguminosarum* strain 3841. Mol Plant Microbe Interact 7: 99–112

Peters N, Frost J, Long S (1986) A plant flavone luteloin induces expression of *Rhizobium meliloti* nodulation genes. Science 233: 977–980

Peters NK, Crist-Estes DK (1989) Nodule formation is stimulated by the ethylene inhibitor aminoethoxyvinylglycine. Plant Physiol 91: 690–693

Petrovics G, Putnoky P, Reuhs B, Kim J, Thorp TA, Noel KD, Carlson RW, Kondorosi A (1993) The presence of a novel type of surface polysaccharide in *Rhizobium meliloti* requires a new fatty acid synthase-like gene cluster involved in symbiotic nodule development. Mol Microbiol 8: 1083–1094

Philip-Hollingsworth S, Hollingswoth RI, Dazzo FB (1991) *N*-acetylglutamic acid: an extracellular nod signal of *Rhizobium trifolii* ANU843 that induces root hair branching and nodule-like primordia in white clover roots. J Biol Chem 266: 16854–16858

Phillips DA, Torrey JG (1971) Studies on cytokinin production by *Rhizobium*. Plant Physiol 49: 11–15

Phillips DA, Joseph CM, Maxwell CA (1992) Trigonelline and stachydrine released from alfalfa seeds activate NodD2 protein in *Rhizobium meliloti*. Plant Physiol 99: 1526–1531

Priefer UB (1989) Genes involved in lipopolysaccharide production and symbiosis are clustered on the chromosome of *Rhizobium leguminosarum* biovar *viciae* Vf39. J Bacteriol 171: 6161–6168

Prinsen E, Chauvaux N, Schmidt J, John M, Wieneke U, Degreef J, Schell J, Vanonckelen H (1991) Stimulation of indole-3-acetic acid production in *Rhizobium* by flavonoids. Ind J Exp Biol 29: 184–186

Recourt K, van Brussel AAN, Driessen AJM, Lugtenberg BJJ (1989) Accumulation of a *nod* gene inducer, the flavonoid narigenin, in the cytoplasmic membrane of *Rhizobium leguminosarum* biovar *viciae* is caused by the pH-dependent hydrophobicity of naringenin. J Bacteriol 171: 4370–4377

Recourt K, Schripsema J, Kijne J, van Brussel A, Lugtenberg B (1991) Inoculation of *Vicia sativa* subsp. *nigra* roots with *Rhizobium leguminosarum* biovar *viciae* results in release of *nod* gene activating flavanones and chalcones. Plant Mol Biol 16: 841–852

Relic B, Perret X, Estrada-Garcia MT, Kopcinska J, Golinowski W, Krishnan HB, Pueppke SG, Broughton WJ (1994) Nod-factors of *Rhizobium* are a key to the legume door. Mol Microbiol 13: 171–178

Reuhs B, Kim J, Badgett A, Carlson R (1994) Production of cell-associated polysaccharides of *Rhizobium fresii* USDA205 is modulated by apigenin and host root extract. Mol Plant Microbe Interact 7: 240–247

Robertson J, Lyttleton P (1982) Coated and smooth vesicles in the biogenesis of cell walls plasma membranes infection threads and peribacteroid membranes in root hairs and nodules of white clover. J Cell Sci 58: 63–78

Rocher P, Debelle F, Maillet F, Lerouge P, Faucher C, Truchet G, Dénarié J, Promé JC (1991) Molecular basis of symbiotic host specificity in *Rhizobium meliloti—nodH* and *nodPQ* genes encode the sulfation of lipo-oligosaccharide signals. Cell 67: 1131–1143

Rohrig H, Schmidt J, Wieneke U, Kondorosi E, Barlier I, Schell J, John M (1994) Biosynthesis of lipooligosaccharide nodulation factors: *Rhizobium* NodA protein is involved in *N*-acylation of the chitooligosaccharide backbone. Proc Natl Acad Sci USA 91: 3122–3126

Rossen L, Shearman C, Johnston A, Downie J (1985) The *nodD* gene of *Rhizobium leguminosarum* is autoregulatory and in the presence of plant exudate induces the *nodABC* genes. EMBO J 4: 3369–3373

Roth L, Stacey G (1989) Cytoplasmic membrane systems involved in bacterium release into soybean nodule cells as studied with two *Bradyrhizobium japonicum* mutant strains. Eur J Cell Biol 49: 24–32

Sadowsky M, Olson E, Foster V, Kosslak R, Verma DPS (1988) Two host-inducible genes of *Rhizobium fredii* and characterization of the inducing compound. J Bacteriol 170: 171–178

Sanjuan J, Grop P, Göttfert M, Hennecke H, Stacey G (1994) NodW is essential for full expression of the common nodulation genes in *Bradyrhizobium japonicum.* Mol Plant Microbe Interact 7: 364–369

Savoure A, Magyar Z, Pierre M, Brown S, Schultze M, Dudits D, Kondorosi A, Kondorosi E (1994) Activation of the cell machinery and the isoflavonoid biosynthesis pathway by active *Rhizobium meliloti* Nod signal molecules in *Medicago* microcallus suspensions. EMBO J 13: 1093–1102

Scheres B, McKhann HI, Zalensky A, Lobler M, Bisseling T, Hirsch AM (1992) The PsENOD12 gene is expressed at two different sites in Afghanistan pea pseudonodules induced by auxin transport inhibitors. Plant Physiol 100: 1649–1655

Scheu A, Economou A, Hong G, Ghelani S, Johnston A, Downie JA (1992) Secretion of the *Rhizobium leguminosarum* nodulation protein NodO by haemolysin-type systems. Mol Microbiol 6: 231–238

Schlaman HR, Spaink HP, Okker RJ, Lugtenberg BJ (1989) Subcellular localization of the *nodD* gene product in *Rhizobium leguminosarum.* J Bacteriol 171: 4686–4693

Schlaman HRM, Okker RJH, Lugtenberg BJJ (1990) Subcellular localization of the *Rhizobium leguminosarum nodI* gene product. J Bacteriol 172: 5486–5489

Schmidt J, John M, Wieneke U, Krussmann H-D, Schell J (1986) Expression of the nodulation gene *nodA* in *Rhizobium meliloti* and localization of the gene product in the cytosol. Proc Natl Acad Sci USA 83: 9581–9585

Schmidt J, Wingender R, John M, Wieneke U, Schell J (1988) *Rhizobium meliloti nodA* and *nodB* genes are involved in generating compounds that stimulate mitosis of plant cells. Proc Natl Acad Sci USA 85: 8575–8582

Schmidt J, Röhrig H, John M, Wieneke U, Stacey G, Koncz C, Schell J (1993) Alteration of plant growth and development by *Rhizobium nodA* and *nodB* genes involved in the synthesis of oligosaccharide signal molecules. Plant J 4: 651–658

Schmidt PE, Broughton WJ, Werner D (1994) Nod factors of *Bradyrhizobium japonicum* and *Rhizobium* sp. NGR234 induce flavonoid accumulation in soybean root exudate. Mol Plant Microbe Interact 7: 384–390

Schu PV, Takegawa K, Fry MJ, Stack JH, Waterfield MD, Emr SD (1993) Phosphatidylinositol 3-kinase encoded by yeast *VPS34* gene essential for protein sorting. Science 260: 88–91

Schwedock J, Long SR (1990) ATP sulphurylase activity of the *nodP* and *nodQ* gene products of *Rhizobium meliloti*. Nature 348: 644–647

Sheaman CA, Rossen I, Johnson AWB, Downie JA (1986) The *Rhizobium leguminosarum* nodulation gene *nodF* encodes a polypeptide similar to acyl carrier protein and is regulated by *nodD* plus a factor in pea root exudate. EMBO J 5: 647–652

Smit G, Puvanesrajah V, Carlson RW, Babour WM, Stacey G (1992) *Bradyrhizobium japonicum nodD1* can be specifically induced by soybean flavonoids that do not induce the *nosYABCSUIJ* operon. J Biol Chem 267: 310–318

Spaink HP, Sheeley DM, van Brussel AAN, Glushka J, York VN, Lugtenberg BJJ (1991) A novel highly unsaturated fatty acid moiety of lipo-oligosaccharide signals determines host specificity of *Rhizobium*. Nature 354: 125–129

Spaink HP, Wijffelman CA, Pees E, Okker RJH, Lugtenberg BJJ (1989) Localization of functional regions of the *Rhizobium nodD* product using hybrid *nodD* genes. Plant Mol Biol 12: 59–73

Stacey G, Luka S, Sanjuan J, Banfalvi Z, Niewkoop AJ, Chun JY, Forsberg LS, Carlson R (1994) *nodZ*, a unique host-specific nodulation gene, is involved in the fucosylation of the lipooligosaccharide nodulation signal of *Bradyrhizobium japonicum*. J Bacteriol 176: 620–633

Staehelin C, Granado J, Müller J, Wiemken A, Mellor RB, Felix G, Regenass M, Broughton WJ, Boller T (1994a) Perception of *Rhizobium* nodulation factors by tomato cells and inactivation by root chitinases. Proc Natl Acad Sci USA 91: 2196–2200

Staehelin C, Schultze M, Kondorosi E, Mellor RB, Boller T, Kondorosi A (1994b) Structural modifications in *Rhizobium meliloti* Nod factors influence their stability against hydrolysis by root chitinases. Plant J 5: 319–330

Stanfield SW, Ielpi L, O'Brochta D, Helinski DR, Ditta GS (1988) The *ndvA* gene product of *Rhizobium meliloti* is required for β-(1 → 2) glucan production and has homology to the ATP-binding export protein HlyB. J Bacteriol 170: 3523–3530

Stephens L, Smrcka A, Cooke F, Jackson T, Sternweis, Hawkins P (1994) A novel phosphoinositide 3 kinase activity in myeloid-derived cells is activated by G protein βγ subunits. Cell 77: 83–93

Stokkermans TJW, Peters NK (1994) *Bradyrhizobium elkanii* lipo-oligosaccharide signals induce complete nodule structures on *Glycine soja*. Planta 193: 413–420

Sturtevant D, Taller B (1989) Cytokinin production by *Bradyrhizobium japonicum*. Plant Physiol 89: 1247–1252

Surin BP, Downie JA (1988) Characterization of the *Rhizobium leguminosarum* genes *nodLMN* involved in efficient host specific nodulation. Mol Microbiol 2: 173–183

Surin BP, Watson JM, Hamilton WDO, Economou A, Downie JA (1990) Molecular characterization of the nodulation gene, *nodT*, from 2 biovars of *Rhizobium leguminosarum*. Mol Microbiol 4: 245–252

Sutton J, Lea E, Downie J (1994) The nodulation-signaling protein NodO from *Rhizobium leguminosarum* biovar *viciae* forms ion channels in membranes. Proc Natl Acad Sci USA 91: 9990–9994

Taller BJ, Sturtevant DB (1991) Cytokinin production by rhizobia. In: Hennecke H, Verma DPS (eds) Advances in molecular genetics of plant–microbe interactions. Kluwer, Dordrecht, pp 215–221

Thimann KV (1936) On the physiology of the formation of nodules on legume roots. Proc Natl Acad Sci USA 22: 511–514

Truchet G, Barker DG, Camut S, Debilly F, Vasse J, Huguet T (1989) Alfalfa nodulation in the absence of *Rhizobium*. Mol Gen Genet 219: 65–68

Valius M, Kazlauskas A (1993) Phospholipase C-γ 1 and phosphatidylinositol 3 kinase are the downstream mediators of the PDGF receptor's mitogenic signal. Cell 73: 321–334

Van Brussel AAN, Recourt K, Pees E, Spaink HP, Tak T, Wijffelman CA, Kijm JW, Lugtenberg BJJ (1990) A biovar-specific signal of *Rhizobium leguminosarum* bv *viciae* induces increased nodulation gene-inducing activity in root exudate of *Vicia sativa* subsp. *nigra*. J Bacteriol 172: 5394–5401

Van Der Meer I, Stuitje A, Mol J (1993) Regulation of general phenylpropanoid and flavonoid gene expression. In: Verma DPS (ed) Control of plant gene expression. CRC Press, Boca Raton, pp 125–155

Vanrhijn PJS, Feys B, Verreth C, Vanderleyden J (1993) Multiple copies of *nodD* in *Rhizobium tropici* CIAT899 and BR816. J Bacteriol 175: 438–447

Vazquez M, Santana O, Quinto C (1993) The NodI and NodJ proteins from *Rhizobium* and *Bradyrhizobium* strains are similar to capsular polysaccharide secretion proteins from Gram-negative bacteria. Mol Microbiol 8: 369–377

Verma DPS (1990) Endosymbiosis of *Rhizobium*: internalization of the "extracellular compartment" and metabolite exchange. In: Gresshoff PM, Roth LE, Stacey G, Newton WE (eds) Nitrogen fixation: achievements and objectives. Chapman and Hall, New York, pp 235–237

Verma DPS (1992) Signals in root nodule organogenesis and endocytosis of *Rhizobium*. Plant Cell 4: 373–382

Verma DPS, Nadler K (1984) Legume-*Rhizobium* symbiosis: host's point of view. In: Verma DPS, Hohn T (eds) Genes involved in microbe–plant interactions. Springer, Wien New York, pp 57–93 [Dennis ES et al (eds) Plant gene research. Basic knowledge and application]

Whitman M, Kaplan DR, Schaffhausen B, Cantley L, Roberts TM (1985) Association of phosphatidylinositol kinase activity with polyoma middle-T competent for transformation. Nature 315: 239–242

Whitman M, Downes PC, Keeler M, Keller T, Cantley L (1988) Type I phosphatidylinositol kinase makes a novel inositol phospholipid, phosphatidylinositol-3-phosphate. Nature 332: 644–646

Wood EA, Butcher GW, Brewin NJ, Kannenberg EL (1980) Genetic derepression of a developmentally regulated lipopolysaccharide antigen from *Rhizobium leguminosarum* 3841. J Bacteriol 171: 4545–4555

Xu P, Lloyd CW, Staiger CJ, Drobak BK (1992) Association of phosphotidylinositol 4-kinase with the plant cytoskeleton. Plant Cell 4: 941–951

Yotsushima K, Nakamura K, Mitsui T, Igaue I (1992) Purification and characterization of phosphatidylinositol-specific phospholipase C in suspension-cultured cells of rice (*Oriza sativa* L.). Biosci Biotech Biochem 56: 1247–1251

Subject Index

T. Boller, F. Meins, Jr. (eds.)

Genes Involved in Plant Defense

1992. 34 figures. X, 364 pages.
Cloth DM 218,– öS 1526,–
ISBN 3-211-82312-3
Plant Gene Research
Basic Knowledge and Application

The co-evolution of plants and microbes has led to an elaborate system of genes involved in recognition, attack and defense. This volume deals with these genes and regulation of their expression. The topic is treated broadly ranging from considerations of population genetics to the identification of defense-related genes and their regulation. The book provides a concise, authoritative review of latest developments in a rapidly developing and active field of agronomic importance.

SpringerWienNewYork

P.O.Box 89, A-1201 Wien • New York, NY 10010, 175 Fifth Avenue
Heidelberger Platz 3, D-14197 Berlin • Tokyo 113, 3-13, Hongo 3-chome, Bunkyo-ku

SpringerPlantGeneResearch

R.G. Herrmann (ed.)

Cell Organelles

1992. 55 figures. XII, 467 pages.
Cloth DM 229,–, öS 1600,–
ISBN 3-211-82264-X
Plant Gene Research
Basic Knowledge and Application

The book presents a comprehensive overview of developments in the biology of plastids, mitochondria, glyoxisomes and peroxisomes. Written by the field's most accomplished authors, the topics of its 12 articles range from organelle biogenesis and heredity to evolution, with coverage of both molecular and classical aspects. It includes a modern treatment of genetic compartmentalization in the plant cell, which is one of the fundamental characteristics of eukaryotes. The treatise serves as an introduction of pertinent literature, is an invaluable resource for research professionals and advanced graduate students in plant molecular and cell biology, and will prove useful to all botanists and teachers who demand a conceptual framework.

SpringerWienNewYork

P.O.Box 89, A-1201 Wien • New York, NY 10010, 175 Fifth Avenue
Heidelberger Platz 3, D-14197 Berlin • Tokyo 113, 3-13, Hongo 3-chome, Bunkyo-ku

Springer-Verlag and the Environment

MIX
Papier aus verantwortungsvollen Quellen
Paper from responsible sources
FSC® C105338

If you have any concerns about our products,
you can contact us on
ProductSafety@springernature.com

In case Publisher is established outside the EU,
the EU authorized representative is:
Springer Nature Customer Service Center GmbH
Europaplatz 3, 69115 Heidelberg, Germany

Printed by Libri Plureos GmbH
in Hamburg, Germany